OCULAR HISTOLOGY
A TEXT AND ATLAS

BEN S. FINE M.D.

Research Associate, Ophthalmic Pathology Branch,
Armed Forces Institute of Pathology;
Associate Research Professor of Ophthalmology,
The George Washington University;
Senior Attending Ophthalmologist,
The Washington Hospital Center,
Washington, D.C.

MYRON YANOFF M.D.

Associate Professor in Ophthalmology and Pathology,
Medical School and Hospital of the University of Pennsylvania;
Chief, Section of Ophthalmic Pathology,
Department of Ophthalmology,
Hospital of the University of Pennsylvania,
Philadelphia

OCULAR HISTOLOGY

A TEXT AND ATLAS

WITH 490 ILLUSTRATIONS, 19 IN FULL COLOR

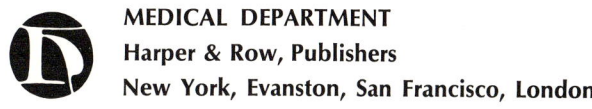

MEDICAL DEPARTMENT
Harper & Row, Publishers
New York, Evanston, San Francisco, London

to our wives and children

OCULAR HISTOLOGY: A Text and Atlas. Copyright © 1972 by Harper & Row, Publishers, Inc.

All rights reserved. No part of this book may be used or reproduced in any manner whatsoever without written permission except in the case of brief quotations embodied in critical articles and reviews. Printed in the United States of America. For information address Medical Department, Harper & Row, Publishers, Inc., 49 East 33rd Street, New York, N.Y. 10016

First Edition
STANDARD BOOK NUMBER: 06–140825–5
LIBRARY OF CONGRESS CATALOG CARD NUMBER: 70–172877

contents

	Foreword by Bruce H. Smith	vii
	Foreword by Lorenz E. Zimmerman	ix
	Preface	xi
1.	Electron Microscopy: Techniques and Interpretation	1
2.	The Cell and Its Contents	15
3.	Cell Interrelations	27
4.	Extracellular Materials	31
5.	Arrangement of Tissues and Cells	41
6.	The Retina	47
7.	The Vitreous Body	109
8.	The Lens	129
9.	The Cornea and Sclera	141
10.	The Uveal Tract	167
11.	The Anterior Chamber Angle	213
12.	The Optic Nerve	233
	Index	249

foreword

This text and atlas of ocular histology is published as a part of the observances of the golden anniversary of the American Registry of Pathology, which had its origins in the Registry of Ophthalmic Pathology in 1921. Fifty years ago, when American ophthalmic pathology was in its infancy, the members of the American Academy of Ophthalmology and Otolaryngology had the foresight to create a histopathology laboratory for the study of ocular tissues at the old Army Medical Museum, which was then under the guidance of Major George Callender, the curator. Since then, the Registry of Ophthalmic Pathology has grown to include over 90,000 specimens, and the Registry itself has expanded to encompass 27 separate registries under the overall name of the American Registry of Pathology, itself one of four departments in the Armed Forces Institute of Pathology. The Registry of Ophthalmic Pathology, now under the guidance of Dr. Lorenz E. Zimmerman, has become a major force in ophthalmic pathology today, and is responsible for a very heavy program in consultation, education, and research in this specialty.

The years of experience and the careful and dedicated work of the authors are clearly demonstrated in this volume. The staff of the Armed Forces Institute of Pathology takes pride in having been able to assist the authors in the preparation and completion of this excellent book.

BRUCE H. SMITH
Captain, MC, USN
Director, AFIP

As this new book by two former Fellows at the Armed Forces Institute of Pathology (AFIP) goes to press, we are celebrating the fiftieth anniversary of the establishment of our Ophthalmic Pathology Laboratory. The necessary organizational agreements for this laboratory were concluded in 1921, and the facility became operational in 1922. Under the guidance of such dedicated pioneers as Callender, Ash, DeCoursey, and Wilder, this laboratory produced many prize-winning exhibits, a number of important research papers based on cases contributed to the Registry of Ophthalmic Pathology, and a series of atlases published by the American Registry of Pathology.

During the period between its founding in 1921 and 1955, the laboratory was housed in the old Army Medical Museum building on the Mall near the Capitol. Space was extremely limited, facilities outmoded, and modern research impossible. By the end of World War II a backlog of several thousand cases had accumulated, and this situation continued during the hectic period of the Korean conflict. Needless to say, the staff in its cramped quarters was nearly overwhelmed.

In 1955, the AFIP moved into its new home on the grounds of the Walter Reed Army Medical Center, a modern research building where more space for a larger staff and excellent facilities for experimental pathology and electron microscopy became available for the first time. It was shortly after we moved into the new building that Ben Fine—still in residency at the District of Columbia General Hospital—began to visit the AFIP to study ophthalmic pathology at every opportunity. Upon completion of his clinical training, he was awarded a full-time NIH fellowship in ophthalmic pathology at the AFIP, and rapidly became one of our most active investigators. Soon after the arrival of the first electron microscope at the AFIP in 1958, Ben Fine teamed with A.J. Tousimis, then a graduate student in biophysics working at the AFIP, and began his electron microscopic studies of ocular tissues.

Thus began not only Ben Fine's career as one of the world's foremost authorities on electron microscopy of the eye, but also the modern training program in ophthalmic histology and pathology at the AFIP. Before our move into the new building, training and research programs were necessarily limited to descriptive pathology, clinicopathologic investigations, and a very limited amount of histochemistry. In our new quarters it gradually became possible to broaden the training and research opportunities, and Ben Fine was the first and certainly one of the most outstanding of a group of about 50 Fellows who have profited by the opportunities made possible through this expanding program.

Myron Yanoff, another former Fellow in Ophthalmic Pathology at the AFIP, is also a pioneer. As the scope of ophthalmic pathology and ophthalmic research is rapidly enlarging, a number of those who have embarked on careers in ocular pathology during the past few years have found it desirable to avail themselves of more than the ordinary formal training, with the expanded objectives of Board certification in both pathologic anatomy and ophthalmology. Myron Yanoff was one of the first to obtain such dual certification.

Together, Drs. Fine and Yanoff are a most interesting team. While both have maintained an active practice of clinical ophthalmology, they are nevertheless completely dedicated to their half-time laboratory work. Both are stimulating teachers who are playing important roles in the formal education of residents and Fellows under their tutelage. The need for a suitable textbook on ocular histology to assist them in their teaching activities provided the incentive for the preparation of this volume. Needless to say, I encouraged them to proceed with their plans, as have the various administrative officers who have been at the AFIP since their work was started. The result is this book—one of the very first to incorporate electron microscopic observations into a detailed presentation of the normal structure of ocular tissues.

I am very proud that the authors of this new book are among my former Fellows, and that the environment of our laboratory, now observing its golden anniversary, was so conducive for their work. I trust that those who use it will find this book a fitting gem to set off this golden crown.

LORENZ E. ZIMMERMAN, M.D.
Chief, Ophthalmic Pathology Branch
Armed Forces Institute of Pathology

preface

This book is an outgrowth of a series of lectures presented to postgraduate students attending the biannual course in ophthalmic pathology at the Armed Forces Institute of Pathology and to those attending the postgraduate course in ophthalmology at the University of Pennsylvania in Philadelphia.

Although a number of references currently are available for ophthalmic histology, the observations that can be made by electron microscopy remain rather widely scattered throughout the literature and therefore are not readily available to the student. We have endeavored to assemble a text that would bridge the gap between conventional light microscopy and electron microscopy. We also have attempted to provide a short introduction to contemporary cytology which may serve as a background for the more extensive presentations that are currently available.

Our aim is to present, in as systematic and simplified a manner as possible, a cytologic *method* or approach for examining the way in which ocular tissues are assembled and arranged, and to point out, wherever possible, any clinical relevance that the anatomic observations may have. We hope that this dual approach will be of value to ophthalmologists, to those who require a contemporary histologic basis from which to investigate a variety of problems in ocular pathology, as well as to others interested in the visual sciences.

It is in this sense of contemporary cytology that the eye, despite its content of tissues with high degrees of specialization, may be considered a sequestered microcosm of the tissues of the body. The eye not only contains examples of most other tissues, but frequently, because of their high degree of functional specialization, may dramatically illustrate a high degree of morphologic exaggeration. For example, the photoreceptor outer segments provide striking illustrations of the degree to which a cilium of a sensory cell may be altered; the pigmented or light-absorbing cells produce melanin granules not in one but in at least two characteristic and rather large sizes; the vitreous body is not only highly transparent, but also one of the most delicate collagenous tissues to be found in man; and the anterior lens capsule is by far the thickest basement membrane in the body. On the other hand, this high degree of specialization does not completely obscure the close relationship of ocular tissues to others: the retina is still part of the central nervous system, the cornea and lens are derivatives of the skin, and the ciliary epithelium is a neuroepithelial gland. From the characteristic and highly specialized morphologic features, therefore, we can, in turn, often infer something about function.

The cytologic examples used are all taken from ocular tissues, primarily human. Most of the preparations are original, as are some of the concepts; for the latter we accept responsibility. In instances where there is a difference of opinion, either in definition or in terminology, or where additional clarification seems warranted, explanatory footnotes are included. The point of view preferred by the authors is presented as the "accepted opinion." No doubt in time some of the concepts and even some of the terminology presented will be further modified.

The authors wish to thank all who have helped in the preparation of this text. In particular, we also wish to dedicate this book in part to Dr. Maximilian Salzmann, who, by 1912, had already stated most of what we have come to say here, and to Dr. Lorenz E. Zimmerman, for creating the environment in which this work was possible.

Our special thanks to Margaret J. Patton for her invaluable and patient assistance in the preparation of the manuscript; to Genevieve C. Overmyer, for all of the related typing of the manuscript; and to Drs. David Barsky, Joseph W. Berkow, Jay L. Helfgott, David M. Kozart, Michael Ramsey, Benjamin Rones, Harold G. Scheie, Mark O. M. Ts'o, Roderick Willis, and Lorenz E. Zimmerman, for encouragement and editorial comments. In addition, we are grateful to John C. Barber, Elizabeth McDonnell, and Frances Schulz, of the Armed Forces Institute of Pathology, for their help with many of the line drawings; and to Drs. Mark O. M. Ts'o, Michael Mund, Merlyn Rodrigues, Peter Wobmann, and Max Helfgott, for their kindness in contributing illustrations. John M. Wehrung, Director of the Sperry Rand Microanalysis Laboratory in Rockville, Maryland, gave us invaluable assistance with the scanning electron microscopy.

Finally, we extend our sincere appreciation to Efrain Perez-Rosario, who for more than a decade has assisted in all of the preparation and sectioning techniques required for our various publications and for this book.

Acknowledgment is given for the support received for much of this work from contract No. DA–49–193–MD–2680, from the Medical Research and Development Command, United States Army, Washington, D.C., and from research grants Nos. EY–00289, EY–00397, and EY–00133 and training grant No. EY–00032 from the National Eye Institute (formerly part of the National Institute for Neurological Diseases and Blindness), National Institutes of Health, Bethesda, Maryland.

B.S.F.
M.Y.

chapter 1

Electron Microscopy: Techniques and Interpretation

The transmission electron microscope
Tissue preparation
    Sectioning, mounting, and embedding
    Fixation
Special methods of examination
    Shadow-casting
    Surface replication
    Scanning electron microscopy
    "Staining" and histochemical methods
    Radioautography
Photography
    Technique
    Interpretation

THE TRANSMISSION ELECTRON MICROSCOPE

The transmission electron microscope (T-E/M) is the electronic counterpart of the inverted compound light microscope (Fig. 1–1).[6] The source of light is replaced by a source of electrons, usually a heated tungsten filament from which the electrons are boiled off. The electrons (at a potential of 50 to 60 kV generally) are fired down an evacuated column (i.e., a vacuum approximately that of a good television tube) through the central apertures of a series of circular electromagnets. Control of current flow through these circular electromagnets collimates, spreads, or contracts this beam, which, like the electromagnetic spectrum, travels in waves. As with the light microscope, the first lens, the one near the illuminating source, is called the *condenser* lens; the next, near the specimen, is called the *objective* lens; the last which is not a lens into which the observer looks but rather one that projects the image onto a fluorescent screen, is called a *projector* lens. The screen fluoresces generally in some shade of green, and the image is viewed as a shadow on this green background. A permanent record is obtained by moving the screen aside temporarily and allowing the imaged beam to fall directly on a photographic emulsion (glass plate or cut film).

A latent image is produced in silver halide crystals and is brought out subsequently by conventional photographic development techniques. Because the various wavelengths within the electron beam cannot be separated out, all the photographs are in shades of black and white. The vacuum of the column generally precludes the examination of living tissue. With the transmission electron microscope direct magnifications of from $\sim 500\times$ to $\sim 150{,}000\times$ can be obtained with resolutions that may exceed 10 A. Greater enlargement can be secured by conventional photographic methods, but the usefulness of this procedure is limited.

TISSUE PREPARATION

Sectioning, Mounting, and Embedding

Because an electron beam of 50 to 60 kV cannot penetrate conventional tissue slices or sections, methods were developed to cut sections thin enough to allow *some* of the electrons to be *transmitted* through them.[8,13] The sections necessarily range from 200 to 600 A in thickness.* Because specimens are difficult to hold rigidly for thin sectioning, plastic embedding materials were introduced (methacrylates at first, now almost universally replaced by epoxy resins).[10] The tissue embedded in plastic and mounted in a special microtome is cut with either a glass or a diamond knife. The diamond knife is most useful for cutting hard or dense tissue (e.g., melanin-bearing cells or dense collagen, as in the cornea and sclera).

Microtomes of special design move the tissue forward in exceedingly small increments using either mechanical or thermal advancing

---

* A conventional section of ocular tissue measures 8 to 10 microns (μ) in thickness; a tissue section for electron microscopy measures 200 to 600 Angstroms (A). This difference can better be appreciated when the following relations are considered:

1 millimeter (mm) = 1,000 microns (μ)
1 micron (μ) = 1,000 millimicrons (mμ) or nanometers (nm)
1 millimicron (mμ) = 10 Angstroms (A)

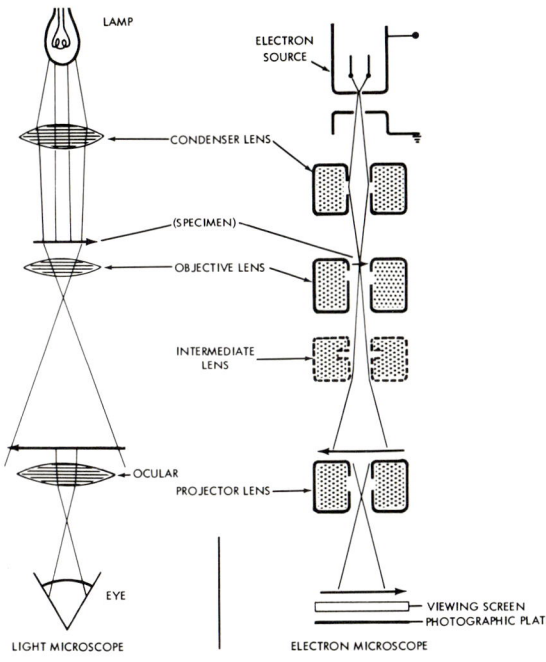

FIG. 1–1. Schematic drawing illustrating analogy between compound light microscope and transmission electron microscope. (From Fine, B. S., et al. Arch Ophthal (Chicago) 62:931, 1959.)

mechanisms. More specialized microtomes automatically advance and section the tissue so that, with care, it can be sectioned serially.

The plastic-embedded tissue sections are then mounted on copper screens (or when necessary for special techniques, nickel or gold screens) called *grids* (Fig. 1–2) which are placed into the microscope column for examination. Only that tissue lying within an aperture in the grid can, of course, be observed. When a small area is to be examined, a number of sections usually are prepared in the event the area of interest is obscured by the metallic supporting gridwork. To minimize such occurrences or to permit proper sequencing when serial sections are examined, a variety of specially designed grids are available.

Fixation

It has long been known that autolytic changes (especially vacuolation of the cytoplasm and chromatin aggregation of the nucleus) can occur in tissues prepared for light microscopy and are caused either by delay in fixation or improper fixation. Proper fixation is even more critical when the higher resolving power of the electron microscope is used for examination. Tissues, therefore, are usually fixed promptly on separation from their blood supply (immersion fixation) or in situ by intraarterial injection of the fixative (perfusion fixation). Because the fixative used most often (buffered osmium tetroxide) penetrates poorly, the tissue usually is cut into small blocks ($\sim$ 1 to 2 cu mm), so that the

FIG. 1–2. Copper screens, or grids, commonly used to support tissue sections in electron microscope. Marker indicates 1 mm.

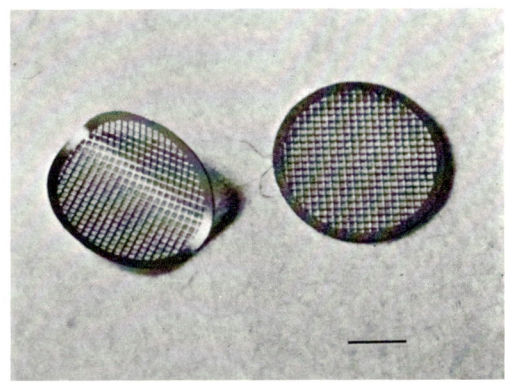

layer of interest lies within 1 to 2 mm of the fixative. A surface layer such as retina, which is less than 1 mm thick, can be suitably fixed by immersion alone while it is still adherent to its backing of adjacent tissues (choroid and sclera). Those tissues which are of no particular interest may be subsequently discarded.

The osmium tetroxide* fixative blackens all the tissue it touches so that topographic orientation may be difficult after fixation and small lesions may be hard to find. A more suitable, all-around fixative, 2 to 4% glutaraldehyde,† can, if used with care, penetrate larger blocks of tissue without blackening them and produce results roughly equivalent to initial fixation with buffered osmium. The osmium tetroxide, however, enhances the contrast of all cell membranes and most of their cytoplasmic particles. For adequate contrast enhancement in tissue sections, therefore, initial fixation in glutaraldehyde must be followed by treatment with osmium tetroxide.

SPECIAL METHODS OF EXAMINATION

Various special examination techniques are available in electron microscopic studies. The topography, or three-dimensional appearance, of material to be examined can be obtained by the use of shadow-casting, surface replication, or scanning electron microscopy (S-E/M). The contrast among cell components can be enhanced by "staining" and histochemical methods, and radioautographic techniques can also be applied when transmission electron microscopy is used.

Shadow-Casting

This old method used in transmission electron microscopy to enhance the three-dimensional appearance of particles or particulate matter entails "spraying" them with a thin molecular layer of metal from an elevated, angled point source (Fig. 1–3). Since the spray comes in from an angle, a zone behind the object (shadow) does not become coated. When the shadow-cast preparation is subsequently examined in the electron microscope,

---

* We have found Dalton's chrome-osmium fixative[3] very useful. Fixation times vary from a short exposure of 35 minutes for a tissue as delicate as the retina to as long as 1 hour for a tissue as dense as cornea.

† Glutaraldehyde (phosphate buffered) has been useful in 2 and 4% solutions.[8,13] The lower concentrations are best for preparing small pieces of tissue directly for electron microscopic study.

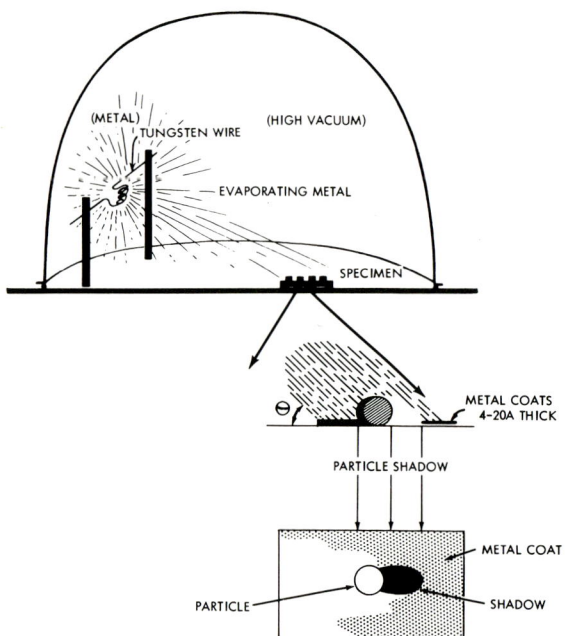

FIG. 1–3. Schematic drawing illustrating technique of shadow-casting. (From Fine, B. S., et al. Arch Ophthal (*Chicago*) 62:931, 1959.)

the region of the shadow allows the electron beam of the microscope to pass through more easily, producing greater exposure to the photographic plate. On the usual positive print (Fig. 1–4A) the shadow is white, but the final illustration is reversed to show the shadow as dark (Fig. 1–4B).

The metallic coating is produced by vaporizing a small amount of metal (e.g., uranium,* palladium, chromium) on a heated tungsten filament in a vacuum chamber (Fig. 1–3). Since the filament is placed at a predetermined distance and height from the object to be coated, the three dimensions of the object can be determined by simple measurements. Tissue sections can also be studied by shadow-casting. For this purpose, they must be embedded in a plastic (e.g., methacrylate) that can be removed to produce some exposed three-dimensional material. The technique is nicely illustrated in the examination of the vitreous body (see Chap. 7) where the structural filaments and ground substances are seen almost in situ.

Surface Replication

In surface replication either a thin or a thick plastic coat is applied to a surface, allowed to harden, and peeled away. The thin coat may be examined directly in the electron microscope or may be shadow-cast for enhancement of the surface detail. The thick coat may be used to obtain a positive replica or it may be examined directly, in the manner of the thin coat.

Scanning Electron Microscopy

The scanning electron microscope (S-E/M) is the electronic counterpart of the stereoscopic (dissecting) light microscope.

In scanning electron microscopy (Fig. 1–5) the electron beam is focused to an exceedingly small spot. The electron beam (or "probe")

* Designated "U-shad" in this text.

plays over (or "scans") the surface of a metal-coated, dried specimen. The beam or scan, controlled by deflecting magnetic fields, traverses the specimen in a series of sequential lines. Bombardment of the metal-coated ("shadowed") specimen produces *secondary* electrons which are collected, translated into light, and subsequently converted into a series of corresponding sequential lines on an oscilloscope, much as is done on a television screen. The picture produced on the screen may be photographed, the photography being generally carried out on a second screen which duplicates the one used for viewing.

Preparation of specimens for scanning electron microscopy is essentially similar to their prepartion for transmission electron microscopy except that upon reaching the stage of final dehydration in 100% propylene oxide, the tissue is not embedded in plastic but allowed to air-dry as slowly as possible. The fixed, air-dried specimen is then mounted whole on a specimen mount ("pedestal" or "stub," Fig. 1–6) by means of an electrically conducting cement. The mounted specimen is then lightly coated (shadowed) in a vacuum chamber with a thin film of metal (e.g., aluminum, gold, gold-palladium) to enhance the production of secondary electrons as well as to carry away charges that may be generated at the point of impact by the electron probe.

By this technique, striking three-dimensional enlargement of surfaces can be readily obtained (Fig. 1–7). Because of the great depth of field, direct magnification of from $\sim 20\times$ to $\sim 50,000\times$ can be obtained, and resolutions to ~ 150 A. Information deep to the surface cannot be obtained unless the surface layer is removed by either chemical or physical means (Fig. 1–8).

For correlation of the surface view with internal structure, the dried specimen can be removed from the mount, reinfiltrated with propylene oxide, and subsequently embedded in epon as for transmission electron microscopic examination. Thin (1.5-μ) sections of this material can be made for correlative light

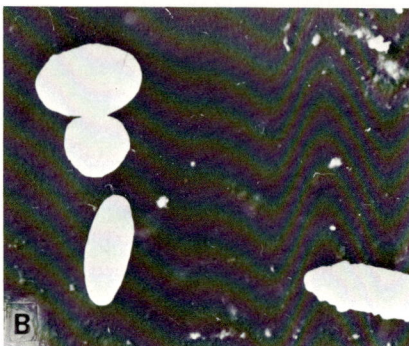

FIG. 1–4. Effect of shadow-casting. **A.** Melanin granules with light shadows. **B.** Reversal of photographic process produces dark shadows but white melanin granules. ×10,000.

FIG. 1–5. Schematic drawing of scanning electron microscope. Transmission electron microscope is modified to scan or play a narrow beam of electrons over surface of specimen. The series of lines ultimately produced on display screen by secondary electrons is photographed with camera attachment.

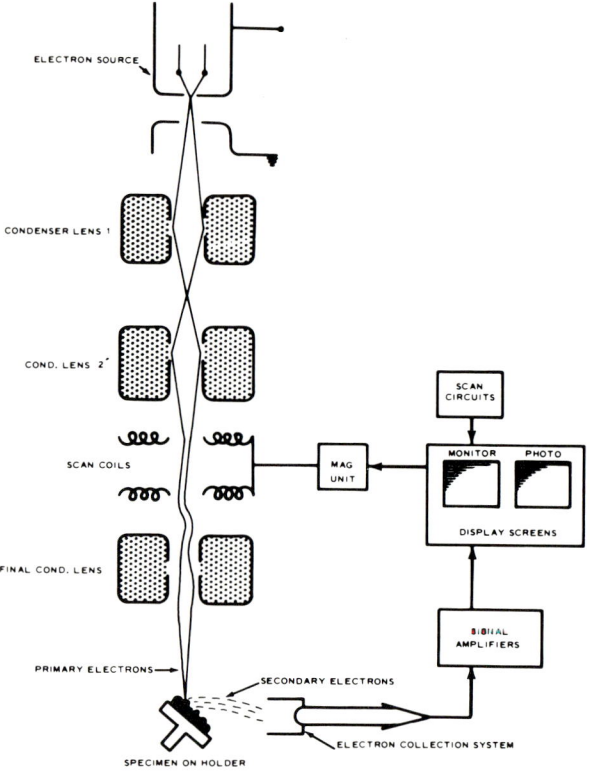

FIG. 1–6. Aluminum specimen mount (stub) for scanning electron microscope, without specimen, showing surface area for mounting (**above**), side view of stub (**left**), and stub with fixed, dried specimen mounted and prepared for examination (**right**). Diameter of mounting area is 12 mm. Specimen mount illustrated here is aluminum pin-type variety designed for Cambridge scanning electron microscope, manufactured by Cambridge Instruments, England. Mounts for other makes of scanning electron microscopes differ in configuration.

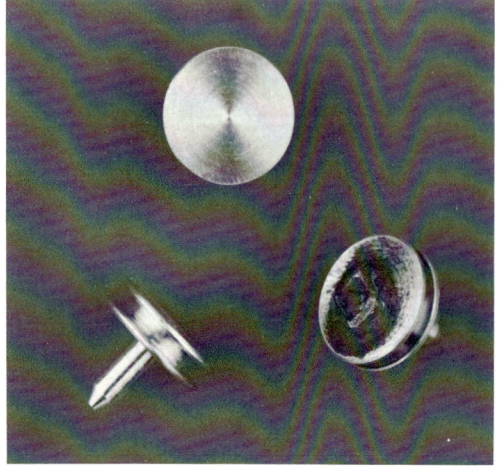

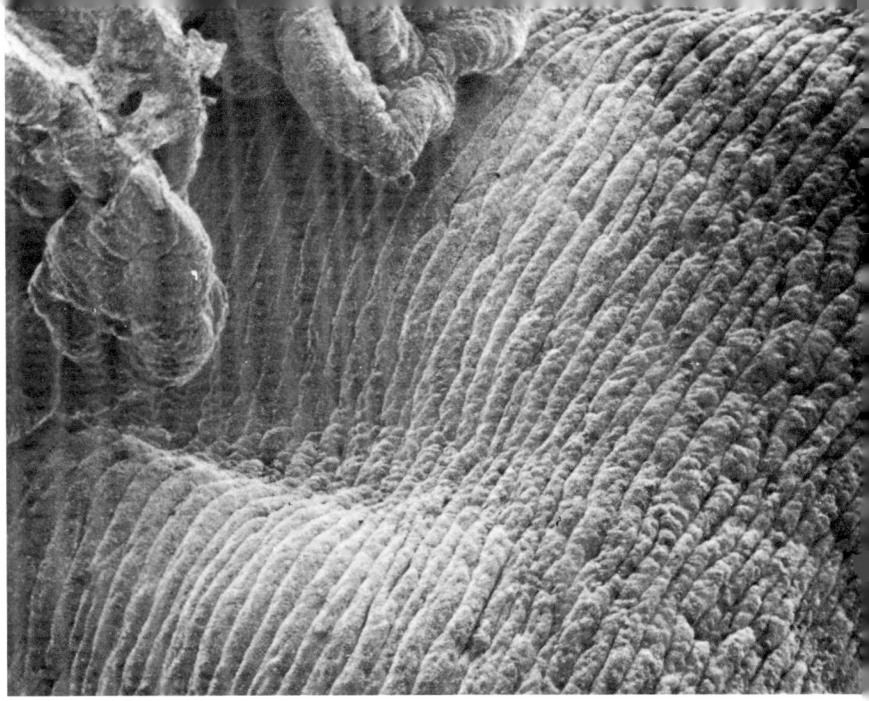

microscopic study (Fig. 1–9) and thinner sections for correlative electron microscopic (i.e., T-E/M) study (Fig. 1–10).

"Staining" and Histochemical Methods

Various compounds are used to enhance the contrast of membranes and particles in a cell or of various extracellular materials. Uranyl acetate, lead citrate, and lead hydroxide are used to treat tissue sections already improved from the native state by fixation with osmium tetroxide. In addition, lead citrate and lead

FIG. 1–7. Scanning electron micrograph of posterior surface of rhesus monkey iris. Circumferential furrows and ridges are clearly displayed. At left, overlying cilary crests are seen and circumferential furrows and ridges continue up anterior declivity of crests. ×165.

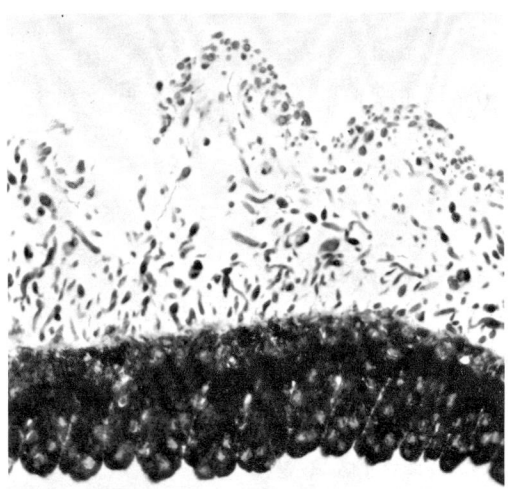

FIG. 1–8. Scanning electron micrograph of melanin granules within cells of iris pigment epithelium. Granules were exposed by tearing away enveloping cytoplasm. ×7,300.

FIG. 1–9. Thin (1.5-μ) section of iris stained with toluidine blue, made by embedding dried specimen in Figure 1–7 in epon. ×210. (AFIP Neg. 70–4570.)

FIG. 1–10. Inset A. Scanning electron micrograph of posterior surface of iris pigment epithelial layer. **Main figure.** Transmission electron micrograph of same material after reinfiltration and embedding, showing that cell cytoplasm between granules is moderately well preserved except for many cytoplasmic holes. Thin basement membrane (*BM*) is poorly outlined. **Inset B.** Electron micrograph of iris stromal melanocyte. Myriad pinholes are present in cytoplasm between melanin granules and within substance of nucleus (*N*). **Main figure,** ×16,500; **inset A,** ×1,500; **inset B,** ×16,500.

hydroxide enhance the contrast of glycogen particles. Seligman and his coworkers[14] have introduced a histochemical procedure, the thiosemicarbazide (TSC) method, for the demonstration of glycogen. This method is the electron microscopic equivalent of the periodic acid–Schiff (PAS) treatment so widely used in light microscopy (see Chap. 4). TSC-positive particles can be even more accurately identified as glycogen when digested with diastase (Fig. 1–11).[2,5]

The contrast of extracellular materials such as collagen and elastin can often be enhanced by treatment of the tissue section with either phosphotungstic acid or with the thiosemi-

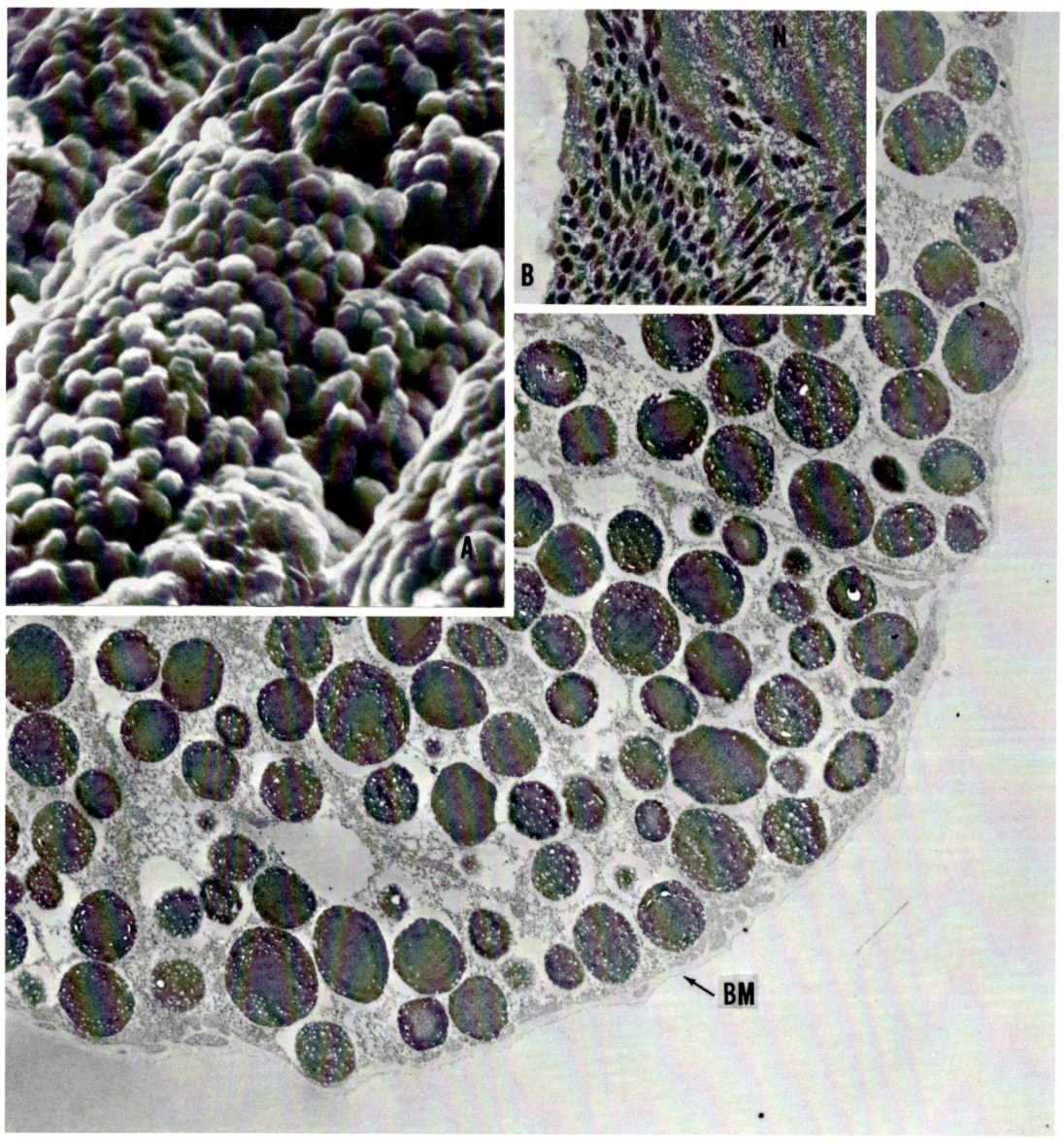

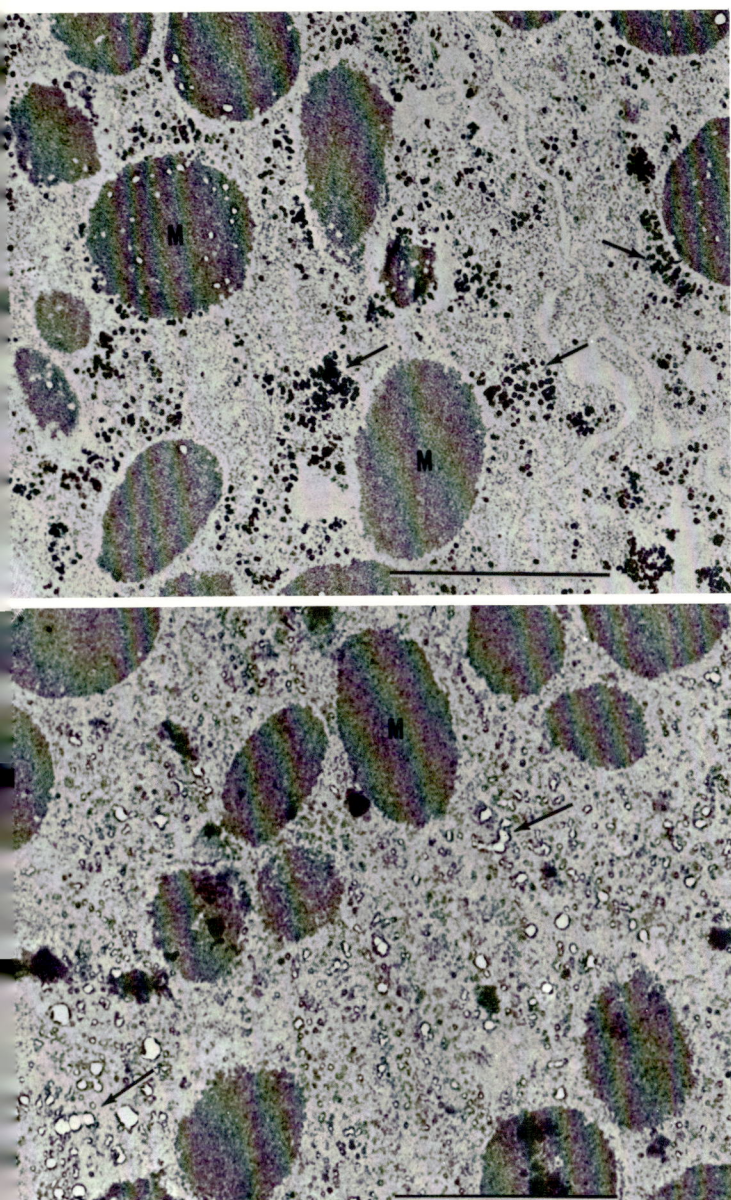

FIG. 1–11. Thiosemicarbazide method for glycogen detection. **A.** In TSC-treated section of iris pigment epithelium, glycogen particles (*arrows*) are exceedingly dense. **B.** Pretreatment of tissue section with diastase digests out particles (*arrows*), identifying them as glycogen. *M*, melanin granules. **A** and **B,** ×33,000. (from Berkow, J. W., and Fine, B. S. *Amer J Ophthal* 69:994, 1970.)

carbazide (TSC) method. More recently, another method[1] (silver tetraphenylporphine sulfonate—Ag-TPPS) has been introduced which stains the elastin of elastic tissue exceedingly densely and with considerable specificity (see Chap. 10).

Radioautography

The technique of light microscopic radioautography has been modified for electron microscopy.[9] Electron microscopic radioautography has been applied to ocular tissues,[7,11,17,18] but no examples are described here.

PHOTOGRAPHY

Technique

The photographic techniques used in electron microscopy are generally those of conventional black-and-white photography.

Interpretation

The photograph contains the information. Unlike the light micrograph, in which the observed detail is *less* than can be seen by the examiner, the detail in the electron micrograph is *greater* than can be seen by the unaided eye of the examiner on the fluorescent screen. This higher information content in the micrograph is due to the combination of the extreme thinness of the tissue sections and the high resolution and depth of field of the electron microscope.

The information in the photograph is in shades of black and white which indicate relative tissue densities to the electron beam. Structures, therefore, are described as electron-dense or electron-lucent, or simply as dense or lucent. In experiments *not* employing contrast enhancements, the relative tissue densities remain quite constant and therefore are believed to reflect the in vivo situation. Thus, a very lucent tissue or cytoplasm (within limits) might be interpreted as "watery." Intracellular particles may vary from being extremely electron-dense to being almost invisible ("negative staining").

Note in Figure 1–12 the ganglion cell cytoplasm (*GCY*) present on the lower left, and a number of bipolar cells (*BIP*) on the right that are surrounded by the lucent cytoplasm of the Müller cells (*MC*). Above are the complex interdigitating neurites of the inner plexiform layer (*IPL*). Two lines (*arrows*) cross the picture diagonally. They are scratches in the tissue section produced by defects in the sharp cutting edge of the knife. These microscratches are analogous to the larger ones of light microscopy. Note also that the nuclei are dense. A nucleus, however, is less dense than its nucleolus (*NCL*). The other nuclei in the photograph do not necessarily lack nucleoli; these may lie in another plane of section in such thin slices. The small dense bodies (*DB*) in the ganglion cell cytoplasm

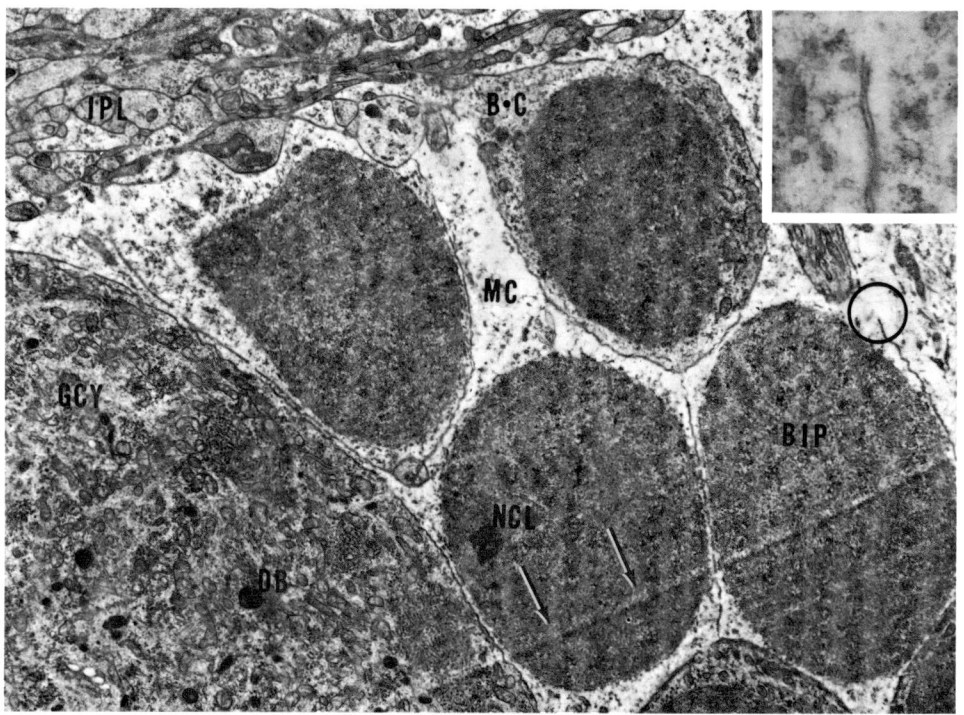

FIG. 1–12. Sample electron micrograph from innermost bipolar cell layer of retina, illustrating some problems in interpretation. See text for explanation of abbreviations. *Arrows* point to knife marks, or microscratches, across tissue section. Note break in plasma membrane of adjacent cells illustrated in *encircled area* and again in enlarged **inset.** Such sharp fractures of surface membranes occur during fixation and/or tissue processing. Lucency of cytoplasm in immediate vicinity of plasma membrane fracture also indicates artifact of preparation. **Main figure,** ×5,500; **inset,** ×25,000.

are more dense than the bipolar cell nucleolus. The cytoplasm of the bipolar cell (B·C) is more dense than the lucent cytoplasm of the adjacent Müller cells. That there are in reality two separate but adjacent cells can only be appreciated at higher magnification of the encircled area (**inset**), which shows the two adjacent cell plasma membranes together with a uniform ("typical") intercellular space.

Misinterpretation of electron micrographs can be easily avoided if the basic principles of cells, cell membranes, intercellular spaces, and planes and thinness of sectioning are constantly kept in mind. For example, in Figure 1–13 continuity of the extracellular collagen (CO) with the cytoplasm of the cell is apparent. Closer examination, however, reveals that (1) the plasma membranes of the cell have been sectioned obliquely at the *free arrows*, whereas they have been sectioned normally at N; (2) the spaces around the collagen fibrils are much less dense than the cytoplasm of the cell, which is quite homogeneous, clearly indicating that these fibrils lie in a different matrix or milieu (in this case extracellular); (3) although it is possible that the extracellular fibrils are being produced by adjacent cells, the evidence must, of necessity, be indirect, since no formed or partially formed fibrous structure is present within the cytoplasm, and no filament or fibril can be found traversing a *normally* sectioned plasma membrane.

The appearance of the cell varies with the type of fixative used.[15] For example, nuclear chromatin aggregation is characteristic of initial exposure to glutaraldehyde (Fig. 1–14A), while a more homogeneous appearance is produced by initial exposure to Dalton's chrome-osmium fixative (Fig. 1–14B). Another example is the appearance of cytoplasmic tubules in some cells treated initially with glutaraldehyde (Fig. 6–54) compared with cytoplasmic filaments in the same cells exposed initially to an osmium tetroxide fixative (see Chaps. 6 and 12).

What can be observed at any one time is limited by the volume of tissue examined (Fig. 1–15). Conventional light microscopic sections with a thickness of 8 to 10 μ have a surface area limited only by the size of the original specimen or the size of the supporting slide. Thin light microscopic sections*

---

* Photographs of thin light microscopic sections are designated throughout this monograph as "1.5-μ sections," and when stained with paraphenylenediamine, are also designated as "PD." All other light microscopic sections are of the 8- to 10-μ thick conventional type, with the staining technique specified: H&E, hematoxylin and eosin; PAS, periodic acid–Schiff; Masson, Masson trichrome. All illustrations are from human material unless otherwise specified.

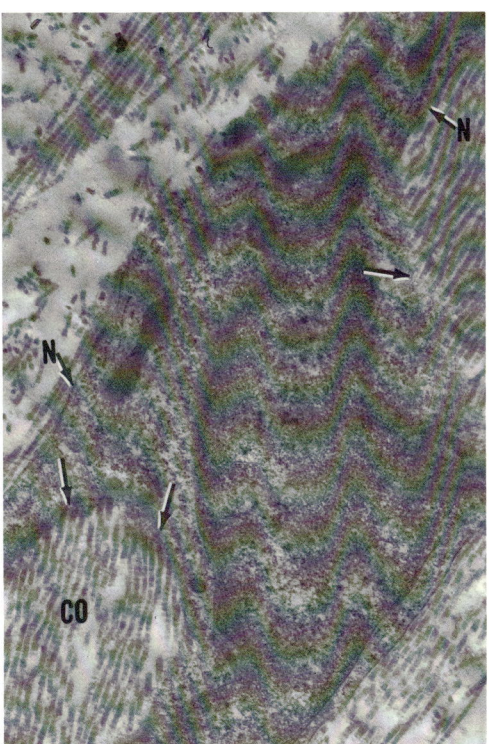

FIG. 1–13. Electron micrograph illustrating problems in interpretation. Oblique sectioning (at *free arrows*) of cell lying in matrix of collagen fibrils (CO) gives false impression that collagen is either within cell or being directly extruded from it. Appearance of correctly sanctioned plasma membrane (N arrows) refutes this interpretation. ×10,800.

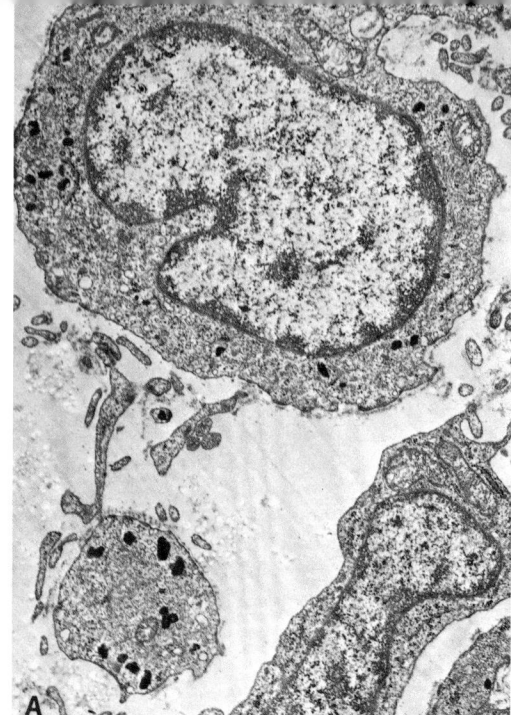
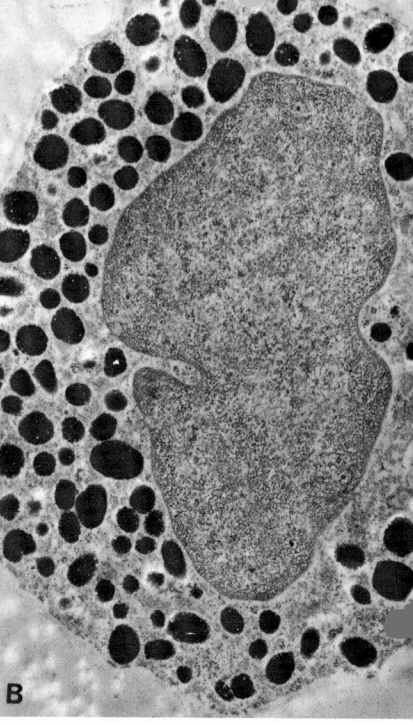

FIG. 1–14. Effect of fixative on cell appearance. Initial fixation with glutaraldehyde (**A**) produces characteristic clumping of nuclear chromatin, while initial fixation in osmium tetroxide (**B**) produces a more homogeneous-appearing nucleus. Iris stroma cells; **A,** ×4,000; **B,** ×12,000.

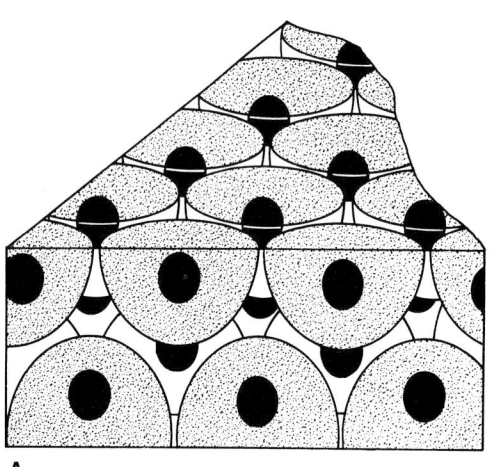
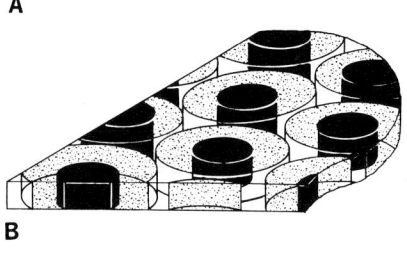
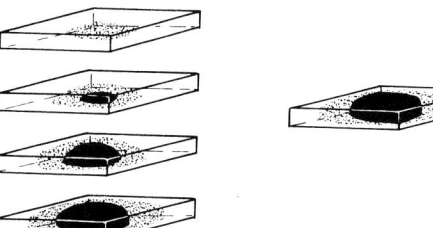

FIG. 1–15. Schematic representation of relative volumes of tissues examined by (**A**) conventional light microscopic section (∼ 8 to 10 μ in thickness), (**B**) thin epon-embedded sections for light microscopy (∼ 1 to 2 μ in thickness), and (**C**) thin epon-embedded sections for electron microscopy (∼ 200 to 600 A in thickness).

with a thickness of 1 to 1.5 μ have a surface area generally limited to the size of the small fragment of embedded tissue. The thin sections for electron microscopy, 200 to 600 A in thickness, have a surface area limited to a few square millimeters at most.

The thin tissue sections embedded in plastic for light microscopic examination may be stained by such dyes as paraphenylenediamine,[4] toluidine blue,[16] methylene blue,[12] and hematoxylin-phloxin.[12] These sections are useful for preliminary evaluation of the tissue prior to electron microscopic examination and may also provide better two-dimensional detail than can be seen in conventional sections for light microscopy.

Further examples of correct interpretation, or of misinterpretation, of electron micrographs are noted throughout the text.

REFERENCES

1. Albert, E. N., and Fleischer, E. A new electron-dense stain for elastic tissue. *J Histochem Cytochem* 18:697, 1970.
2. Berkow, J. W., and Fine, B. S. Glycogen in normal human iris pigment epithelium. *Amer J Ophthal* 69:994, 1970.
3. Dalton, A. J. A chrome osmium fixative for electron microscopy. *Anat Rec* 121:281, 1955.
4. Estable-Puig, J. F., Bauer, W. C., and Blumberg, J. M. Paraphenylenediamine staining of osmium-fixed, plastic-embedded tissue for light and phase microscopy. *J Neuropath Exp Neurol* 24:531, 1965.
5. Fine, B. S. Free-floating pigmented cyst in the anterior chamber. *Amer J Ophthal* 67:494, 1969.
6. Fine, B. S., Tousimis, A. J., and Zimmerman, L. E. Some general principles of electron microscopy. *Arch Ophthal (Chicago)* 62:931, 1001, 1959.
7. Herron, W. L., Riegel, B. W., Myers, O. E., and Rubin, M. L. Retinal dystrophy in the rat: A pigment epithelial disease. *Invest Ophthal* 8:595, 1969.
8. Kay, D. (ed.). *Techniques for Electron Microscopy,* ed. 2. Philadelphia, Davis, 1965.
9. LeBlond, C., and Warren, K. B. (eds.). *The Use of Radioautography in Investigating Protein Synthesis.* New York, Academic Press, 1966.
10. Luft, J. H. Improvements in epoxy resin embedding methods. *J Biophys Biochem Cytol* 9:409, 1961.
11. Magalhaes, M. M., and Coimbra, A. Electron microscope radioautographic study of glycogen synthesis in the rabbit retina. *J Cell Biol* 47:263, 1970.
12. Munger, B. L. Staining methods applicable to sections of osmium-fixed tissue for light microscopy. *J Biophys Biochem Cytol* 11:502, 1961.
13. Pease, D. C. *Histological Techniques for Electron Microscopy,* ed. 2. New York, Academic Press, 1964.
14. Seligman, A. M., Hanker, J. S., Wasserkrug, H., Dmochowski, H., and Katzoff, L. Histochemical demonstration of some oxidized macromolecules with thiocarbohydrazide (TCH) or thiosemicarbazide (TSC) and osmium tetroxide. *J Histochem Cytochem* 13:629, 1965.
15. Trump, B. F., and Ericsson, J. L. E. The effect of the fixative solution on the ultrastructure of cells and tissues. *Lab Invest* 14:1245, 1965.
16. Trump, B. F., Smuckler, E. A., and Benditt, E. P. A method for staining epoxy sections for light microscopy. *J Ultrastruct Res* 5:343, 1961.
17. Young, R. W., and Bok, D. Participation of the retinal pigment epithelium in the rod outer segment renewal process. *J Cell Biol* 42:392, 1969.
18. Young, R. W., and Bok, D. Autoradiographic studies on the metabolism of the retinal pigment epithelium. *Invest Ophthal* 9:524, 1970.

chapter 2
The Cell and Its Contents

**Nucleus, cytoplasm, and plasma
  membrane
Cytoplasmic organelles
  Endoplasmic reticulum
  Mitochondria
  The Golgi complex
  Centrioles and cilia
  Filaments, microtubules, and ground
    substance
Cytoplasmic inclusions
  Ribosomes
  Glycogen
  Pigment granules
  Lipoidal bodies
  Lysosomes (cytosomes)
  Secretion granules or vacuoles**

Electron microscopy is a method of studying histology at the cytologic level.[3,10] In applying this technique to ocular tissues, therefore, we are concerned not only with histologic terminology but also with cytologic terminology. Where applicable, light microscopic cytologic terminology has been carried over without modification into electron microscopic cytology. On occasion some exclusively electron microscopic terms have been added.

NUCLEUS, CYTOPLASM, AND PLASMA MEMBRANE

When a cell is fixed and stained for light microscopy with such dyes as hematoxylin and eosin (H&E),[11,17] it is clearly shown to contain a nucleus, the dense blue-staining (basophilic or hematoxylinophilic) rather centrally located structure, and a pink-staining (acidophilic or eosinophilic) cytoplasm. The cytoplasmic portion of the living cell appears delicately structured except for a rather narrow uniform zone around the periphery. The nonstructured periphery is called the *ectoplasm* (Fig. 2–1). The greater volume of ap-

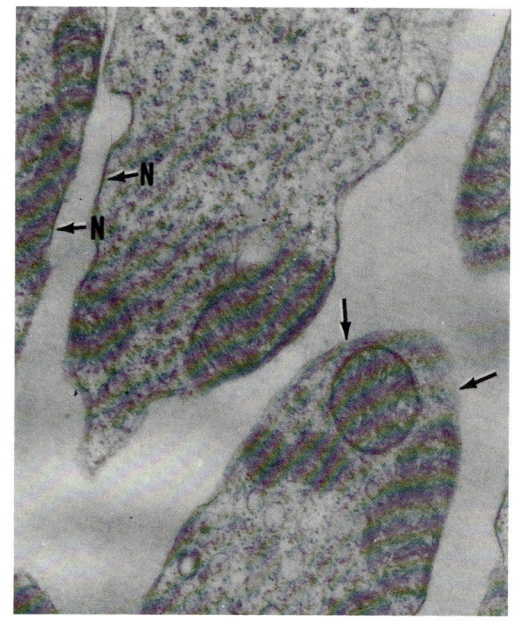

FIG. 2–2. Plasma membrane, sectioned perpendicularly at *N arrows*, appears as a single dense line. Membrane, sectioned obliquely at *free arrows*, appears blurred or indistinct. ×14,000.

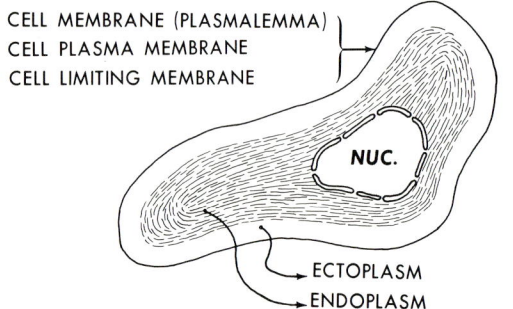

FIG. 2–1. The cytoplasm.

parently structured cytoplasmic protoplasm is called the *endoplasm*. The entire cell is bounded by the plasma membrane (also called cell membrane, cell limiting membrane, or plasmalemma), which cannot be visualized with the light microscope.

With proper fixation and sectioning at right angles to the cell surface, the sharp profile of the plasma membrane appears in an electron micrograph (Fig. 2–2) as a continuous dense line ∼ 75 A in thickness. At higher magnification, however, three further subdivisions of this single membrane are recognizable: a central lucent zone lying between two denser zones, each ~ 25 A in thickness. Each of the three parts is called a *leaflet*. The composite trilaminar membrane is known as a *unit membrane*.[21] With few exceptions, the micrographs presented in this monograph demonstrate these membranes at *low* magnification, and therefore the unit membrane is seen only as a single line (see Chap. 3).

Oblique sectioning of a unit membrane produces a blurred zone of moderate density. If the membrane undulates as it passes in and out of the plane of the thin section, it may seem to appear and disappear (Fig. 2–2).

CYTOPLASMIC ORGANELLES

Endoplasmic Reticulum

Before the development of thin-sectioning techniques a method had to be found which would permit an electron beam to pass through cells. In 1945, Porter et al.[18] examined the attenuated periphery of a monolayer of cultured cells in the electron microscope (Fig. 2–3). They noted that in addition to the already well recognized discrete, thread-like structures, the mitochondria, the peripheral cytoplasm contained a system of interconnecting lines and irregular enlargements forming a cytoplasmic network or reticulum. The network, therefore, was called the *endoplasmic reticulum*. Subsequent development of suitable thin-sectioning techniques revealed that the network was not limited to the endoplasm, that it was not connected to the surface plasma membrane except in unusual circumstances, and that its membrane-bound tubules and cisterns permeated much of the cytoplasm of most cells. In cells engaged actively in the synthesis of certain proteins (e.g., acinar cells of the pancreas),[22] this system appeared "highly organized" or layered, and the membranes were coated on one side with small, dark particles ~ 150 A in diameter (Fig. 2–4).[12,13] The particles (once known as "Palade granules" but now termed *ribosomes*)

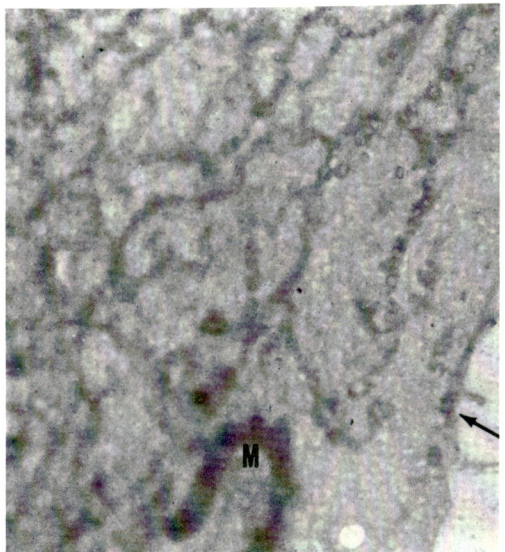

FIG. 2–3. Transmission electron micrograph, showing interconnected bead-like structures (reticulum) within attenuated cytoplasm of a monolayer of cultured cells. *Arrow* points to free surface of cell. M, mitochondrion. (From Porter, K. R., et al. *J Exp Med* 81:233, 1945.)

were demonstrated to contain ribonucleoprotein, the material that accounts for cytoplasmic basophilia. Where the particles occur bound to the cytoplasmic membranes of the endoplasmic reticulum, the reticulum is known as *granular* (rough-surfaced) endoplasmic reticulum (e.g., ganglion cells of the retina, discussed in Chap. 6, and nonpigmented epithelial cells of the ciliary body pars plana, discussed in Chap. 10). The remaining reticulum is known as the *agranular* (smooth-surfaced) reticulum (e.g., retinal pigment epithelium, discussed in Chap. 6).

The system of endoplasmic reticulum ends blindly at the nucleus as several double-membraned arcuate expansions encircling the nucleus (Fig. 2–15) to form the nuclear double membrane (Fig. 2–5). Apertures or *pores* lying between these expansions maintain continuity between the nuclear and cytoplasmic compartments. The nucleus (within the nuclear membrane) divides into two portions: the larger, generally homogeneous part contains the network of chromatin identified as desoxyribonucleic acid (DNA); the smaller well-localized, intensely basophilic part, the nucleolus, has been characterized as mostly ribonucleic acid (RNA). A close relation exists between nuclear RNA and cytoplasmic RNA via small clusters of RNA particles that travel to the cytoplasm via the nuclear pores, the "messenger" (mRNA).

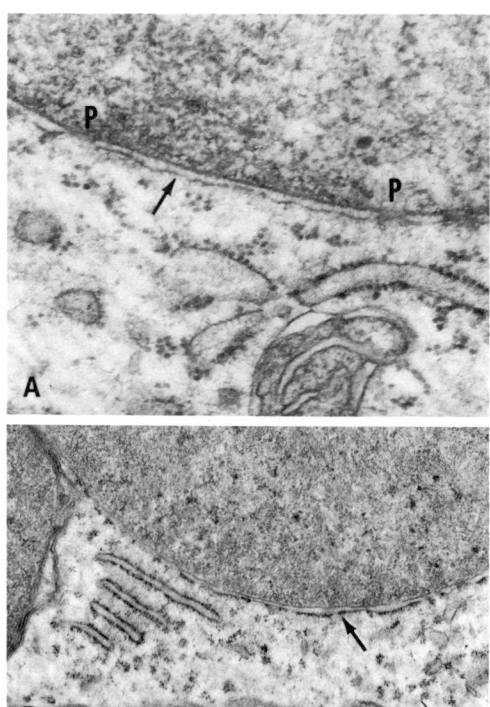

FIG. 2–5. Nuclear double membrane formed by endoplasmic reticulum. In **A** double membrane is agranular (*arrow*); in **B** outer membrane is studded with ribosomes (*arrow*). Continuity between nucleus and cytoplasm is present between expansions of reticulum as nuclear pores (*P*). **A,** ×42,000; **B,** ×16,000.

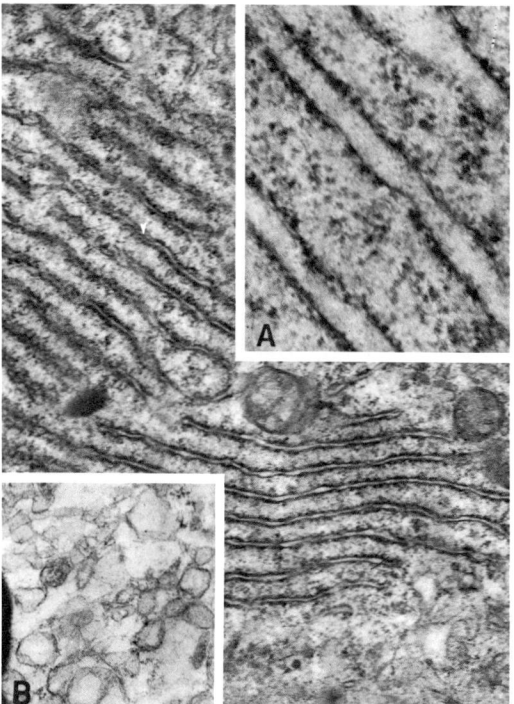

FIG. 2–4. Stacks of organized granular endoplasmic reticulum in cell from nonpigmented ciliary epithelium. This highly organized appearance indicates high degree of synthetic activity by cell. **Inset A** shows ribosomes aligned along one surface of parallel cytoplasmic membranes (Nissl substance of ganglion cell). **Inset B** shows portion of endoplasmic reticulum which is free of granules (agranular or smooth-surfaced) and less well arranged. **Main figure,** ×15,000; **inset A,** ×45,000; **inset B,** ×30,000.

Mitochondria

In addition to the widely distributed endoplasmic reticulum, the cytoplasm contains a number of discrete organelles (Fig. 2–3) previously described by light microscopists as slender thread-like granules—mitochondria. These organelles move about freely in the living cell and can be stained supravitally

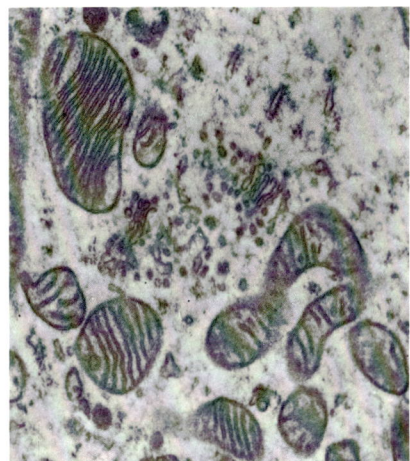

FIG. 2–6. Mitochondria with their characteristic internal double-membraned cristae. ×22,000.

FIG. 2–8. Continuity (at *arrow*) of mitochondrial inner wall and a crista. ×100,000.

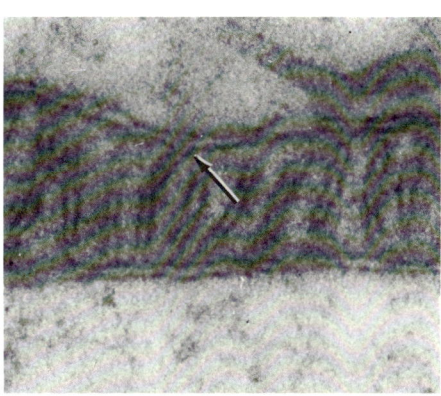

with Janus green. By electron microscopy, in tissue sections[3,10] they generally appear (Fig. 2–6) as elongated, spherical and sometimes branching structures bounded by a double membrane, each of which is a unit membrane (Figs. 2–7 and 3–5; e.g., retinal photoreceptors, Chap. 6). The inner membrane is thrown into folds (Fig. 2–8), forming the cristae mitochondriales or simply *cristae*. The rather uniform *intermembrane space* between the outer membranes is continued into the cristal folds as the *intracristal space*. The homogeneous mitochondrial ground substance or *matrix* occupies all the space between the double-walled cristae and lies therefore in the *intercristal spaces*.[16] A number of discrete, very dense granules (300 to 500 A in diameter) are frequently present within the mitochondrial matrix. They are called simply *mitochondrial* or *matrix* granules.

Mitochondria may differ considerably in appearance from tissue to tissue (compare corneal endothelium with neural retina), varying in both cristal arrangement and length, to density of the intercristal matrix.

The mitochondria function as the main source of cell energy[16] by converting adenosine diphosphate (ADP) to adenosine triphosphate (ATP) by oxidative phosphorylation. Biochemical assay of isolated mitochondrial fragments indicates that the phosphorylating and respiratory enzymes are in general bound to the membranes, while the enzymes of the Krebs citric acid cycle are within the matrix. The mitochondria are therefore sometimes called the "powerhouses" of the cell.

There is evidence that the membrane-bound enzymes are not haphazardly arranged but are properly (morphologically as well as biochemically) sequenced along the membranes. "Elementary particles"[16] (with globular heads 80 to 100 A in diameter), which can be observed by rapid freezing and

MITOCHONDRIA INTERNAL ORGANIZATION

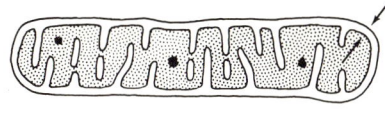

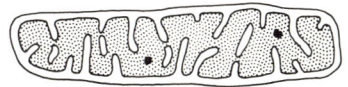

FIG. 2–7. Schematic drawing of a sectioned mitochondrion. Double membrane of outer wall lies between *arrows*.

high-resolution electron microscopy to be attached exclusively to the matrix side of the inner mitochondrial membrane, are also considered possible foci in which enzymes may be assembled to perform specific functions.

The matrix granules are believed to be binding sites for various cations, especially calcium.

Mitochondria readily swell or contract with osmotic changes, and therefore their variable morphologic appearance in this regard is sometimes physiologic, sometimes artifactitious.

The Golgi Complex

Unconnected with the endoplasmic reticulum, the Golgi apparatus or complex is a second, smaller system of agranular, membrane-bound, flattened vesicles (Fig. 2–9).[5] Since the apparatus lacks granules, it does not stain with the usual methods for light microscopy, and in H&E-treated preparations occasionally may be seen as a "negatively stained" area (an unstained area in a region otherwise well stained, e.g., the juxtanuclear "halo" in a plasma cell). In highly oriented or polarized cells such as those in various epithelia, the Golgi apparatus generally is found on the apical side of the nucleus and forms a "crown" to the apical pole of the nucleus. It is recognized in sections as several layers of agranular, flattened vesicles surrounded by clusters of large and small rounded vesicles (e.g., nonpigmented epithelial cells of the ciliary body pars plana, Chap. 10).

Centrioles and Cilia

Within each cell are two short, cylindrical structures (centrioles) usually located in the vicinity of the Golgi apparatus. The region they occupy is known in light microscopy as the centrosome. The centrioles are best observed as two small blue dots when stained with iron hematoxylin. They generally are

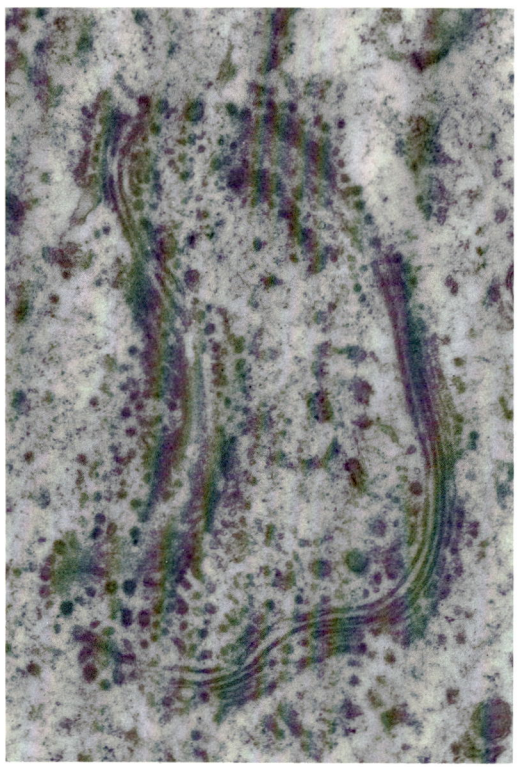

FIG. 2–9. Golgi complex, showing several stacks of agranular membranes and associated free vesicles of varying sizes. ×24,000.

composed of nine short microtubules ("filaments") arranged as a short cylinder.

More detailed investigation has shown that the microtubules are arranged in twos or threes (doublets or triplets). In many epithelia one centriole comes to lie near the apical cytoplasm of the cell (Fig. 2–10), and its tubules become enormously elongated, protruding from the apical end and pushing the apical plasma membrane outward. From the basal end of the centriole one or two rootlets project deep into the apical cytoplasm. These rootlets consist of exceedingly delicate filaments with cross-striations of ~ 600 A. This composite structure is known as a *cilium* (Fig. 2–11) and the *fixed centriole* is known as a *basal body*. The second centriole is not modified and re-

FIG. 2–10. Two centrioles. Centriole at right is unmodified. Centriole at left has become modified into a basal body for a cilium (*arrow*). ×27,000.

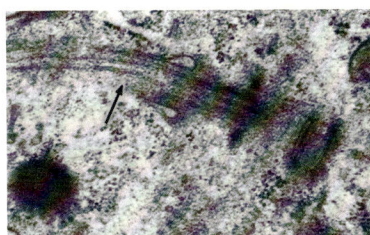

mains as a *free* centriole. Cross-sections of the shafts of cilia generally show nine peripheral filaments. In "motile" cilia two additional filaments are present axially, to produce what is called a 9 + 2 pattern. A few special "non-motile" cilia (e.g., photoreceptors) lack central or axial filaments and have the 9 + 0 pattern when viewed in cross-section.

Filaments, Microtubules, and Ground Substance

Many delicate filaments may be found intracellularly. They range from 50 to 80 A in diameter and vary in concentration from one cell type to another (Fig. 2–12). Epithelial cells, especially those of epidermis or its ocular counterpart, the corneal epithelium (Figs. 3–3 and 9–4), contain large numbers of such delicate intracellular filaments. The filaments are sometimes considered to be a type of "endoskeleton" of the cell, and their relation to other structural configurations (*terminal web, terminal bar,* and *desmosomes*) is noted later. Smooth and striated muscle cells contain special varieties of intracytoplasmic filaments which are so characteristic that the cells can be identified from small fragments.

In addition to filaments that may be present under all conditions of fixation, *microtubules* (~ 200 to 250 A in diameter) often may be found, especially in certain cells which have been *initially* fixed in an aldehyde

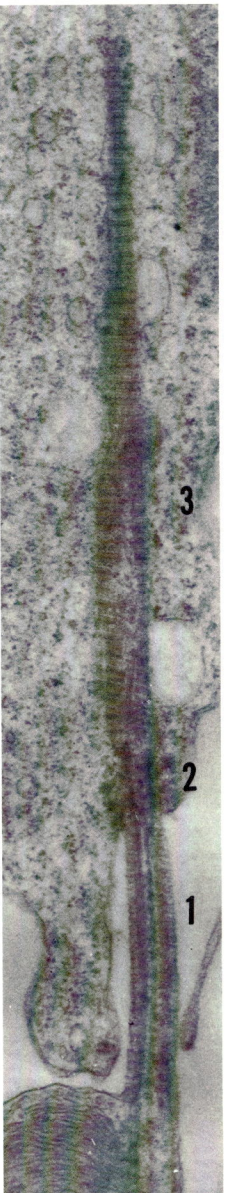

FIG. 2–11. Typical cilium with ciliary extension (*1*), fixed basal body (*2*), and cross-striated rootlets (*3*). Rabbit retina, ×25,000.

(e.g., glutaraldehyde or formaldehyde; Fig. 2–12). The "tubular" structures may therefore be present in either filament form (~ 50 to

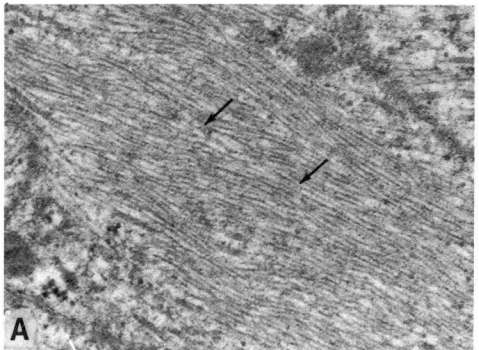 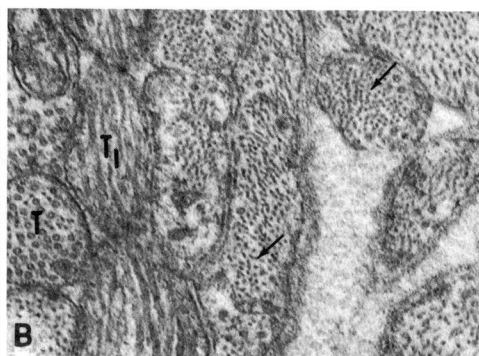

FIG. 2–12. Intracellular filaments and tubules. In **A,** *arrows* indicate longitudinal appearance of filaments. In **B,** *arrows* indicate cross-sectional appearance of filaments. Cells containing microtubules are axons. Tubules are cut in cross-section at T and longitudinally at T_1. Optic nerve, glutaraldehyde fixation; **A,** ×30,000; **B,** ×36,000.

80 A in material initially fixed in osmium tetroxide) or microtubule form (∼ 200 to 250 A in material initially fixed in glutaraldehyde) depending upon the fixative initially used. In certain ocular tissues (e.g., the retina) such variations can be produced readily.

Excluding the free ribonucleoprotein particles (mRNA, ribosomal rosettes or polysomes) and the glycogen particles (∼ 300 A in diameter), the remainder of the cytoplasmic substance generally appears as a watery (i.e., lucent) material containing a very diffuse grayish substance. This "leftover" watery region is referred to as the cytoplasmic *ground substance.*

CYTOPLASMIC INCLUSIONS

Ribosomes

Free clusters of ribonucleoprotein particles (ribosomes-RNP), measuring ∼ 150 A, may be present in the ground substance of the cytoplasm, sometimes in the form of small rosettes (polysomes). Cytoplasmic basophilia is due to the presence of the ribosomes.

Glycogen

The appearance of glycogen in an electron micrograph varies with the cell type and with the method of fixation.[2,15,19] With good fixation, the irregularly shaped particles average 200 to 300 A in diameter (thus generally larger than ribosomes) and are sometimes called *beta particles.* The beta particles are distinguished from clusters of such particles, which are called *alpha particles* or rosettes.

If unstained, aggregates or clusters of the particles may be recognized by their "negative" appearance. Treatment of a tissue section with a lead stain (e.g., lead hydroxide) accentuates the density of these particles and facilitates their identification. They can also be demonstrated by the thiosemicarbazide (TSC) method (Fig. 1–11*A*). Diastase digestion may be necessary at the electron microscopic level to identify individual or widely scattered particles of glycogen (Fig. 1–11*B*).

Pigment Granules

In normal ocular tissue, two varieties of pigment granules can be seen by light microscopy. The most obvious granule is melanin, which is extremely dense to both light and electron microscopy. The less obvious granule is lipofuscin. Melanin granules are found within many of the cells of the neuroepithelial layers and within the pigment-bearing cells of the uveal tract. The melanin granules of these two different ocular coats are easily distinguished from each other when viewed by light microscopy because those of the neuroepi-

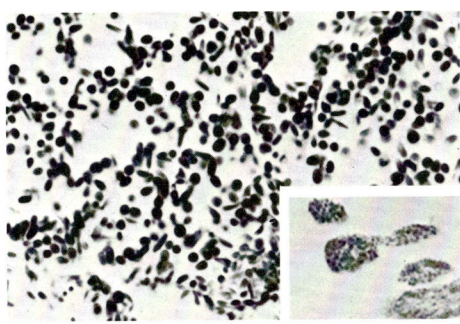

FIG. 2–13. Light micrographs of melanin granules of neuroepithelium (**main figure**) and of uveal tract (**inset**). Note that neuroepithelial granules are larger and appear more dense. Both sections 1.5 μ, ×1,000. (AFIP Neg. 68–9944.)

thelium are larger and therefore appear darker than those of the uveal tract (Fig. 2–13). In fact, the neuroepithelial granules are large enough (∼ 1 μ) to be observed as discrete by oil-immersion light microscopy. Those of the uveal tract lie close to the limits of resolution (∼ 0.3 to 0.5 μ) of the light microscope. Melanin granules are also easily recognized by electron microscopy (Fig. 2–14) because of their exceedingly dense appearance, their regular contours, and, with glass knife sectioning, their often "fractured" appearance. Diamond knives generally cut the melanin granules more smoothly. The second variety of granule (lipofuscin) may be confused with the melanin granule because of its slightly brownish color by light microscopy and its proximity to the melanin granule. By electron microscopy the lipofuscin granules are more easily sectioned and are less dense (Fig. 6–9). They appear more numerous in certain cells as the organism ages (i.e., few are found in infant eyes, more in young adult eyes, and most in aged eyes).

Other pigments may be present, such as hemoglobin in red blood cells (Fig. 6–64). Products of hemoglobin (hemosiderin, ferritin) or iron deposits of foreign origin, often found in pathologic tissue, are not considered here.

Lipoidal Bodies

Normal cells may contain fat droplets (lipoidal bodies) which appear by electron microscopy to be greatly varied. Some are homogeneous while others show lamellar configurations. Some have less density; others have greater density; some may even appear to lose their central substance in the process of fixation and staining.

Lysosomes (Cytosomes)

A heterogeneous group of "dense bodies" (actually variable in appearance and density) is found in many cells throughout the body. Attempts have been made to characterize these bodies according to their function. One variety, originally fractionated from rat liver cells, was found on biochemical assay to contain acid hydrolases or "lytic" enzymes; hence the term "lysosome." Although originally proposed as the place where cell autolysis[6] (e.g., "suicide bags") begins, the dense

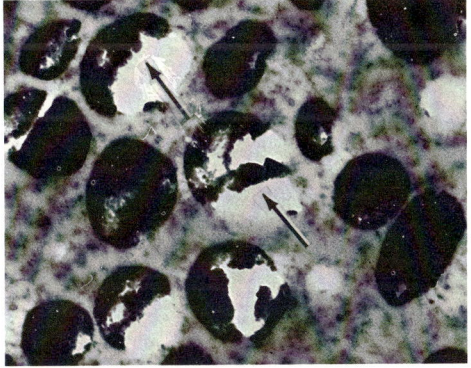

FIG. 2–14. Electron micrograph of neuroepithelial melanin granules. Note extreme density of granules, their circular and oval profiles, and their often "fractured" appearance (*arrows*). ×10,000.

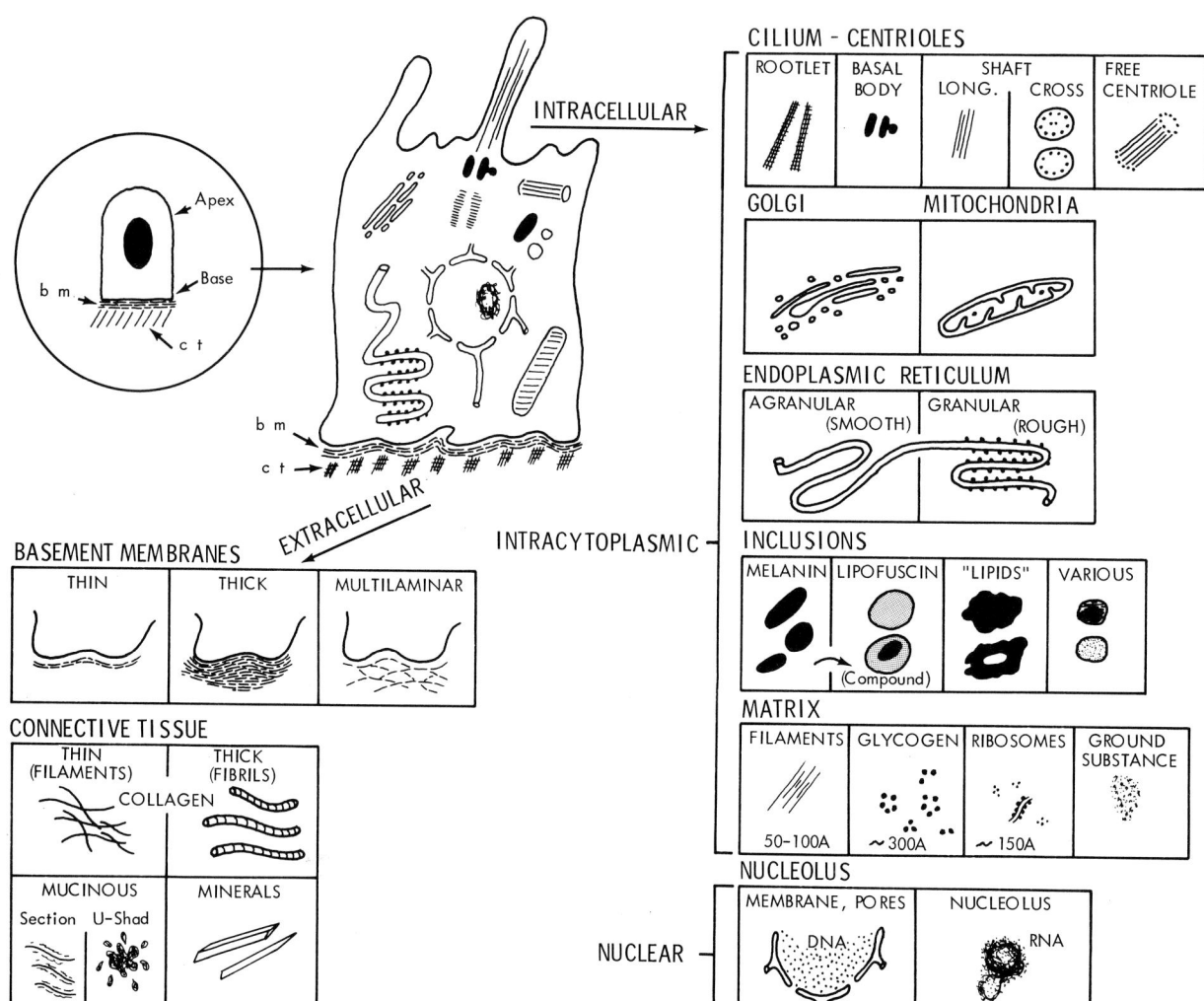

FIG. 2–15. Schematic drawing of the cell, its components (organelles), its contents (inclusions) and relationship to extracellular materials.

bodies are now thought, on morphologic grounds, to be a heterogeneous group and unlikely to function in all cells in the same way; hence the nonspecific term, *cytosome*.[7,23] A lysosome can be identified in a tissue section only by histochemical means.[9] Morphologic identification in situ, therefore, is generally limited to the histochemical demonstration of preferably two or more hydrolases in the same location.[9] Biochemical assay is generally carried out on relatively homogeneous preparations of the bodies collected by either differential or zonal centrifugation.

The dense bodies or "primary" lysosomes generally have a single limiting membrane and are usually PAS positive, exhibit metachromasia, and often stain for acid mucopolysaccharides.

"Secondary" lysosomes is a term generally applied to *digestive* vacuoles which may be present in a cell. Such vacuoles contain material which may be recognized as having been taken in from outside the cell (phagocytosis or heterophagocytosis) or as material actually derived from parts of its own cytoplasm, e.g., mitochondria (autophagocytosis).

In many ocular tissues (especially ganglion

cells of the retina) dense cytoplasmic bodies of unknown composition appear more dense with age and are neither PAS-positive by light microscopy nor the equivalent TSC-positive by electron microscopy. The dense bodies in the ganglion cells may possibly function as "wastebaskets" for materials that cannot be removed in any other way. Some of the increase in density with age may therefore be due to increasing polymerization of these sequestered materials.

The granules of such cells as blood neutrophils contain a variety of enzymes and therefore are generally considered good examples of "true," i.e., primary, lysosomes.

Other dense cytoplasmic bodies, differing in appearance, can be observed in various tissues and are sometimes named "microbodies."[1,8,20] Their functions are not known.

Secretion Granules or Vacuoles

Secretory cells[14] synthesize a "useful" material to be extruded to the exterior, via a lumen as in exocrine cells or directly into the tissue fluids or blood stream as in endocrine cells. The cells produce materials that range from mucous vesicles of the salivary glands to the denser, enzyme-packed granules of the pancreas. The secretory materials are thought to be synthesized at least partially within the highly organized forms of granular endoplasmic reticulum.[4] They are transferred to the Golgi apparatus, presumably by isolated vesicles, where they are either concentrated or chemically altered, or both. Subsequently, they are extruded to the exterior, generally through the plasmalemma of the cell via the apical cytoplasm.

Figure 2–15 summarizes schematically the structure of the cell and its relation to the various extracellular materials (Chap. 4).

REFERENCES

1. Afzelius, B. A. The occurrence and structure of microbodies: A comparative study. *J Cell Biol* 26:835, 1965.
2. Biava, C. Identification and structural forms of human particle glycogen. *Lab Invest* 12:1179, 1963.
3. Bloom, W., and Fawcett, D. W. *A Textbook of Histology*, ed. 9. Philadelphia, Saunders, 1968.
4. Caro, L. G. Electron microscopic radioautography of thin sections: The Golgi zone as a site of protein concentration in pancreatic acinar cells. *J Biophys Biochem Cytol* 10:37, 1961.
5. Dalton, A. J. Golgi Apparatus and Secretion Granules. In Brachet, J., and Mirsky, E. A. (eds.). *The Cell: Biochemistry, Physiology, Morphology*, vol. 2. New York, Academic Press, 1961, pp. 603–619.
6. deDuve, C. Lysosome, A New Group of Cytoplasmic Particles. In Hayashi, T. (ed.). *Subcellular Particles*. New York, Ronald Press, 1959, pp. 128–159.
7. Ericsson, J. L. E., Trump, B. F., and Weibel, J. Electron microscopic studies of the proximal tubule of the rat kidney. II. Cytosegrosomes and cytosomes: Their relationship to each other and to the lysosome concept. *Lab Invest* 14:1341, 1965.
8. Ericsson, J. L. E., and Trump, B. F. Electron microscopic studies of the epithelium of the proximal tubule of the rat kidney. III. Microbodies, multivesicular bodies, and the Golgi apparatus. *Lab Invest* 15:1610, 1966.
9. Gahan, P. D. Histochemistry of lysosomes. *Int Rev Cytol* 21:1, 1967.
10. Ham, A. W., and Leeson, T. S. *Histology*, ed. 4. Philadelphia, Lippincott, 1961.
11. Luna, L. G. (ed.). *Manual of Histologic and Special Staining Technics*, ed. 3. New York, McGraw-Hill, 1968.
12. Palade, G. E. A Small Particulate Component of the Cytoplasm. In Palay, S. L. (ed.). *Frontiers in Cytology*. New Haven, Yale University Press, 1958.
13. Palade, G. E., and Siekevitz, P. Pancreatic microsomes: An integrated morphological and biochemical study. *J Biophys Biochem Cytol* 2:671, 1956.
14. Palay, S. L. The Morphology of Secretion. In

Palay, S. L. (ed.). *Frontiers in Cytology.* New Haven, Yale University Press, 1958.

15. Paluello, M., and Rosati, G. The influence of fixation and dehydration on the isolated glycogen. *J Micr* 7:275, 1968.
16. Parsons, D. F. Recent advances correlating structure and function in mitochondria. *Int Rev Exp Path* 4:1, 1965.
17. Pearse, A. G. E. *Histochemistry, Theoretical and Applied,* ed. 2. Boston, Little, Brown, 1960.
18. Porter, K. R., Claude, A., and Fullam, E. F. A study of tissue culture cells by electron microscopy: Methods and preliminary observations. *J Exp Med* 81:233, 1945.
19. Revel, J. P. Electron microscopy of glycogen. *J Histochem Cytochem* 12:104, 1964.
20. Rhodin, J. *Correlation of Ultrastructural Organization and Function in Normal and Experimentally changed Proximal Convoluted Tubule Cells of the Mouse Kidney.* Stockholm, Godvil, 1954.
21. Robertson, J. D. Unit Membranes: A Review with Recent New Studies of Experimental Alterations and a New Subunit Structure in Synaptic Membranes. In Locke, M. (ed.). *Cellular Membranes in Development.* New York, Academic Press, 1964, p. 1.
22. Siekevitz, P., and Palade, G. E. A cytochemical study on the pancreas of the guinea pig. V. In vivo incorporation of leucine-1-C^{14} into the chymotrypsinogen of various cell fractions. *J Biophys Biochem Cytol* 7:619, 1960.
23. Trump, B. F. An electron microscopic study of the uptake, transport, and storage of colloidal materials by the cells of the vertebrate nephron. *J Ultrastruct Res* 5:291, 1961.

chapter 3

Cell Interrelations

The Intercellular Space
The Desmosome
The Terminal Bar

THE INTERCELLULAR SPACE

In electron micrographs the intercellular space is seen as a rather uniform (~ 150 A) lucent space between the dense plasma membranes of adjacent cells (Fig. 3–1).[3] Except for some modification (e.g., nonpigmented ciliary epithelium) in normal tissues and much more gross exaggerations in pathologic tissues,[9] this space is remarkably uniform and probably represents the in vivo situation closely.

THE DESMOSOME

In any plane from apex to base (Fig. 3–2), focal densities (*desmosomes*)[5,7] may connect adjacent cells, especially in certain epithelia

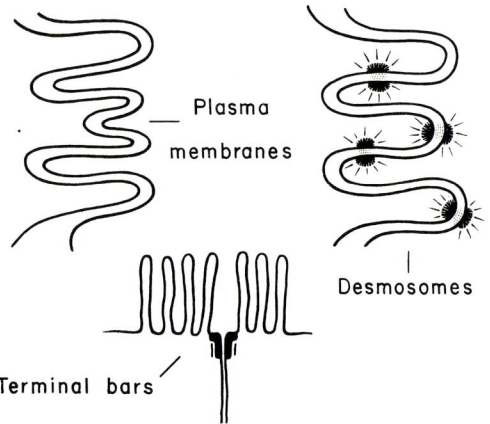

FIG. 3–2. Schematic drawing showing a typical intercellular space, desmosomes distributed along the intercellular space, and an apical terminal bar attachment in cross-section.

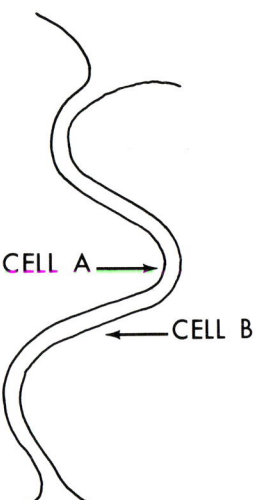

FIG. 3–1. Typical uniform intercellular space bounded by limiting plasma membranes of adjacent cells A and B.

(e.g., epidermis,[6] corneal epithelium, neuro-epithelial layers). Long appreciated by light microscopists as small attachment foci on the intercellular bridges of the epidermal prickle or squamoid cell layer (as seen with iron hematoxylin stain), the desmosomes partially occlude the intercellular space and function as strong focal attachments between adjacent cells. The attachments are so strong that the cell generally ruptures long before the desmosome can be made to separate by mechanical means.

On electron microscopic examination, a desmosome in cross-section consists of apposing plasma membranes that are of greatly increased density and separated by a slightly widened intercellular space containing a *cement substance* of moderate electron density (Fig. 3–3). The adjacent cell cytoplasm is

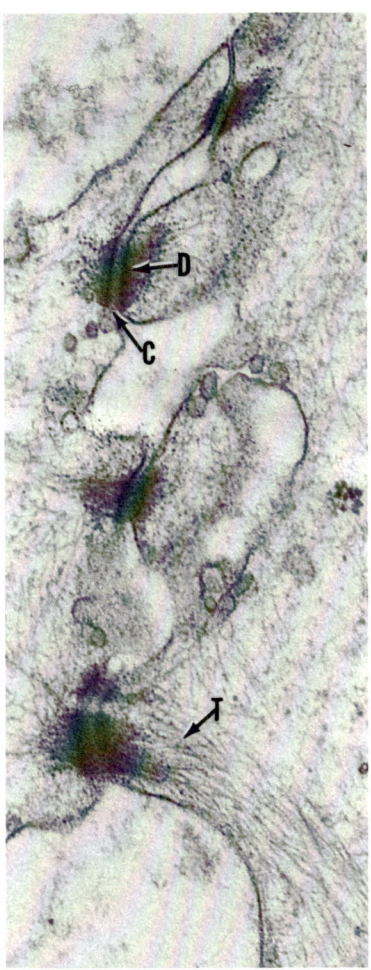

FIG. 3–3. Typical desmosomes (focal attachments or maculae adherentes). *C*, cement substance; *D*, plasmalemmal densities; *T*, tonofibrils. Rabbit corneal epithelium, ×39,000.

testinal epithelium, retinal pigment epithelium) have long been appreciated by light microscopists, especially in sections stained with iron hematoxylin. The attachment zones form complete encircling girdles around the epithelial cells so that on surface view they often resemble "chickenwire" and were once called *fenestrated membranes*. The outer limiting membrane of the retina (Figs. 6–32 and 6–33) is a good example of such a "fenestrated membrane." The girdle-like attachments resemble desmosomes in that they attach adjacent cells to each other, but they differ somewhat from desmosomes in their location, which is always apical, and in their electron microscopic structural appearance (Fig. 3–4).

The terminal bar seen by light microscopy to be characteristic of various epithelia can be shown by electron microscopy to consist of two parts.[2] The basilar portion (*zonula adherens*) closely resembles a desmosome in cross-section except for its slightly narrower intercellular space and its encircling arrangement. The apical portion (*zonula occludens*)

focally increased in density. Clusters of intracytoplasmic filaments, known as *tonofibrils*, radiate toward these focal areas. In three dimensions, the desmosome appears as a small oval plaque. More recent terminology designates these foci of attachment as *maculae adherentes*.[2]

THE TERMINAL BAR

Attachment densities lying in a single plane near the lumen of various epithelia (e.g., in-

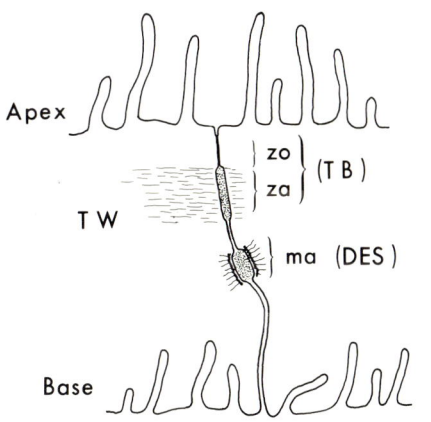

EPITHELIUM - JUNCTIONAL COMPLEX

FIG. 3–4. A junctional complex consisting of a complete terminal bar (*TB*) composed of zonula occludens (*Zo*) and zonula adherens (*Za*), together with associated desmosome (*DES*), i.e., macula adherens (*ma*). *TW*, terminal web.

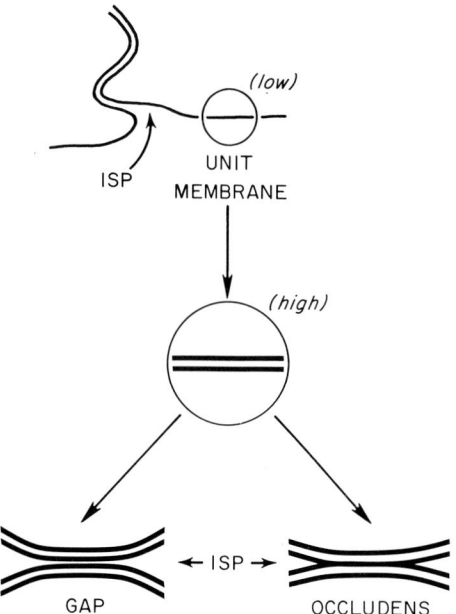

FIG. 3–5. Schematic drawing showing trilaminar appearance of unit membrane at high magnification. An occludens junction is pentalaminar; a gap junction septalaminar. *ISP,* intercellular space.

consists of dense adjacent cell membranes so closely apposed that the intercellular space appears more or less obliterated.* No adjacent cytoplasmic densities accompany this apical part of the terminal bar.

The nonspecific term, *junctional complex,* occasionally is used for the apical attachment zones and may include a third part, a typical desmosome or *macula adherens,* on the basilar side of the bipartite arrangement. Then the complex becomes a tripartite arrangement.

These arrangements may have some variability (compare the fenestrated membrane of the retinal pigment epithelium and the external limiting membrane of the retina; Chap. 6).

In locations where such attachments appear morphologically similar but do not form complete girdles, the terms *fascia occludens* and *fascia adherens* have been proposed (e.g., ciliary epithelium, Chap. 10).

REFERENCES

1. Brightman, M. W., and Reese, T. S. Junctions between intimately apposed cell membranes in the vertebrate brain. *J Cell Biol* 40:648, 1969.
2. Farquhar, M. G., and Palade, G. E. Junctional complexes in various epithelia. *J Cell Biol* 17:375, 1963.
3. Fawcett, D. W. Structural Specializations of the Cell Surface. In Palay, S. L. (ed.). *Frontiers in Cytology.* New Haven, Yale University Press, 1958.
4. Goodenough, D. A., and Revel, J. P. A fine structural analysis of intercellular junctions in the mouse liver. *J Cell Biol* 45:272, 1970.
5. Kelly, D. E. Fine structure of desmosomes, hemidesmosomes and an adepidermal globular layer in developing newt epidermis. *J Cell Biol* 28:51, 1966.
6. Odland, G. F. The fine structure of the interrelationship of cells in the human epidermis. *J Biophys Biochem Cytol* 4:529, 1958.
7. Overton, J. Desmosome development in normal and reassociating cells in the early chick blastoderm. *Develop Biol* 4:532, 1962.
8. Robertson, J. D. Unit Membranes: A Review with Recent New Studies of Experimental Alterations and a New Subunit Structure in Synaptic Membranes. In Locke, M. (ed.). *Cellular Membranes in Development.* New York, Academic Press, 1964, p. 1.
9. Zimmerman, L. E., and Fine, B. S. Production of hyaluronic acid by cysts and tumors of the ciliary body. *Arch Ophthal (Chicago)* 72:365, 1964.

---

* On the basis of electron microscopic appearance, a further distinction can be made for this apical part of the terminal bar. If the outer leaflets of adjacent cell *unit membranes*[8] (Fig. 3–5) can be demonstrated to fuse (i.e., to form a pentalaminar configuration), the junction is truly occlusive. If, however, the outer leaflets of adjacent cell unit membranes come very close together but do not fuse, the intercellular space is not completely obliterated (i.e., a gap of 20 to 30 A persists). The term *gap junction*[1,4] is then applied to this seven-layered configuration, and the space that is observed morphologically also represents some functional continuity with the adjacent intercellular space. Gap junctions have been observed between ependymal cells, between astrocytes, and between some electrically coupled neurons.[1] Junctions which persist with more than one method of fixation are considered to be more "real" than "tight" junctions seen with only one method of fixation. Some of the spurious tight junctions are considered to be mere appositions of adjacent cell membranes and are called *labile appositions.*[1]

chapter 4
Extracellular Materials

Fibrous materials: Collagen and reticulin
Mucinous materials: Polysaccharides and glycoproteins
Crystalline materials: Minerals
Elastic tissue: Microfibrils, Elastin
Basement membranes: Thick, Thin, Multilaminar
Effect of cell growth and differentiation on configuration of cell membrane

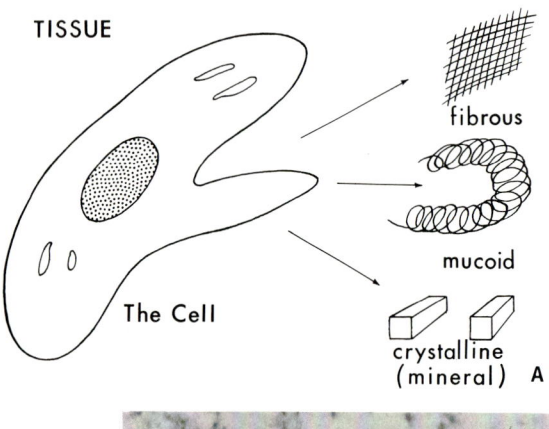
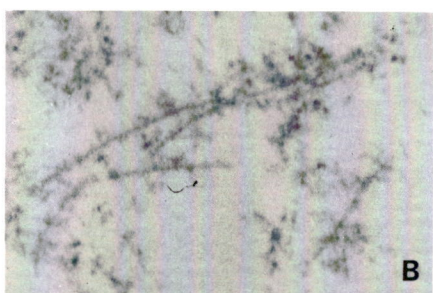

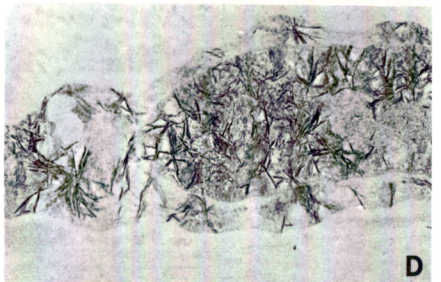

An aggregate of cells together with their extracellular materials constitutes a tissue (Fig. 4–1). The extracellular materials are separable morphologically into at least five groups: fibrous materials, mucinous materials, crystalline materials, elastic tissue, and basement membranes.

FIBROUS MATERIALS: COLLAGEN AND RETICULIN

The fibrous components of ocular tissues consist of bundles of macromolecules composed of linear arrangements of amino acids. Amino acids, having rather rigid "backbones" to their molecules, produce in turn rigid macromolecules which, in aggregates, form rigid filaments, fibrils, and fibers (Fig. 4–2).

Collagen is the most ubiquitous fibrous component (Gr. *kolla* glue plus Gr. *gennan* to produce). It is characterized by its morphology, its x-ray diffraction pattern, and its amino acid composition (especially its content of hydroxyproline and hydroxylysine).[5]

X-ray diffraction studies show that "native" collagen (i.e., that which is derived from

FIG. 4–1. Components of tissue. **A.** Tissue consists of cells surrounded by extracellular materials. Electron microscopy visualizes fibrous extracellular materials as filaments (**B**) or fibrils (Fig. 4–3), mucinous materials as fluffy aggregates in thin sections (**B**) or as irregular drying patterns in shadow-cast preparations (**C**), and minerals as crystalline structures (**D**). **B**, ×32,000; **C**, ×16,000; **D**, ×21,000.

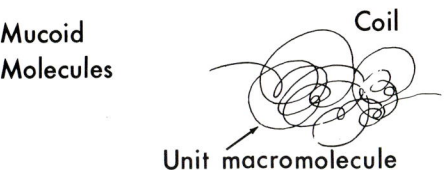
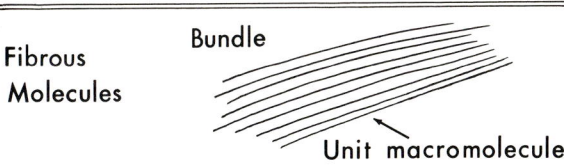

FIG. 4–2. Schematic drawing of unit macromolecules and their aggregate forms.

natural sources such as tendons in contradistinction to collagen produced in vitro) produces a repeating pattern of ~ 640 A. The pattern is almost identical with a periodicity (*banding*) observed in the fibrils by electron microscopy (Fig. 4–3), a periodicity which has come to be considered characteristic of collagen. Further subdivision of the periodic bands (intraperiod banding) can also be observed.

Fibrous structures of smaller diameter often are seen in fetal tissues, in cultures of young fibroblasts, and in the vitreous body. Although they all possess a smaller periodicity (~ 220 A), their amino acid composition is appropriate for collagen. The smaller diameter structures may be termed *filaments* of embryonic, fetal, or "young" collagen.

The smallest basic unit of a collagen fibril or filament is the tropocollagen molecule. It measures ~ 12 to 14 A in diameter and 2,800 A in length. The molecule is composed of three polypeptide chains arranged in a spiral. Two of the amino acids, hydroxyproline and hydroxylysine, are considered highly characteristic of the collagen molecule, as they are not found elsewhere in significant concentration. The collagen molecule is intimately associated with a small amount of sugar, a glycoprotein from which it cannot be completely separated, suggesting that this small amount of sugar is also an integral part of the collagen molecule. The tropocollagen molecules can either go into solution or align themselves to produce increasingly visible filaments and fibrils of varying diameters and lengths (see Chap. 11 for discussion of in vitro preparations).

In ocular tissues, the term *fibril* has been reserved for those collagens that are of sufficient diameter to bear the 640-A periodicity (e.g., in the cornea, uvea, and sclera). The term *filament* is used for the finer structures characteristic of vitreous framework, and the term *fiber* is used for the more grossly visible structures such as the zonule of the lens.

"Reticulin" is a term for the hypothetical structural component of a *reticulum* (e.g., liver reticulum),[9] a delicate network observed around a variety of cells. Currently, a reticulum or its reticulin is demonstrated by the deposition of silver salts on the delicate extracellular network. By electron microscopy, the fibrous component of the network is seen to vary from "typical" 640-A period collagen *fibrils* (e.g., iris stroma) to "atypical" 220-A

FIG. 4–3. Collagen fibrils as seen by electron microscopy in thin section. Cross-striation or periodicity of ~ 640 A is characteristic of most collagen fibrils. ×25,000.

period collagen *filaments* (e.g., vitreous framework). That the densely packed typical collagen fibrils in the corneal stroma which stain poorly for reticulin can be made to stain well if foci of the stroma are first mechanically disrupted suggests that reticulin is but a mechanically looser arrangement of various collagen fibrils or filaments.

MUCINOUS MATERIALS: POLYSACCHARIDES AND GLYCOPROTEINS

Considerable confusion exists in the literature concerning the exact meaning of the terms mucins, mucoids, mucinous substances, mucosubstances, mucoproteins, mucopolysaccharides, acid mucopolysaccharides, and glycoproteins. All these substances are amino-sugar-containing compounds. Except for glycogen and the nucleic acids, all carbohydrate polymers of higher animals contain amino sugars. For the sake of simplicity, the following terminology will be used in this discussion:

Polysaccharide. A macromolecule composed of monosaccharide units with or without protein attachments.

Homopolysaccharide. A macromolecule composed of many monosaccharide units generally not linked to proteins, e.g., glycogen.

Heteropolysaccharide. A macromolecule composed of a mixture of constituent building blocks, e.g., amino sugars and hexuronic acids in *acid mucopolysaccharides*. The heteropolysaccharides may be weakly linked to proteins.

Glycoprotein.[12] A saccharide, usually an oligosaccharide, that is strongly attached to a protein and forms an integral part of the protein molecule. Examples are basement membranes (lens capsule, Descemet's membrane) and collagen (reticulin).

Acid mucopolysaccharides (AMP) are heteropolysaccharides containing amino sugars and hexuronic acids or sulfate esters or both.[3,7,8,13] In their native state AMP are thought to be associated with proteins by weak ionic linkages or labile covalent bonds. They differ, therefore, from *glycoproteins* in containing hexuronic acids or sulfate esters or both, and in not having oligosaccharide units or firm linkage by covalent bonds to protein. Since AMP are composed of amino sugars and hexuronic acids, they have little intrinsic strength and so lie coiled up within a jacket of water. In tissue, therefore, they form a slippery, loose, watery network between collagen fibrils and filaments, best exemplified in the vitreous body. These networks appear morphologically as indistinct "fluffy" patches on fibrous structures when seen in thin electron microscopic sections (Figs. 4–1B and 4–2). On the surface of shadow-cast, thicker sections from which only the water and various ions have been removed, they appear as "drying patterns" (Fig. 4–1C).

Another characteristic of AMP is that they contain a large number of negatively charged groups or ions. They are, therefore, *polyanionic* and have to carry an equivalent number of cations, usually sodium, called *counterions*. Use is made of the polyanionic nature of AMP in a number of histochemical techniques. The colloidal iron method utilizes a trivalent ferric cationic complex that attaches to the anions of AMP and then is made visible by staining with ferrocyanide (the Prussian blue reaction). Alcian blue is a tetravalent cationic complex that attaches to polyanionic substances and directly colors the material blue. Another histochemical technique which utilizes the polyanionic nature of AMP with the additional prerequisite of sulfate esters is the metachromatic staining* achieved with certain cationic dyes (e.g., toluidine blue, methyl violet, crystal violet). Since hyaluronic

---

* Metachromasia may be defined as the staining of a tissue component so that the absorption spectrum of the resulting tissue dye complex differs sufficiently from that of the original dye, and from its ordinary tissue complexes, to give a marked change in color.

acid does not contain sulfate esters, it does not exhibit metachromasia.

Acid mucopolysaccharides do not react with the PAS stain.[12] For a substance to be PAS-positive, it must have unsubstituted vicinal glycol groups (CHOH-CHOH). The periodic acid oxidizes the groups to dialdehydes (CHO-CHO). The Schiff reagent (leucofuchsin) then reacts with the dialdehydes. The Schiff reagent–dialdehyde complexes are colored red to magenta and are easily identified in tissue sections. The homopolysaccharides (e.g., glucose) have abundant vicinal glycol groups along the course of their long-chain monosaccharide units and are, therefore, PAS-positive. Glycoproteins (e.g., lens capsule) also contain many vicinal glycol groups in the large number of unsubstituted sugars located at the ends of their oligosaccharide chains and are also PAS-positive. Glycogen can be differentiated from glycoprotein by applying diastase to the tissue under consideration. The diastase digests out the glycogen so that the PAS method no longer stains glycogen but continues to color glycoprotein. Other groups of compounds which give a PAS-positive reaction include glycolipids (e.g., gangliosides and cerebrosides) and unsaturated lipids and phospholipids (e.g., sphingomyelin and lipofuscins).

The AMP found in the eye are listed in Table 4–1. Hyaluronic acid is the only AMP found in the vitreous body. The exact composition of the AMP associated with rods and cones is as yet undetermined, but may contain sialic acid.[4]

CRYSTALLINE MATERIALS: MINERALS

Calcification does not occur in the normal eye (except in aging tissues), but it may occur in pathologic tissues. Figure 4–1D shows hydroxyapatite crystals within a cell.

ELASTIC TISSUE: MICROFIBRILS, ELASTIN

Elastic tissue is best observed when it occurs in quantity (i.e., ligamentum nuchae or the internal elastic lamina of larger arteries). It is generally identified in light microscopy by special staining methods (e.g., Verhoeff's or Weigert's stains).

Chemically it differs from collagen by its lower content of hydroxyproline, by the absence of hydroxylysine, by its higher valine content and by its relative chemical inertness (i.e., resistant to hydrolysis by mild acid or alkali). By electron microscopy (see Figs. 10–65, 12–20) it is characterized by two components,[6,10] the fibrillar part or *microfibrils* (~ 110 angstrom diameter) and the homogeneous part *elastin*. It appears to originate early in life in a manner similar to collagen, as recognizable extracellular filaments or *microfibrils* lying adjacent to the plasma membranes of the synthesizing cells (e.g., fibroblasts, smooth muscle cells). The homogeneous component, *elastin*, appears later, within the bundle of microfibrils, producing the composite structure *elastic tissue*. The microfibrils and the elastin differ considerably from each other in their chemical composition. Elastin contains the amino acids desmosine and isodesmosine which are responsible for cross-linking the polymers to produce a special elastic syncytium. The microfibrils lack these two amino acids and are considered to act as a sort of scaffolding which serves to shape the homogeneous deposits of elastin into fibrous or sheetlike arrangements (e.g., in ligaments or in elastic arteries).

BASEMENT MEMBRANES

Associated with, but lying outside, the basal plasma membranes of various epithelia (or endothelia) is the *basement membrane*, a continuous layer of material of moderate electron density. The layer may be so thin (Fig. 4–4) as to be inadequately observable by light microscopy or so thick (Fig. 4–5) that it can be seen with the naked eye. *Thin* basement membranes are separated from the cell plasma membrane by a lucent zone (Figs. 6–59 and 6–60). When the basement membrane

Table 4-1. Acid Mucopolysaccharides of the Eye

Acid mucopolysaccharide	Characteristic						
	Amino sugar	Uronic acid	Sulfate groups	Alcian blue after testicular hyaluronidase	Alcian blue after streptococcal hyaluronidase	Metachromasia	Main location
Hyaluronic acid	Glucosamine	Glucuronic acid	0	Negative	Negative	Negative	Vitreous, cornea, sclera
Chondroitin sulfate A	Galactosamine	Glucuronic acid	1	Negative or Weakly Positive	Positive	Positive	?Cornea, sclera
Chondroitin sulfate B	Galactosamine	Iduronic acid	1	Positive	Positive	Positive	?Cornea, sclera
Chondroitin sulfate C	Galactosamine	Glucuronic acid	1	Negative or Weakly Positive	Positive	Positive	?Cornea, sclera
Chondroitin	Galactosamine	Glucuronic acid	0	Negative	Negative	?	Cornea
Keratosulfate	Glucosamine	(Galactose)	1	?	?	Positive	Cornea
Heparin	Glucosamine	Glucuronic acid, iduronic acid	2½	Positive	Positive	Positive	Mast cells
Heparin monosulfuric acid	Glucosamine	Glucuronic acid, iduronic acid	1	Positive	Positive	Positive	Mast cells
Retinal receptor AMP	?	?	?	Positive	Positive	Positive	Space between retinal receptors

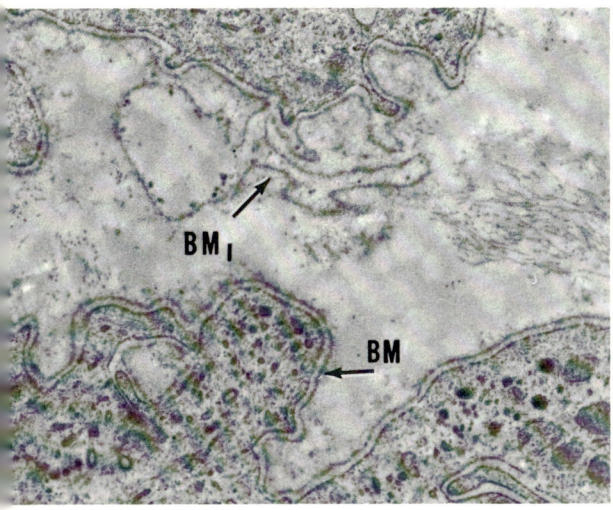

FIG. 4-4. Typical thin basement membrane (BM) normally following contour of cell plasma membrane and separated from it by uniform zone of lucency. In regions of abnormality, thin basement membrane (BM_1) may be markedly folded or convoluted. ×13,500.

is *thick* (Fig. 6–56), this zone is absent, suggesting that the lucent zone consists of a watery material, that some juxtacellular material is lost consistently and uniformly during the processing of the tissue, or that this membrane retracts from the cell during processing.

Some thick ocular basement membranes are formed by an interweaving of multiple thin basement membranes, the composite structure being a *multilaminar* basement membrane. The multilaminar basement membranes may themselves be composed of thickened lamellas (e.g., peripheral corneal epithelium) or of thin lamellas (e.g., internal limiting membrane of the ciliary epithelium).

The thicker ocular basement membranes have long been appreciated by histologists and known by the synonyms "glass membranes" or "cuticular membranes." Many of their chemical and physical properties can be determined from light and electron microscopic studies of the normal and abnormal basement membranes as well as from direct clinical observation in vivo.

Because basement membranes are glycoproteins[2] and have abundant vicinal hydroxyl groups (see above), they stain vividly with the PAS method (Fig. 4–6). The membranes are believed to represent a two-phase system[11] in which the fibrous component consists of collagen in exceedingly fine filamentous to microfilamentous form (perhaps even to the tropocollagen level) embedded in a matrix of various polysaccharides. Some basement membranes (e.g., lens capsule) show a distinct filamentous composition, while others (e.g., kidney glomerulus) are less distinct.

From histopathologic and clinical observations it is clear that the larger basement membranes are transparent and remain so, are relatively indigestible, possess the elastic properties of retracting and folding when disrupted (Fig. 4–6), and are elaborated by the cells that lie upon them (i.e., epithelium or endothelium). When the known properties of thick basement membranes are extrapolated to the thin basement membranes, the properties of the latter can be better understood—their relative indigestibility, their

FIG. 4-5. Scanning electron micrograph of broken edge of posterior lens capsule (arrow) showing thick, sheet-like basement membrane. ×57.

folding (Fig. 4–4) and elastic-like retraction when broken, and their elaboration by the cells that lie upon them.

The free surfaces of such thick glass-like basement membranes are smooth (see Figs. 8–7 and 8–10) and highly reflecting. This accounts for the shining surfaces of such membranes as Descemet's, the lens capsule, or the normal internal limiting membrane of the human retina. As the structural macromolecules slowly alter with age (e.g., increasing polymerization), some of the physical and morphologic properties of these basement membranes presumably are altered. They therefore may become less reflecting, as in the lens capsule or the internal limiting membrane of the retina, or may show foci of rarefaction and densification, as in the internal limiting membrane of the retina or the basement membranes of retinal blood vessels.

Effect of Cell Growth and Differentiation on Configuration of Cell Membrane

With growth, a cell increases in volume and its ratio of volume to surface area changes (Fig. 4–7). However, since the surface area of the cell plasma membrane limits the relations,[1] the cell has only a few alternatives: (1) it may cease to grow larger, (2) it may divide into two daughter cells, (3) it may increase its surface area by developing surface infolding and/or *villi*, (4) it may increase its surface area by simple elongation

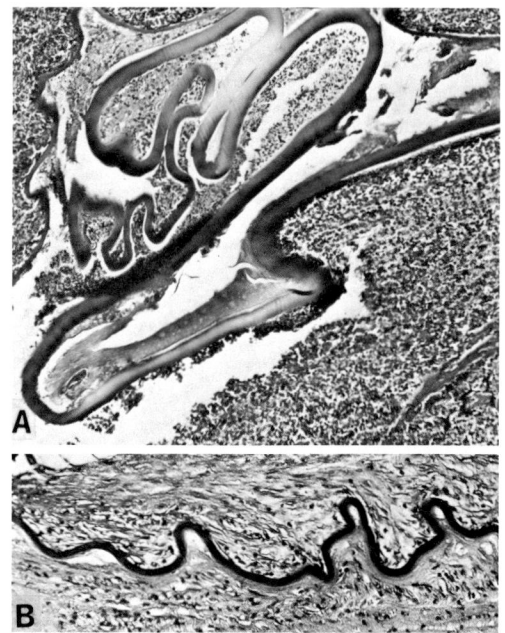

FIG. 4–6. Marked folding of disrupted thick basement membranes. **A.** Membrane from rabbit lens capsule. **B.** Descemet's membrane from rabbit cornea. **A**, PAS, ×100 (AFIP Neg. 66–8918); **B**, PAS, ×100 (AFIP Neg. 66–8919).

into a "strap-like" flattened cell, (5) it may combine the last two alternatives so that the inner half of the cell consists of a stout elongated portion with many small delicate side projections and the outer half consists of

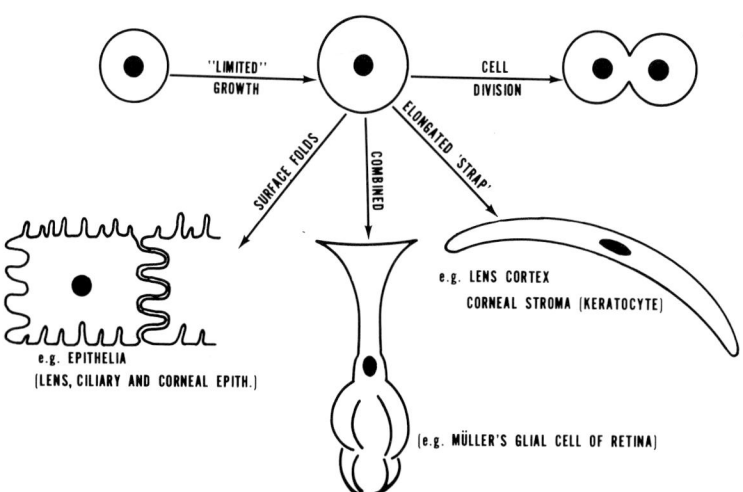

FIG. 4–7. Schematic drawing illustrating effect of cell growth on configuration of plasma membrane.

large delicate cytoplasmic leaflets with much surface area. This last response is best exemplified by the Müller (glial) cell of the retina (see Chap. 6). Examples of other responses to cell growth are epithelial cells that show marked infoldings of their surfaces, lens and corneal stroma cells that undergo remarkable flattening and elongation into strap-like formations, vascular endothelia that flatten into wide sheets, and neurons that produce enormous elongated cell extensions.

REFERENCES

1. Asimov, I. *The Well Springs of Life*. New York, New American Library, 1960.
2. Dische, Z. The Glycans of the Mammalian Lens Capsule—A Model of Basement Membranes. In Siperstein, M. (ed.). *Small Blood Vessel Involvement in Diabetes Mellitus*. Washington, D.C., American Institute of Biological Science, 1964.
3. Dorfman, A. Polysaccharides of connective tissue. *J Histochem Cytochem* 11:2, 1963.
4. Font, R. L., Zimmerman, L. E., and Fine, B. S. Adenoma of the pigment epithelium: Histochemical and electron microscopic observations. Submitted, *Amer J Ophthal.*
5. Gould, B. S. (ed.). *Treatise on Collagen*: vol. 2, *Biology of Collagen*, Parts A and B. New York, Academic Press, 1968.
6. Greenlee, T. K., Jr., Ross, R., Hartman, J. L. The fine structure of elastic fibers. *J Cell Biol* 30:59, 1966.
7. Meyer, K. Chemistry and biology of mucopolysaccharides and glycoproteins. *Cold Spring Harbor Quant Biol* 6:91, 1938.
8. Meyer, K., Davidson, E., Linker, A., and Hoffman, P. The acid mucopolysaccharides of connective tissue. *Biochim Biophys Acta* 21:506, 1956.
9. Puchtler, H. On the original definition of the term "reticulin." *J Histochem Cytochem* 12:552, 1964.
10. Ross, R., Bornstein, P.: The elastic fiber. I: The separation and partial characterization of its macromolecular components. *J Cell Biol* 40:366, 1969.
11. Slayter, G. Two-phase materials. *Sci Amer* 206:124, 1962.
12. Spiro, R. G. Glycoproteins: Structure, metabolism, and biology. *New Eng J Med* 269:566, 1963.
13. Zugibe, F. T. Mucopolysaccharides of the arterial wall. *J Histochem Cytochem* 11:35, 1963.

chapter 5
Arrangement of Tissues and Cells

Epithelia, endothelia, and mesothelia
Connective tissue
Muscle
Nerve tissue
Conventions used in describing the eye
   The three-coat (trilaminar) arrangement
   Directions
   Orientation
   The three-tissue arrangement

EPITHELIA, ENDOTHELIA, AND MESOTHELIA

When an epithelium or endothelium lines a lumen or a free surface, each cell has a free end and an attached end (Fig. 5–1). The free end of the cell is called the *apex*, and the attached end the *base*. In highly oriented or *polarized* cells there are *apical* plasma membranes, *basilar* plasma membranes, and *lateral* plasma membranes.

Lying outside the basilar plasma membranes of epithelia or endothelia is a continuous thick or thin layer of homogeneous-appearing extracellular material, the basement membrane (see Chap. 4). On the other side of the basement membrane is the *connective tissue*, usually collagenous fibrils which also may be *thick* or *thin*. Other interfibrillar materials may be present. As in retinal blood vessels, the connective tissues occasionally may be replaced by another basement membrane and its associated cell.

Ocular epithelia are derived from surface *ectoderm* or *neuroectoderm*. The epithelia may undergo the usual modifications of epithelia elsewhere and may produce apical villi and cilia, glands that secrete mucinous materials, as well as melanin granules. The cells may vary in shape from the flattened *squamous* cell of the corneal epithelial surface through the *cuboidal* cells of the lens epithelium or retinal pigment epithelium to the *columnar* cells of the nonpigmented epithelial layer of the posterior pars plana.

The cells that form the innermost lining of blood vessels and lymphatics are called *endothelial* cells. They are derived from mesoderm.

The cells that line body cavities other than those of the vascular lymphatic system (i.e., pleural, pericardial, peritoneal, and anteriormost boundary of the aqueous compartment) are called *mesothelial cells*.

According to one authority,[2] the term epithelium is used in the context of its *appearance* and *function*. The terms endothelium and mesothelium refer to their mesodermal origin.

Morphologically, endothelial cells are but a modification of epithelial cells. The cell body is the site of the nucleus and therefore protrudes with the apical cytoplasm into the lumen. To form an endothelial tube, the lateral cytoplasm is drawn out into large sheets which adhere to neighboring cells by an attachment girdle (terminal bars) common to various epithelia (see Fig. 6–65). A basement membrane layer outside this elongated

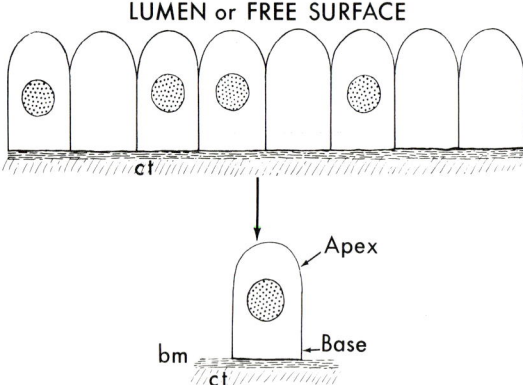

FIG. 5–1. Schematic drawing of highly oriented or polarized epithelia or endothelia lining a lumen or a free surface, showing basic relation of a cell to its basement membrane (*bm*) and adjacent connective tissue (*ct*).

Color Plate I

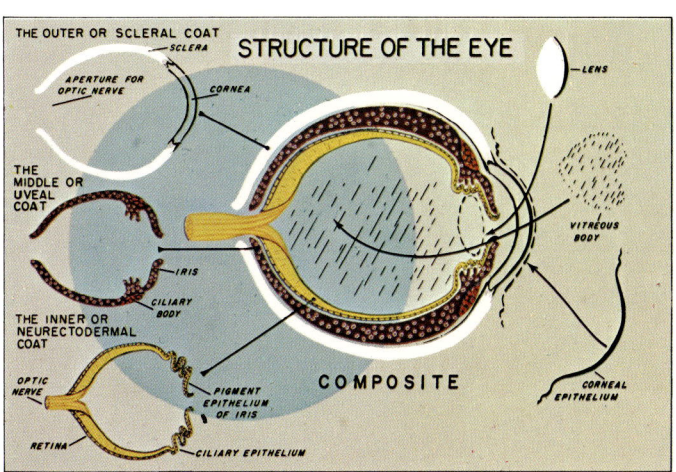

FIG. 5–2. The three-coat (trilaminar) arrangement of the eye.

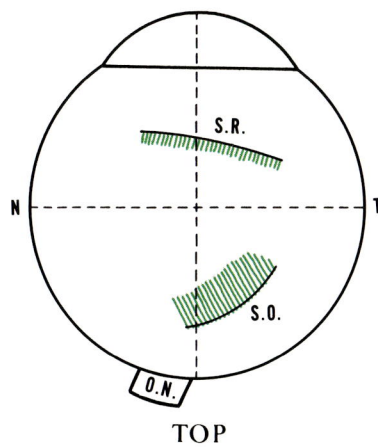

TOP

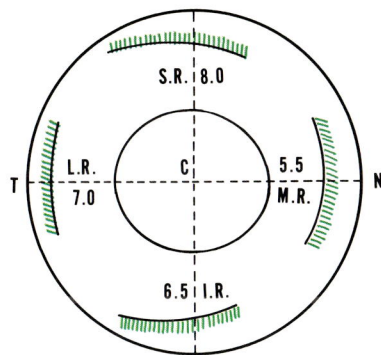

FRONT

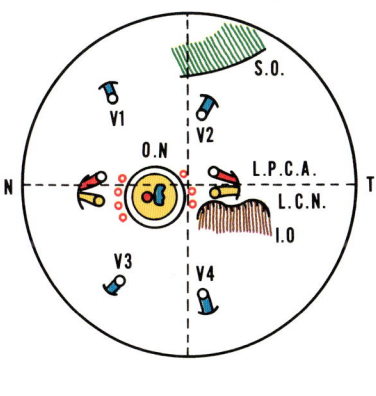

BACK

FIG. 5–5. Landmarks of front, back, and top of enucleated right eye. Tendons are shown in *green,* muscles in *brown,* nerves in *yellow,* arteries in *red,* veins in *blue.* M.R., I.R., L.R., and S.R. refer to medial, inferior, lateral, and superior rectus muscles, respectively; *IO* to inferior oblique muscle; *SO* to superior oblique muscle; *V* to various vortex veins; *ON* to optic nerve. *LPCA,* long posterior ciliary artery; *LCN,* long ciliary nerves. (Modified from Salzmann, M. *The Anatomy and Histology of the Human Eyeball in the Normal State: Its Development and Senescence.* Chicago, University of Chicago Press, 1912.)

basilar plasma membrane completely envelops the endothelial tube.

The term "mesothelium" refers to cells having the ability to act like connective tissue cells in various pathologic conditions.[2] Mesothelia closely resemble epithelia, are generally cuboidal and one cell layer thick, interdigitate laterally with neighboring cells, and may have an apical junctional complex and a moderate to very thick basement membrane (e.g., corneal "endothelium" with its Descemet's membrane).

CONNECTIVE TISSUE

A connective tissue consists of cells embedded in a matrix of extracellular materials. (These extracellular substances are described in Chap. 4.) Connective tissues vary in density from the extreme delicacy of the vitreous body to the marked density of cornea and sclera.

The most common cell type in connective tissue is the fibroblast, which, as the name implies, produces much of the fibrous as well as mucinous extracellular materials. In ocular tissues some of these cells have been given special names (e.g., the *keratocyte* of the cornea). *Histocytes*, which also may be present, have large, ovoid nuclei and are not readily identified until they have engulfed some foreign material. Then they are known as *macrophages* (as distinguished from the *microphages* of the blood stream, more commonly known as *polymorphonuclear leukocytes*). The composition of the material engulfed (e.g., blood-filled macrophages, melanin-filled macrophages, lysed lens-material-filled macrophages) determines the cytologic appearance of macrophages.

Mast cells are large cells that may be found in various connective tissues. In the eye they have been well demonstrated in the loose vascular connective tissue at the limbus and in the uvea. Their cytoplasm is almost filled with large *metachromatic* and PAS-positive granules which contain *heparin*.

Among other cells that may be found in the connective tissue or may invade from the blood stream are the plasma cells, which may be identified by their enormous amount of granular endoplasmic reticulum; the red blood cells themselves; the polymorphonuclear leukocytes, with cytoplasmic granules that differ in appearance from all the others; the eosinophils with their characteristic crystal-containing granules; the basophils; the monocytes; the lymphocytes; and the platelets.

MUSCLE

For discussion of the structure of striated muscles (of which the extraocular muscles are a good example), the reader is referred to standard texts.[1,2] Smooth muscles (ciliary muscles, dilator and sphincter muscles of the iris) are discussed with the ciliary muscle of the ciliary body region (Chap. 10).

NERVE TISSUE

The structure of typical and atypical neurons is described in Chapter 6. Current concepts of the morphology of the glia are discussed in Chapters 6 and 12.

CONVENTIONS USED IN DESCRIBING THE EYE

The Three-Coat (Trilaminar) Arrangement

Classically, the globe is described as three-layered (Fig. 5–2 on Color Plate I). The outer, tough, collagenous, protective, or "skeletal" coat is opaque sclera, except anteriorly where it becomes the transparent cornea. The innermost coat, the neuroectodermal layer, is a double layer of cells derived from an outpouching of the floor of the primitive forebrain. It consists of the sensory neural retina with its pigment epithelium posteriorly and the anterior extensions to form the double-layered epithelium of the ciliary body and iris. The primitive neural epithelium is sometimes called the medullary epithelium. Between the outer and inner layers of the eye lies the vascular and heavily pigmented uveal tract, which is subdivided topographically into three parts: the iris stroma anteriorly, the choroid posteriorly, and the ciliary body in-between.

Directions

The two principal planes in the eye are the equatorial and the meridional (Fig. 5–3).[4] The *equatorial plane* lies midway between the anterior and posterior poles and is perpendicular to the meridional plane. Planes parallel to the equatorial plane (i.e., anterior or posterior to it) are called transverse, coronal, frontal, or radial planes. The *meridional plane* passes through the anterior and posterior poles of the eye. Meridional planes may be horizontal, vertical, or oblique. Planes parallel to the meridional plane but *not* passing through the anterior and posterior poles are called *sagittal* planes.

Orientation

The anatomic (ophthalmologic) orientation of the eye for descriptive purposes is such

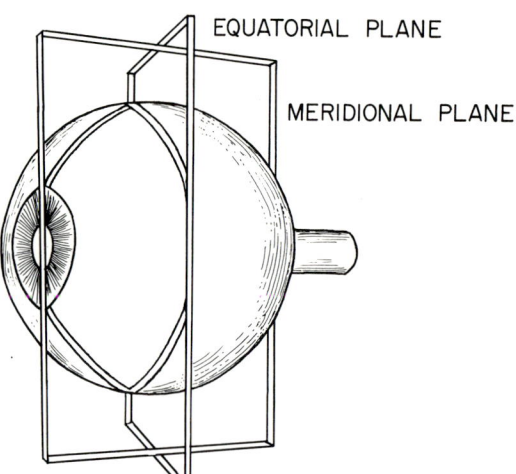

FIG. 5–3. Diagram of equatorial and meridional planes of the eye. (From Scheie, H. G., and Albert, D. M. *Adler's Textbook of Ophthalmology.* Philadelphia, Saunders, 1969.)

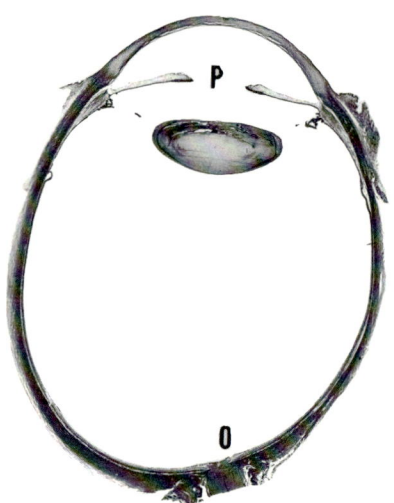

FIG. 5–4. Horizontal meridional section of an eye oriented in ophthalmic manner, with cornea *up* and optic nerve *down*. Note that section passes through both pupil (*P*) and optic nerve head (*O*). Such a section is sometimes called a PO section. H&E, ×3. (AFIP Neg. 70–7292.)

that the cornea is up and the optic nerve down (Fig. 5–4). The directional terms *anterior* and *posterior* are derived from this orientation. The terms *internal* and *external* (or lateral) are applied to directions proceeding *toward* and *away from* the geometric center of the globe.

The posterior view of the globe is the most useful in determining whether an enucleated eye (globe) is right or left (back view, Fig. 5–5, on Color Plate I). The optic nerve and the adjacent fleshy attachment for the inferior oblique muscle are two of the most important landmarks in this view.

The optic nerve exits from the posterior aspect of the globe through a 3-mm diameter scleral canal 3 mm nasal to, and 1 mm superior to, the posterior pole. It can be readily observed lying within its meningeal sheaths.

The temporal side of the eye from optic nerve to limbus features a longer arc than does the nasal side. The horizontal meridian is easily identified by two prominent blue lines radiating anteriorly in opposite directions from points near the optic nerve. The lines are the pathways of the long posterior temporal and long posterior nasal ciliary arteries and the accompanying long ciliary nerves. Once this horizontal meridian is identified, the superior, inferior, temporal, and nasal quadrants of the posterior pole of the eye can be easily determined from the insertions of the oblique muscles. The inferior oblique is the only one of the six extraocular muscles that inserts into the globe via a muscular (or at the most a very short tendinous) attachment. The attachment is easily identified temporal to the optic nerve and is inserted almost precisely along the horizonal meridian. Because the fibers of the inferior oblique muscle course downward, the inferior part of the globe can be identified. The macula lies within the eye anterior to the nasal portion of the muscular insertion. Thus, by examining the posterior aspect of the globe, aligning the silhouettes of the two ciliary vessels into the horizontal meridian, and finding the muscular insertion of the inferior oblique, a right or left globe can be identified. The identification may be confirmed by seeking the tendinous insertion of the superior oblique muscle which attaches to the sclera in the upper temporal quadrant (Fig. 5–5, top view).

Frequently, the vortex veins, usually four in number, are prominent. The position of these veins may vary considerably. In general, two lie above the optic nerve and two below. Each pair straddles the vertical meridian with the upper some 7 or 8 mm behind the equator and the lower ~ 2 mm closer to it. Although the remnant insertions of the four rectus muscles are in a regular but varying relation to the limbus, they generally are not as useful as other landmarks for identifying a right or left eye. The short curved lines which mark the attachments of the rectus muscles do, however, spiral round the cornea in such a way that the medial rectus lies closest to the corneal margin (limbus) and the superior rectus lies the farthest away (Fig. 5–5, front view). A line connecting the insertions of the four rectus muscles forms the *spiral of Tillaux*. Many short posterior ciliary arteries and short ciliary nerves enter the sclera in a cluster around the optic nerve, but they are difficult to identify without the aid of a dissecting microscope. Because the cornea is oval with the horizontal meridian larger than the vertical one, the horizontal meridian of the eye sometimes can be identified from the anterior view of the globe.

The Three-Tissue Arrangement

The parts of the globe may also be described in terms of their predominant tissue component: those composed almost entirely of cells (e.g., lens, retina and its pigment epithelium, and anterior extensions of ciliary and iris epithelium); those composed almost entirely of extracellular materials (e.g., corneosclera, vitreous body); and those composed of an even mixture of cells and extracellular materials (e.g., uveal tract). This classification is particularly appropriate to descriptions of ocular tissue studied by electron microscopy, and it is the one used in the succeeding chapters of this book.

REFERENCES

1. Bloom, W., and Fawcett, D. W. *A Textbook of Histology*, ed. 9. Philadelphia, Saunders, 1968.
2. Ham, A. W., and Leeson, T. S. *Histology*, ed. 4. Philadelphia, Lippincott, 1961.
3. Salzmann, M. *The Anatomy and Histology of the Human Eyeball in the Normal State: Its Development and Senescence.* Tr. by E. V. L. Brown. Chicago, University of Chicago Press, 1912.
4. Scheie, H. G., and Albert, D. M. *Adler's Textbook of Ophthalmology.* Philadelphia, Saunders, 1969.

chapter 6

The Retina

Embryology
The pigment epithelium
The neural (sensory) retina
    The neuronal system
        Rods and cones
        External limiting membrane
        Outer nuclear layer
        Outer plexiform layer
        Inner nuclear layer
            Bipolar cells
            Accessory neurons
        Inner plexiform layer
        Ganglion cell layer
        Nerve fiber layer
        Internal limiting membrane
    The glial system
        The Müller cell
        Accessory glia
    The vascular system
    Topographic variations
        Macula
        Fovea
        The subretinal space
            Ora serrata and anterior subretinal cul-de-sac
            Posterior subretinal cul-de-sac

In terms of the three-tissue classification described in Chapter 5, the retina is a tissue composed chiefly of cells.

EMBRYOLOGY

The *Neuroectodermal cells* lining the floor of the primitive forebrain proliferate outward as a blind tubular diverticulum (Fig. 6–1). The proliferation continues until the diverticulum lies beneath the laterally extended *ectoderm* (the *surface* ectoderm), where it balloons out to form a large single-layered vesicle, the (primary) *optic vesicle.* Invagination of this vesicle produces a two-layered structure, the *optic cup* (or secondary optic vesicle), which almost obliterates the lumen of the optic vesicle. The cells contained in these two primitive neuroectodermal layers

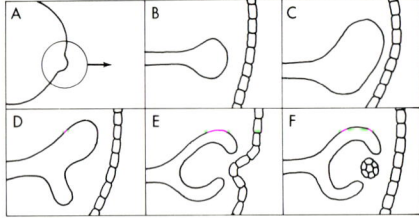

FIG. 6–1. Embryonic development of retina. Invaginated outpouching of floor of primitive forebrain (*A*) extends outward to lie beneath surface ectoderm (*B*) as a vesicle (*C*) which subsequently invaginates (*D*) to form optic cup (*E*). Lens vesicle forms from surface ectoderm (*E, F*).

therefore come to lie in apposition, apex-to-apex (Fig. 6–2, on Color Plate II). The outer layer remains upright and develops heavy pigmentation. It continues into maturity as a single layer of epithelium and becomes the definitive pigment epithelium of the retina and ciliary body and the anterior pigment epithelial layer of the iris. The cells of most of the inverted *inner* neuroectodermal layer multiply repeatedly to produce the final mature, many-layered neural (or sensory) retina (Fig. 6–3, on Color Plate II, and Fig. 6–4). Anteriorly, beyond the *ora serrata*, the inner neuroectodermal layer continues as a single layer of epithelium, the nonpigmented epithelium of the ciliary body and the posterior layer of iris pigment epithelium (see Chap. 10).

THE PIGMENT EPITHELIUM

Like all epithelia and endothelial cells, the pigment epithelial cells have apexes and bases (Fig. 6–5). Lying apposed to the basal plasma membranes of these cells is a thin basement membrane (Fig. 6–6).

The thin basement membrane as observed by electron microscopy is interpreted perhaps most usefully as the equivalent of the cuticular portion ("lamina vitrea") of Bruch's membrane,[48,54] as seen either grossly or by light microscopy. Bruch originally described a continuous shiny cuticular surface which could be best observed grossly by careful removal of the pigment epithelial cell layer. With subsequent examination by light microscopic techniques, the full thickness of tissue lying between pigment epithelium and the plane of the choriocapillaris appeared homogeneous and assumed the designation of "the membrane of

Color Plate II

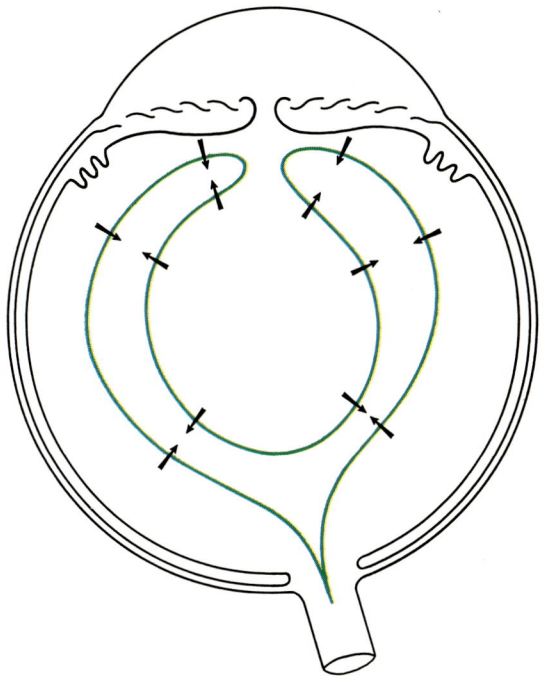

FIG. 6–2. Development of retina. Embryonic invagination to produce optic cup results in two epithelial layers (green) applied to one another with cells arranged apex-to-apex (arrows).

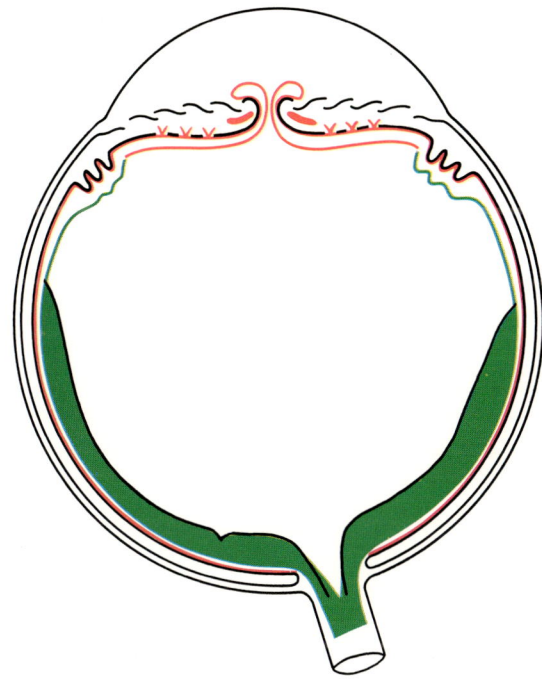

FIG. 6–3. Development of retina. With maturation, specialization takes place in various regions of this two-layered neuroepithelium. *Red* lines indicate those parts that develop pigmentation or neuroepithelial muscles of iris (dilator and sphincter). *Green* regions represent persisting nonpigmented cell layers.

THE RETINA   49

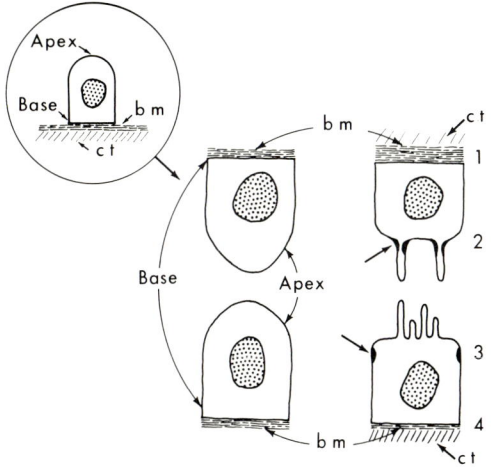

FIG. 6–4. Development of retina. With inversion, neural retina develops basement membrane (*bm*) and associated connective tissue (*ct*) on its anterior surface. Apposing apical regions develop terminal bar attachment girdles (*2, 3*). Further modifications occur. One basement membrane (*1*) becomes thick (internal limiting membrane), while the other basement membrane (of pigment epithelium) remains thin (*4*). Connective tissue associated with (*1*) is composed of thin collagenous filaments (vitreous), while that associated with (*4*) is composed of thick collagenous fibrils (lamina elastica). (Modified from Fine, B. S. *Arch Ophthal (Chicago)* 66:847, 1961.)

Bruch," which was then synonymous with *lamina vitrea*. This zone was later subdivided into two parts: an inner cuticular portion and a more fibrillary outer portion. The latter portion stained prominently with orcein or Weigert's elastic tissue stain, hence the name *lamina elastica*. On electron microscopic examination, this bipartite region may be subdivided into several parts, depending upon the age and location of the tissue examined.

The structures between the *choriocapillaris* and the pigment epithelium in young tissue form a trilaminar arrangement consisting of two basement membranes, one belonging to the pigment epithelium and the other to the endothelium of the choriocapillaris, separated by a zone of connective tissue. Between adjacent segments of choriocapillaris, however, the basement membrane of the endothelium no longer participates in this trilaminar arrangement, and it becomes bilaminar. The connective tissue portion ("lamina elastica") is exaggerated into "pillars" between the segments of the choriocapillaris. The connective tissue pillars are continuous with the connective tissue of the choroid.

With age, portions of the connective tissue in the posterior of the eye become more dense and accumulate fragments of other materials. This occurs rather diffusely in the connective tissue pillars, but is more highly localized in a midzonal plane within the trilaminar regions. This

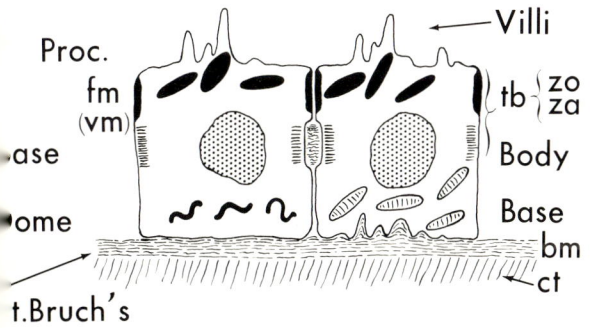

FIG. 6–5. Schematic drawing of pigment epithelial cells, their attachments, associated basement membrane, and connective tissues. Terms commonly used in light microscopy are shown at left and equivalent electron microscopic terms on right.

Proc., apical cell processes	*tb*, terminal bar
fm, fenestrated membrane	*zo*, zonula occludens
vm, Verhoeff's membrane	*za*, zonula adherens
Cut. Bruch's, cuticular portion of Bruch's membrane	*bm*, basement membrane
	ct, connective tissue

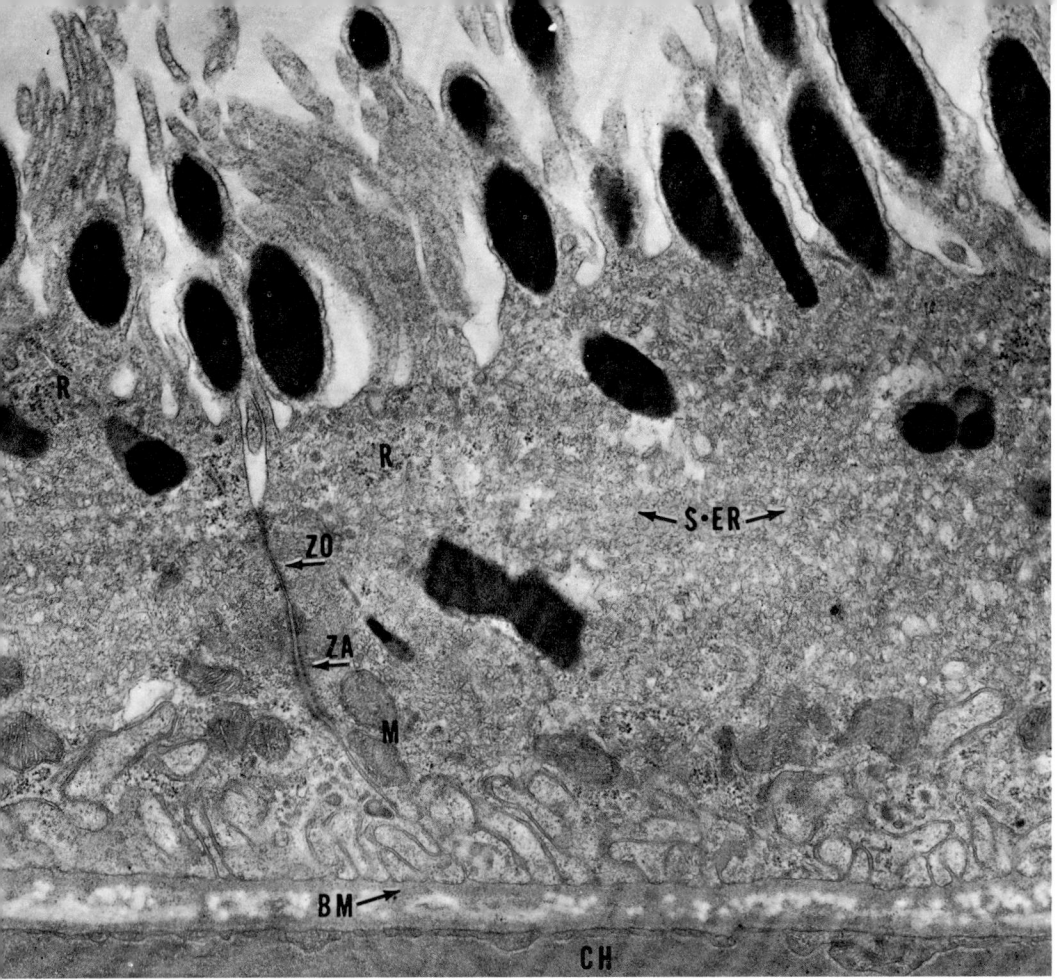

FIG. 6–6. Pigment epithelium from rabbit retina. Thin basement membrane (*BM*) is present along base of cell. Most melanin granules are localized to apical cytoplasm and apical villi. Present between adjacent cells is terminal bar attachment girdle consisting of both a zonula occludens (*ZO*) and a zonula adherens (*ZA*) portion. Mitochondria (*M*) are present in basilar cytoplasm. A few clusters of free and attached ribosomes (*R*) are present in apical cytoplasm. Entire cell is permeated by agranular endoplasmic reticulum (*S·ER*). *CH*, choriocapillaris. ×22,500. (From Fine, B. S. In McPherson, A. (ed.). *New and Controversial Aspects of Retinal Detachment.* New York, Hoeber, 1968.)

midzone, which often becomes basophilic in H&E-stained sections, may then be said to further subdivide the trilaminar zones. It is this midzone which appears to take up the Weigert elastic tissue stain. More anteriorly, clearly recognizable segments of elastic tissue are present in the midzone (see Chap. 10.).

For descriptive and practical reasons it is useful to retain the concept of separating this zone (i.e., Bruch's "membrane") into two *continuous* layers: the lamina vitrea and the lamina elastica.

As in other epithelia, the basal plasma membranes often are irregularly infolded (Fig. 6–7). The apexes of the pigment epithelial cells are modified into villi, which generally envelop the impinging tips of the photoreceptor outer segments. Whether these villi are stout or delicate depends upon the location of the cells within the eye and the type of photoreceptor to which they are related (see under Rods and Cones and under Fovea, later in this chapter).

The limiting plasma membranes of the villi are separated from those of the adjacent photoreceptor outer segments by a typical intercellular space.[23,27] No specialized attachments are present between the retinal pigment epithelium and the photoreceptors. Near their apexes, the adjacent pigment epithelial cells are attached to one another by terminal bars (Figs. 6–6 and 6–8).[1,2,9,23] These terminal bars possess both zonula occludens and zonula adherens portions.

The cytoplasm of the pigment epithelial cell may be subdivided into three zones or layers.[28] *All three* are permeated by the interconnecting membranous system of tubules, cisterns, and channels known as the agranular (smooth-surfaced) endoplasmic reticulum. The outer one-third of the cell contains the mitochondria and the prominent infoldings of the basal plasma membrane. The inner one-third contains most of the melanin (pigment) granules and projects from the apical surface as many villous processes. This zone also contains a few ribonucleoprotein particles (RNP particles, ribosomes) lying freely in small clusters ("rosettes") or attached to short segments of granular (rough-surfaced) endoplasmic reticulum. The intermediate zone is relatively free of cell organelles except for the all-pervading agranular reticulum and the nucleus. When granules of lipofuscin are present, they occupy most of this intermediate zone. Glycogen particles are absent.

The pigment (melanin) granules are *spherical* or *ovoid*. The spherical granules are present in the apical cell cytoplasm, while the ovoid granules generally are present within the apical villi. The approximately equal ratio of spherical to ovoid granules is in sharp contrast to the predominance of the spherical form in the pigment epithelia of the ciliary body and iris.*

* From an examination of human embryonic material,[5,45] development of the large melanin granule of the neuroepithelium can be characterized by the

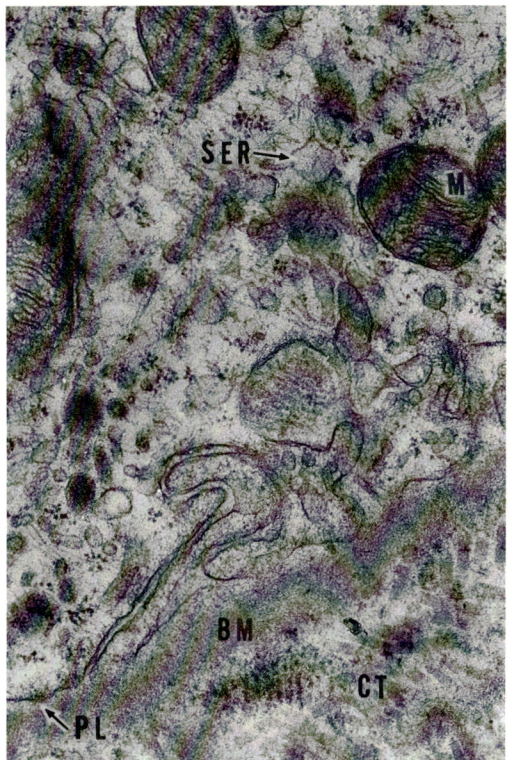

FIG. 6–7. Markedly infolded basal plasma membrane (*PL*) of pigment epithelial cell. A basement membrane (*BM*) and adjacent connective tissue (*CT*) are present. *M*, mitochondria; *SER*, smooth endoplasmic reticulum. ×27,000. (From Fine, B. S. In McPherson, A. (ed.). *New and Controversial Aspects of Retinal Detachment.* New York, Hoeber, 1968.)

following morphologic classification (modified from others)[31,44,56] into the *premelanosome*, which can be observed only by electron microscopy, and the several stages of melanosome (defined here as a developing granule with *any* content of melanin), most of which can be observed but not classified by light microscopy (Fig. 6–12).

Premelanosome. Generally a membrane-enclosed group of highly oriented, weakly electron-dense filaments

Melanosome

   Early immature (filaments). Appearance of densifi-

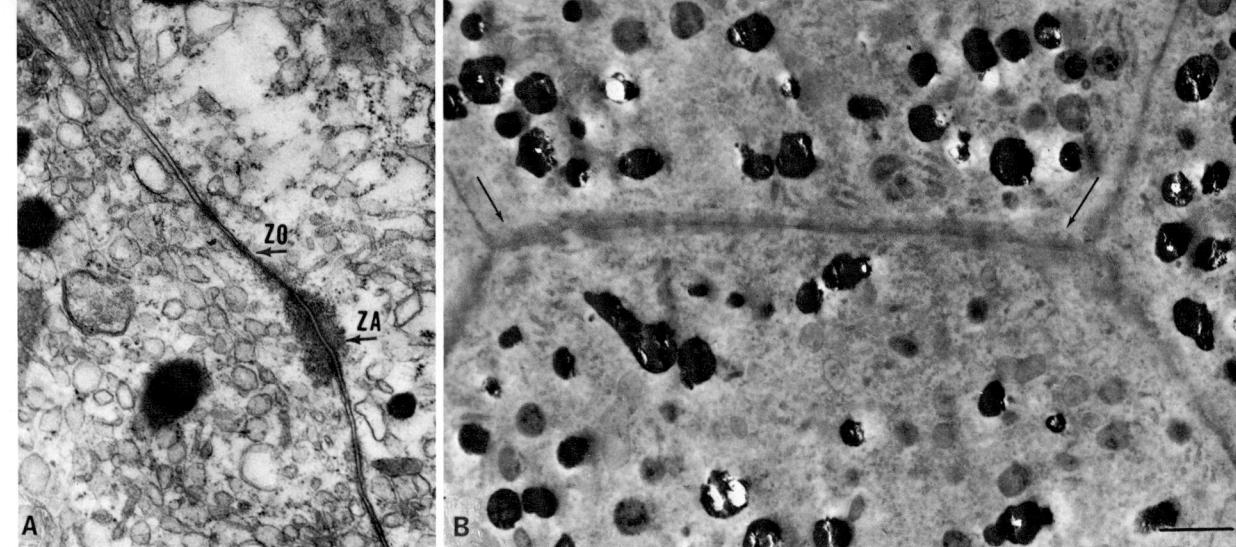

FIG. 6–8. Pigment epithelium. **A.** Terminal bar attachment girdle in human pigment epithelium. *ZA,* zonula adherens; *ZO,* zonula occludens. **B.** Transverse section through retinal pigment epithelium at level of zonula adherens which has been cut along its length (*arrows*), illustrating its circumferential arrangement. **A,** ×27,000; **B,** ×12,300. (**B** from Fine, B. S. *Arch Ophthal (Chicago)* 66:847, 1961.)

The differences in size and shape of the melanin granules throughout the ocular layers can be readily appreciated by oil-immersion light microscopy (Fig. 2–13). The uveal melanin granules are much smaller, mainly ovoid in form, and therefore easily differentiated. Knowledge of such variation is sometimes useful to the pathologist who wishes to identify the source of the granules in an altered tissue or in a pathologic accumulation.

> cation along filaments, varying from a few random spots of densification to complete densification of entire filament
> *Late immature* (*rodlets*). Increase in densification by thickening of filaments into rodlets
> *Mature.* Compaction of rodlets into an almost homogeneously dense granule

These stages are only morphologic guideposts since many variations are seen and many so-called mature granules show regions of incomplete rodlet fusion and small peripheral foci of lucency when examined in very thin section.

The spherical and ovoid melanin granules are uniformly dense except for a number of small denser particles or "holes" in their periphery. Many bodies of less dense material are often present adjacent to the melanin granules. These latter bodies represent lipofuscin (Fig. 6–9). A number of smaller dense melanin granules are frequently found to be embedded in the lipofuscin bodies (Fig. 6–10). The latter composite or *compound* granules are commonly observed in the apical and mid-cytoplasm of older pigment epithelial cells, and together with the lipofuscin granules are prominent in the macular region.

The prominence of the lipofuscin and compound granules within the pigment epithelial cells of the macular region (Fig. 6–10) as contrasted with their relative sparseness in cells of the same layer elsewhere suggests an anatomic basis for the observed clinical finding, during fluorescein angiography, that the macular area is considerably darker than the rest of the retina, namely that the yellow-orange lipofuscin effectively acts as a filter in the macular area and screens out the underlying excited fluorescein in the choroid.

Increase in the number of lipofuscin and/or compound granules can occur readily in the region of the macula not only from aging alone but also from mild acute or chronic insult (e.g., exposure to various forms of irradiation, diseases of the choroid, etc.)

Other intracytoplasmic structures which may be present include discrete dense bodies and lamellated bodies (Fig. 6–11). The latter are believed to represent engulfed photoreceptor outer segment material.[64,73,74] *Myeloid bodies*, as observed in amphibian retinal pigment epithelium,[50,69] are lacking in the human. The Golgi complex of the pigment epithelium is generally not prominent, although on occasion it may be observed dilated with a lucent material (Fig. 6–13), suggesting that this cell may also contribute at least in small measure to the *interreceptor* mucoid.[30]

The pigment epithelial cells vary from the narrow, tall, highly uniform structures in the region of the macula (Figs. 6–10A, 6–14) to the wide, low, very irregular structures in the region of the ora serrata (Figs. 6–10B, 6–15). These latter cells may be less densely pigmented and may be multinucleated.

The pigment epithelial layer regains some of its cuboidal stature at the ora serrata, where it continues as the pigmented ciliary epithelium and finally as the anterior layer of iris epithelium (see Chap. 10).

THE NEURAL (SENSORY) RETINA

The neural retina (pars optica retinae) is the innermost layer of the neuroectodermal coat. In the living eye it is a thin, delicate transparent sheet of tissue containing the special sensory cells (photoreceptors) which convert photic stimuli into nerve impulses and subsequent visual sensations.

The neural retina is ~ 0.4 mm thick at the margin of the optic nerve, tapering to ~ 0.15 mm on the nasal side in the region of the ora serrata. Temporally the retina remains ~ 0.4 mm thick to the periphery of the macula (except for the *fovea centralis*, where it abruptly thins to ~ 0.2 mm for a width of ~ 0.2 mm).[54]

Traditionally, the neural retina is subdivided into nine "layers." These are indicated in Figure 6–16.

The neural retina may also be conveniently and usefully classified into *three systems:* the neuronal, the glial, and the vascular. The neuronal and glial are derived from the primitive neuroectoderm, whereas the vascular is derived from later invasion by mesoderm in the region of the *optic papilla*. The optic papilla generally undergoes atrophy or flattening to become the *nerve head* or *optic disc* (see Chap. 12).

The neuronal and glial systems traverse the entire thickness of the neural retina (Fig. 6–17). The three primary connected nerve cells —the neuroepithelial photoreceptor, the bipolar cell, and the ganglion cell—form the basic arrangement of the neuronal system.

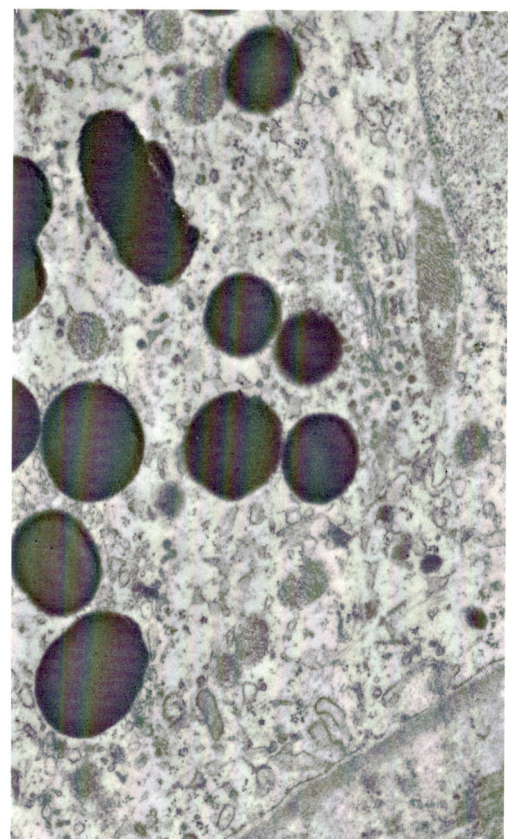

FIG. 6–9. Lipofuscin granules within body of pigment epithelial cell. They are often prominent in region of macula or in older tissue. ×21,000.

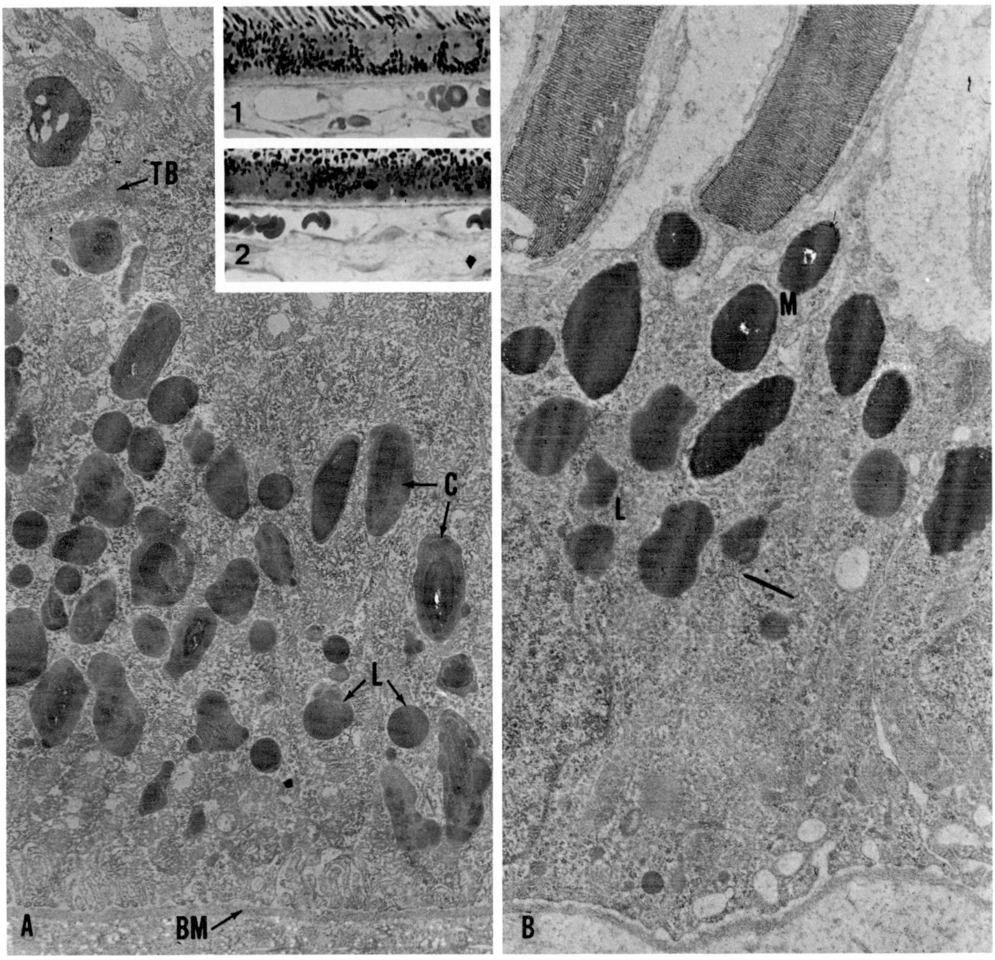

FIG. 6–10. A. Sample of parafoveal pigment epithelium from region similar to inset 1. The body and base of the cell are occupied by lipofuscin (L) and compound (C) (i.e., melanin embedded in lipofuscin) granules. BM, basement membrane of pigment epithelium; TB, terminal bar (zonula adherens portion). ×6,000. **Inset 1,** region of parafoveal pigment epithelium (1.5 micron PD, ×400, AFIP Neg. 71–4896), **inset 2,** pigment epithelium, nasal retina (slightly oblique), (1.5 micron, PD ×400, AFIP Neg. 71–4897). **B.** Extramacular pigment epithelial cells contain fewer granules located mostly within the apical cytoplasm. Ratio of lipofuscin granules (L) to melanin granules (M) in the plane of section is also lower here than in the macula. ×12,300.

FIG. 6–11. Lamellar structure present within basal cytoplasm of pigment epithelial cell. ×16,000.

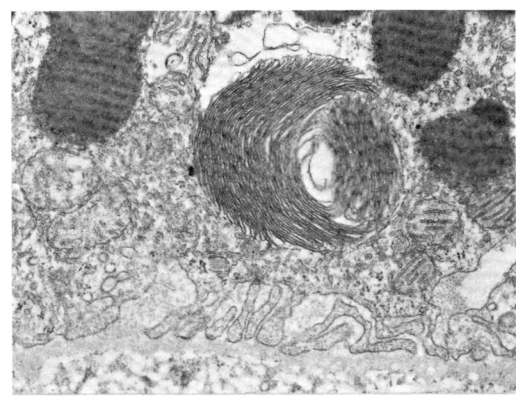

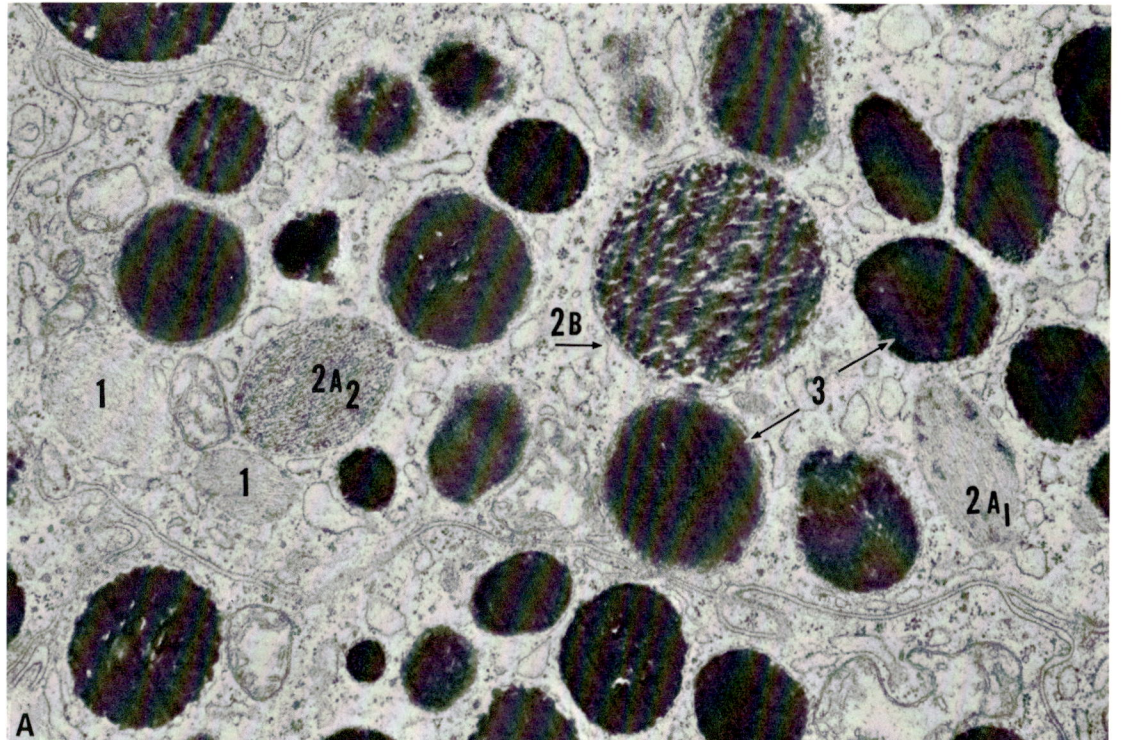

FIG. 6–12. Development of melanin granules in human pigment epithelium. **A.** Melanin granules develop prenatally from premelanosome (*1*) through *immature* stages of melanosome (*2A₁*, beginning of early stage; *2A₂*, later stage; *2B*, rodlet stage) to *mature* melanosome (*3*). Iris pigment epithelium, 16 weeks. **B.** Schematic drawing showing sequence of melanin granule development. **A,** ×8,000. (From Mund, M. L., Rodrigues, M. M., and Fine, B. S.[45])

The best known and largest of the cells in the glial system is the Müller cell.

The basic pattern of the neural retina therefore can be most simply understood by superimposing these systems. For completeness this schema also includes the smaller glia present in the inner layers (fibrous and protoplasmic astrocytes, microglia, and "oligodendrocytes"), the horizontally interconnecting neurons in the bipolar cell layer, the *amacrine* cells on the inner side, and the *horizontal* cells on the outer (Fig. 6–18).

THE NEURONAL SYSTEM

Rods and Cones

The photoreceptor cells are highly specialized neuroepithelial cells analogous to other sensory receptor cells and their special endings (e.g., Pacini's corpuscles for pressure, Krause's end bulbs for cold, or Meissner's corpuscles for light touch). They, like the remainder of the neural retina, are inverted (Fig. 6–19). The photoreceptor cell outer and inner segments make up the layer of rods

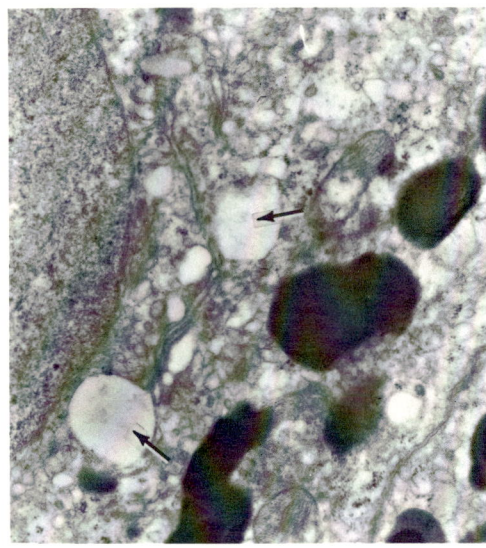

FIG. 6–13. Example of Golgi complex, some of which is dilated with flocculent material (*arrows*). Golgi complex is not prominent in pigment epithelium. ×16,000.

FIG. 6–14. Flat preparation of pigment epithelium in region of macula. H&E, ×700. (AFIP Neg. 70–730. Courtesy of Dr. M. O. M. Ts'o.)

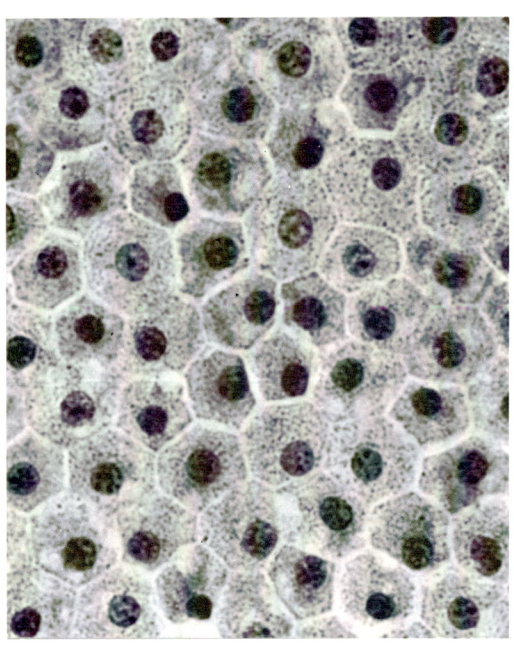

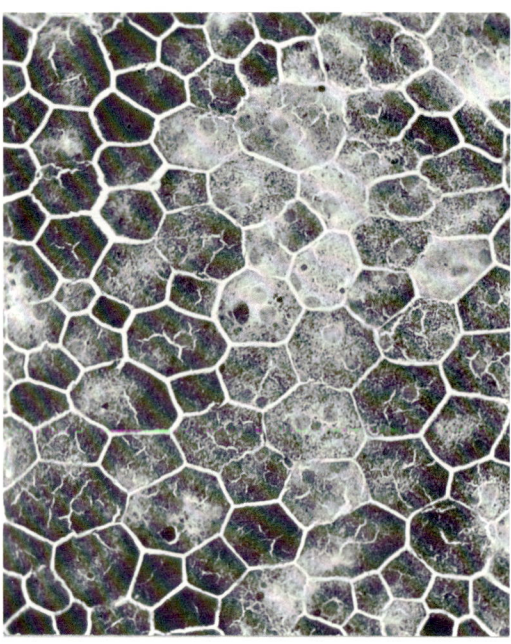

FIG. 6–15. Flat preparation of pigment epithelium in region of ora serrata. H&E, ×210. (AFIP Neg. 70–731. Courtesy of Dr. M. O. M. Ts'o.)

and cones. The cell bodies or *perikarya* are located just internal to the plane of terminal bar attachments, known in light microscopy as the external limiting membrane (Fig. 6–16). The apical extensions, *inner segments*, are located just external to this plane. The rod inner segments are cylindrical; those of the cones contain glycogen (Figs. 6–20 and 6–27) and are larger and conical due mainly to the great number of mitochondria aggregated at their outer ends. The inner segments of the photoreceptors contain the prominent Golgi complex (Figs. 6–20 and 6–21), whose vesicles usually are filled with a lucent material. The material closely resembles that of the large extracellular space (Fig. 6–22) which extends from the external limiting membrane to the apical surface of the pigment epithelium between the photoreceptor cells. The lucent material has been identified histochemically[30,77] as containing one or more acid mucopolysaccharides (Figs. 6–23 through 6–25) whose exact nature is as yet unknown.

Agranular endoplasmic reticulum and free clusters of ribosomes are also present within the photoreceptor inner segments. Myofilaments are not in evidence, so apparently the segment lacks the myoid portion, unlike the case in certain amphibia.

Projecting from the apex of the inner segment is a highly modified cilium (Fig. 6–24). It lacks central (axial) filaments or microtubules of the more usual $9 + 2$ arrangement of cilia observed elsewhere (e.g., ependyma or respiratory epithelium). The cilium here consists of a $9 + 0$ arrangement: 9 filaments

FIG. 6–16. Typical section of retina showing its layered arrangement. *ILM*, internal limiting membrane; *NFL*, nerve fiber layer; *GC*, ganglion cell layer; *IPL*, inner plexiform layer; *INL*, inner nuclear layer; *OPL*, outer plexiform layer (*P*, plexiform; *H*, Henle fibers); *MLM*, middle limiting membrane; *ONL*, outer nuclear layer; *XLM*, external limiting membrane; *RC*, layer of rods and cones (*IS*, inner segments; *OS*, outer segments.) Macula, H&E, ×390. (Modified from Fine, B. S., and Zimmerman, L. E. *Invest Ophthal* 2:446, 1963.)

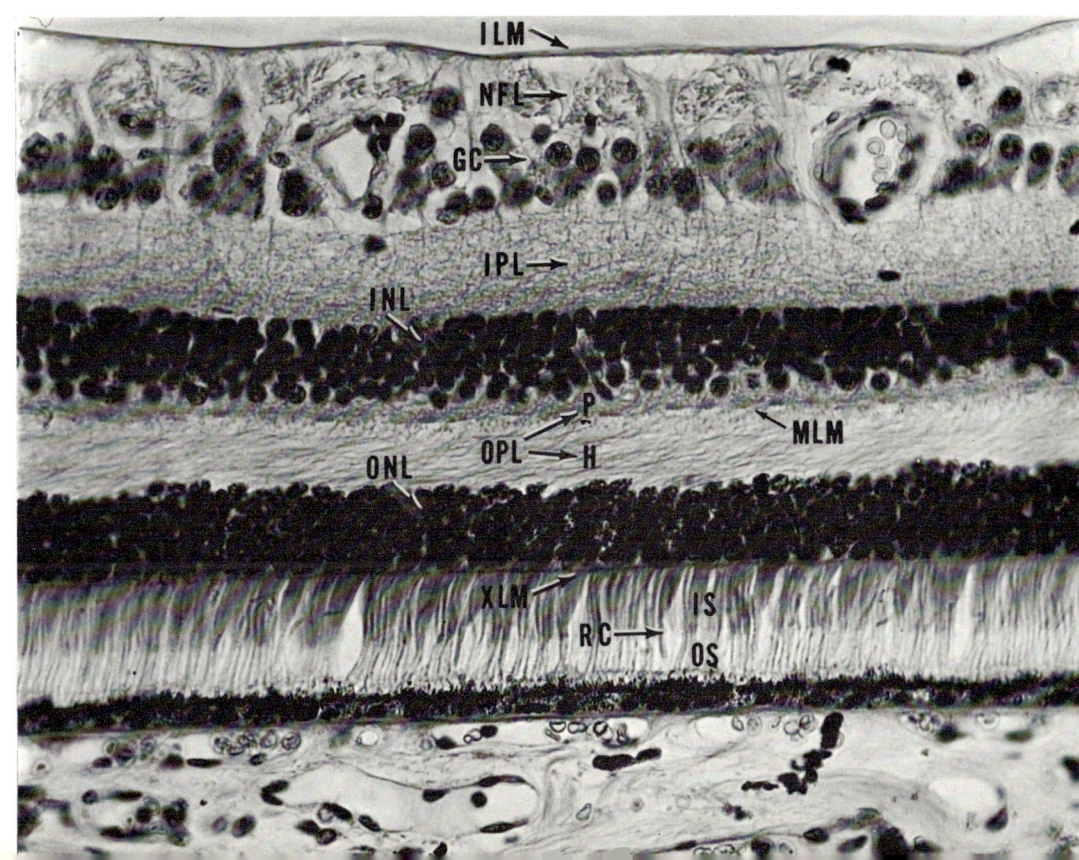

FIG. 6–17. Schematic drawing of neuronal and glial systems of neural retina. Neuronal system of three serially connected cells is at right; large glial system (Müller's cell) is at left. *ILM,* internal limiting membrane; *XLM,* external limiting membrane. (From Fine, B. S. In McPherson, A. (ed.). *New and Controversial Aspects of Retinal Detachment.* New York, Hoeber, 1968.)

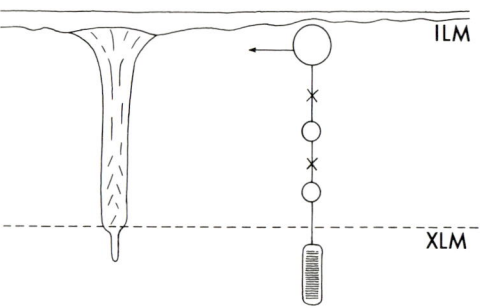

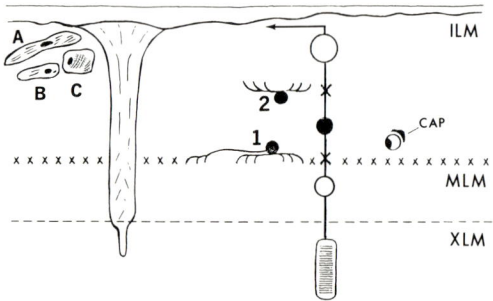

FIG. 6–18. More complete scheme of retinal structure. Included are middle limiting membrane (*MLM*), vascular system represented by its outermost capillary bed (*CAP*), horizontal cells of bipolar layer (*1*), amacrine cells of bipolar layer (*2*), and small glia of inner retinal layers. *A,* fibrous astrocytes; *B,* protoplasmic astrocytes; *C,* (?) oligodendrocytes.

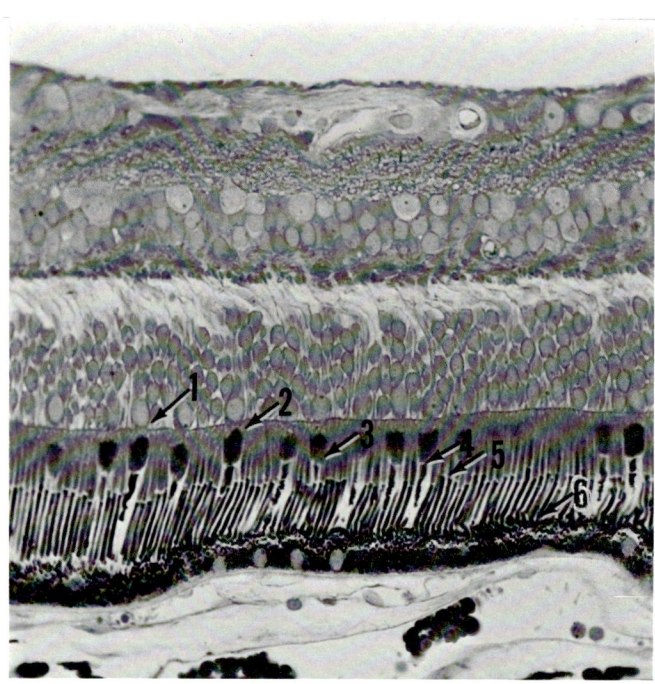

FIG. 6–19. Light micrograph of thin (1.5-μ) section of retina. Figures indicate approximate regions illustrated in subsequent micrographs: *1,* Fig. 6–20; *2,* Fig. 6–21; *3,* Fig. 6–22; *4,* Fig. 6–24; *5,* Fig. 6–26; *6,* Fig. 6–30. PD, ×305. (AFIP Neg. 64–6650. Modified from Fine, B. S. In McPherson, A. (ed.). *New and Controversial Aspects of Retinal Detachment.* New York, Hoeber, 1968.)

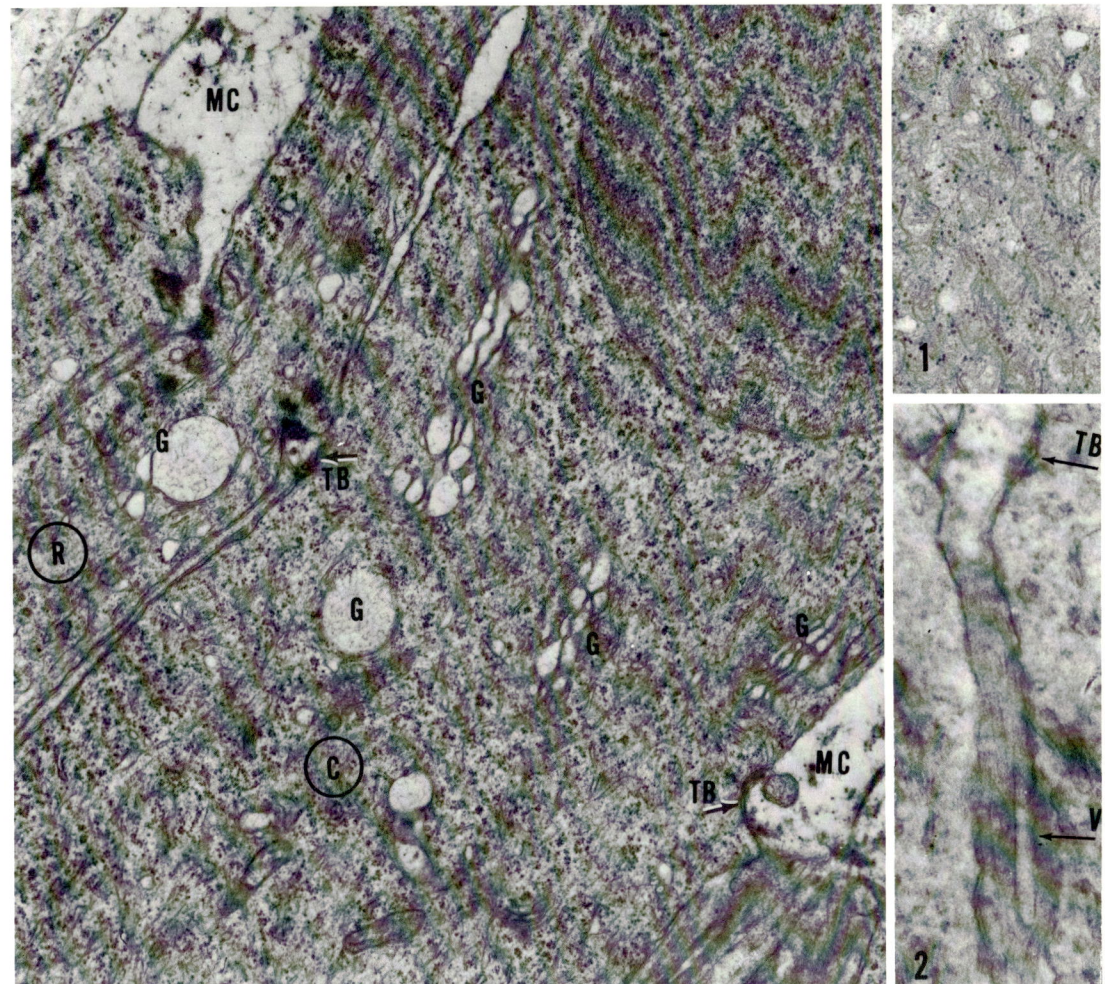

(doublets)* forming a ring, but lacking the 2 "usual" centrally located filaments (Fig. 6–26).

The plasma membrane of the cell extends from the apex of the inner segment to form a cup or *calyx* (Fig. 6–28). The shaft of the cilium covered with plasmalemma projects from the floor of the calyx. Beyond the shaft of the cilium the cell plasmalemma is ballooned outward into a cylindrical or slightly conical sac, the *outer segment*. In the rod cell, this sac is filled with a stack of superimposed,

* In cross-section the *basal body* of the cilium has been described as containing 9 fiber *triplets*, the connecting cilium or shaft as containing 9 fiber *doublets*; within the outer segment the cilium becomes 9 fiber *singlets*.[10]

FIG. 6–20. Large cone (C) inner segment and adjacent rod (R) inner segment containing their respective Golgi complexes (G). External limiting membrane consists of sectioned terminal bar attachment girdles (TB) which unite adjacent photoreceptor and Müller cells (MC). In **inset 2** Müller cell villi (V) project beyond external limiting membrane (TB) for a short distance. **Inset 1** shows random distribution of glycogen between mitochondria in parafoveal cone inner segment (TSC-treated). **Main figure,** ×19,000; **inset 1,** ×16,000; **inset 2,** ×28,700. (Modified from Fine, B. S., and Zimmerman, L. E. *Invest Ophthal* 2:446, 1963.)

FIG. 6–21. Oblique section of photoreceptor inner segments near external limiting membrane. Dilated Golgi cisternae (*arrows*) are present within segment, as are aggregates of agranular reticulum and clusters of ribosomes (*R*). Villi (*V*) of Müller cells are numerous between inner segments near membrane. ×18,700.

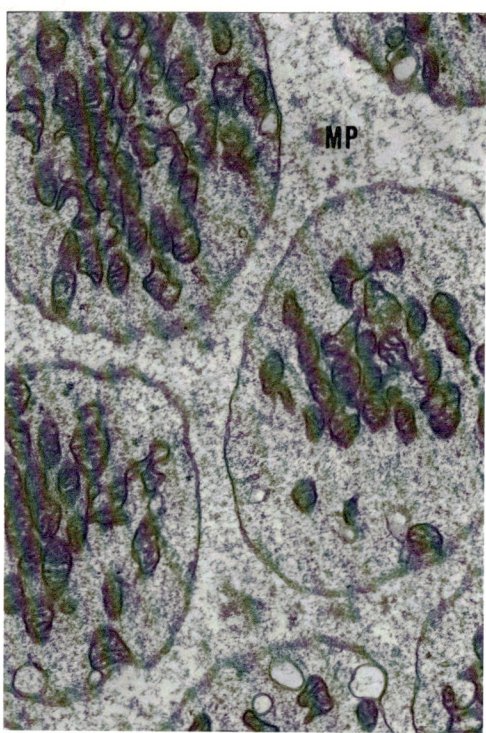

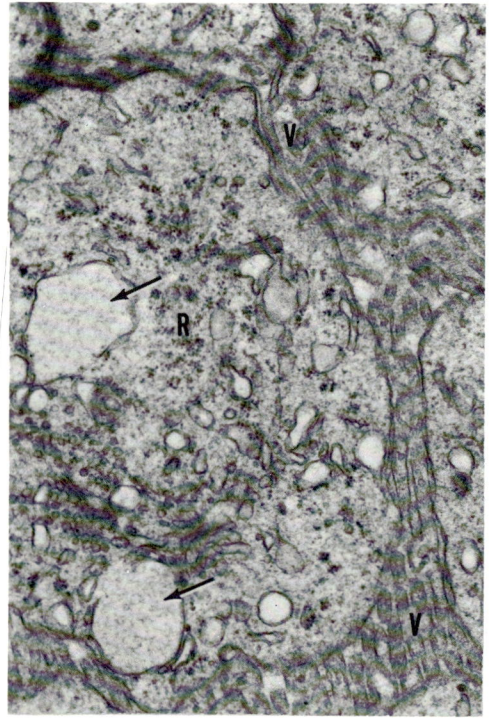

FIG. 6–22. Cross-section of inner segments more distal to Figure 6–21. Note interreceptor mucoid (*MP*), lack of Müller cell villi, and larger aggregates of mitochondria within cell. ×14,000.

flattened "hollow" discs or double lamellas (membranes).[7,8,14,19,22,39,58,60,61,67,70] The lamellas are easily separated from one another and are free of attachment to the surface plasma membrane. Each double-membraned disc is also free from attachment to an adjacent disc. The discs possess several clefts at their periphery where the membranes of the discs are attenuated into tubules, possibly a vestige of their once completely tubular nature.* The

---

* The continuity of lamellas with the surface plasma membrane in the cone[12,28] implies derivation from this surface membrane. That the rod discs (lamellas or sacs) are similarly derived from the surface plasma membrane has been inferred from developmental studies.[16,47] The PAS-positivity of both rod and cone outer segments (Fig. 6–23, on Color Plate III) is due to their high content of lipoprotein membranes, as they contain no glycogen (Fig. 6–27A).

The discs or lamellas can easily break down into tubules, an alteration frequently observed as a fixation artifact. Because some of these intracellular membrane alterations are relatively nonspecific, their dissolution into tubules may also indicate early or chronic *pathologic* change. Proper interpretation then depends on other and/or additional criteria. Bends, loops, bulges, and hook-like formations of the entire photoreceptor outer segments also often occur as artifacts.

Color Plate III

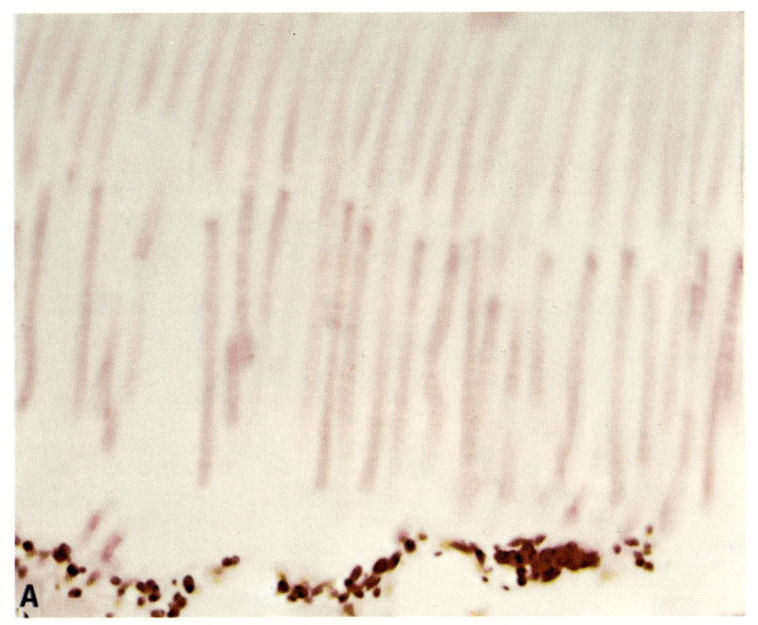

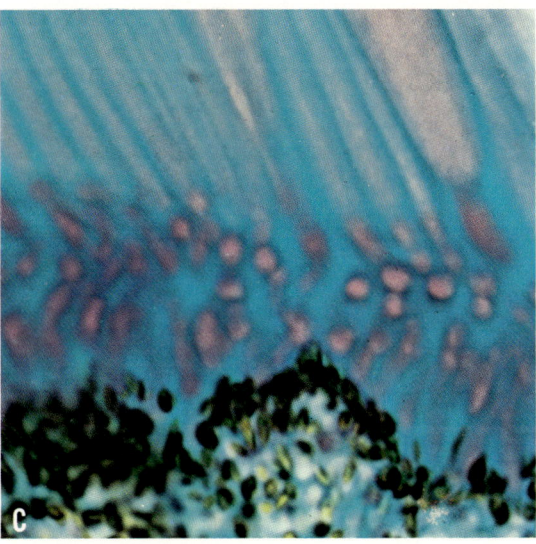

FIG. 6–23. Composition of layer of rods and cones. **A,** photoreceptor outer segments are PAS-positive; **B** and **C,** spaces between photoreceptor outer segments are positive for acid mucopolysaccharides. Cross-sections of PAS-positive outer segments (in **C**) show size of mucoid-filled extracellular spaces. Note diffuse PAS positivity of cone inner segments, due in part to distribution of glycogen (see inset 1 of Fig. 6–20 and Fig. 6–27). **A,** PAS-hematoxylin, ×1,500; **B,** PAS, AMP, ×1,080; **C,** PAS, AMP, ×2,160. (Modified from Fine, B. S., and Zimmerman, L. E. *Invest Ophthal* 2:446, 1963.)

clefts lie in register forming *longitudinal furrows* along the surface of the rod cell outer segments (Figs. 6–26 and 6–28).

Differing in many respects from the rod cell outer segment is the cone cell outer segment (Fig. 6–29).[13,18] Although apparently similar to rod cell lamellas, cone cell lamellas are less easily separated (i.e., the double membranes of each "disc" are tightly bound to each other), they possess no peripheral clefts containing groupings of tubules and are in continuity with the surface plasma membrane at least over short segments of their circumference (*inset,* Fig. 6–24). Their connecting cilia, however, have the 9 + 0 filament arrangement of the rods, pass similarly through a narrow constriction (the isthmus) between inner and outer segments, and possess a basal body and at least two striated rootlets like other cilia. A free centriole generally is observed in the cytoplasm near the basal body.

The rounded, completely enclosed tip of the rod outer segment impinges on the apex of the pigment epithelial cell, which surrounds it with numerous delicate villi as previously mentioned (Fig. 6–30). The attachment between pigment epithelium and rod outer segments appears to be no stronger than that of a usual intercellular space.

The relation of the *extrafoveal cone to the pigment epithelium* generally differs from that of the rod in that the *tapered* outer end of the cone falls short of the usual plane of the pigment cell apexes. Long, heavy pigment epithelial villi extend inward to envelop the cone outer segment (Fig. 6–31). The cones are greatly modified in the fovea (see later in this chapter), and toward the ora serrata they are shorter and plumper than elsewhere, have poorly developed outer segments, and increase in number relative to rod outer segments.

The large extracellular space between photoreceptor outer segments represents the remains of the lumen of the embryonic optic vesicle. The space is filled with a mixture of acid mucopolysaccharides (Figs. 6–23 through 6–25), few, if any, of which are sensitive to hyaluronidase.

Separation of the neural retina from its pigment epithelium in the postnatal eye creates, in effect, a gross reappearance of the lumen of the optic vesicle. This is clinically termed a retinal "detachment."

External Limiting Membrane

The external limiting membrane (XLM); (X in Fig. 6–32) is not a true membrane, but rather attachment sites of the adjacent photoreceptor and Müller cells. These attachments are terminal bars (Fig. 6–33) which lack the zonula occludens portion. Thus the intercellular spaces probably are limited fuctionally but are not absent or totally obstructed.

Outer Nuclear Layer

The photoreceptor cell bodies ("outer nuclear layer" or ONL) lie internal to the plane of the external limiting membrane (Fig. 6–32). Occasionally a photoreceptor cell nucleus (especially of a cone because of their proximity) is found lying *external* to this membrane. The cone cell body or perikaryon is much larger than that of the rod. This difference in size is also carried over into the rod and cone inner segments and axons (Henle fibers).

The cytoplasmic connections which extend from the photoreceptor inner segments at the level of the external limiting membrane to the cell body often are referred to as the *connecting fibers* (Fig. 6–32).[54] Because most of the cone cell bodies lie nearer to the external limiting membrane than do the rod cells, these connecting fibers are necessarily short and stout for the cone cells but long and narrow for the rod cells.

Outer Plexiform Layer

The outer plexiform layer (OPL) is truly plexiform only in its inner third; its outer two-thirds is composed of the axonal extensions

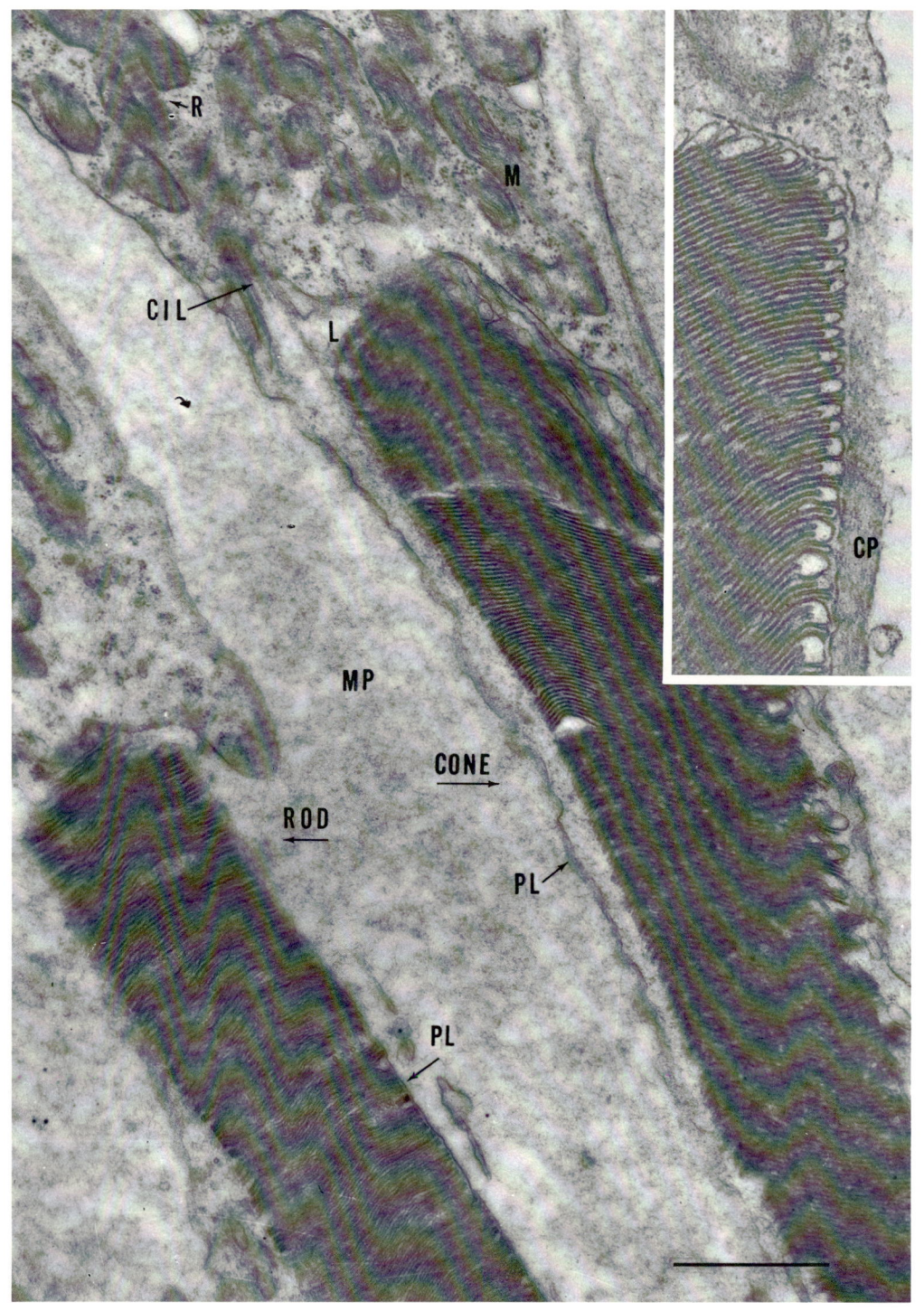

FIG. 6–24. (*Opposite page.*) Rod and cone outer segments. Connecting cilium (*CIL*) of cone is shown here. Intercellular mucopolysaccharides (*MP*) are present. Double lamellas of tapering cone outer segment are strongly attached to each other and are connected to others in different plane (see **inset**) as well as to surface plasma membrane. *CP,* calyceal process; *L,* loops of cone outer segment lamellas; *PL,* plasma membrane; *R,* ciliary rootlets. **Main figure,** human nasal macula, ×23,800; **inset,** monkey cone, parafovea, glutaraldehyde fixation, ×40,000. (Modified from Fine, B. S., and Zimmerman, L. E. *Invest Ophthal* 2:446, 1963.)

(Henle fibers) of the photoreceptors enveloped by the lucent outer cytoplasm of the Müller cells (Figs. 6–34 and 6–35).[29] The axons contain a few segments of agranular reticulum, a small scattering of individual (apparently glycogen, i.e., TSC-positive) particles, a few elongated mitochondria, and uniform packing of neurotubules embedded in a moderately dense ground substance (see Ganglion Cell Layer and Nerve Fiber layer, later in this chapter).

These thick and thin axons of cone and rod photoreceptor cells together with their surrounding "packing" of Müller cells (Fig. 6–36) make up the Henle fiber layer of the outer plexiform layer.*

Near the dendrites of the bipolar cells (Fig. 6–37), the photoreceptor axons dilate into large *synaptic expansions*. The cone expansion is much the larger of the two and the more inwardly placed (Fig. 6–38). With the metal impregnation techniques of light microscopy, the cone expansion appears as a *pedicle* or foot, with small branches. The rod axon expands into a somewhat teardrop-shaped expansion termed the rod *spherule* (Fig. 6–40).[49]

* The term "Henle fiber layer" in recent years has become synonymous with the *oblique* fibers present in the macular region. Except for this obliquity, and consequent greater length, the fibers do not differ substantially from axonal extensions elsewhere, which once were also called "Henle fibers." The outer plexiform layer presumably is so named because it resembles the inner plexiform layer superficially under certain conditions of tissue sectioning (Fig. 6–43). The inner plexiform layer, however, is always plexiform in appearance regardless of the plane of section.

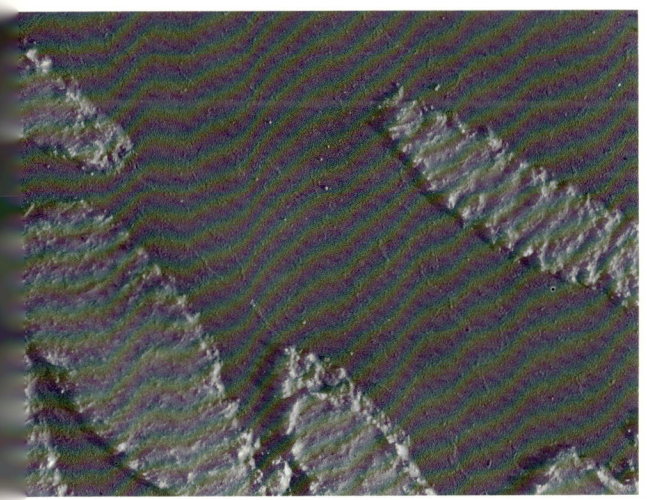

FIG. 6–25. Shadow-cast preparation of photoreceptor outer segments. Interreceptor mucoid is seen as drying patterns between receptor segments. U-shad, ×12,000.

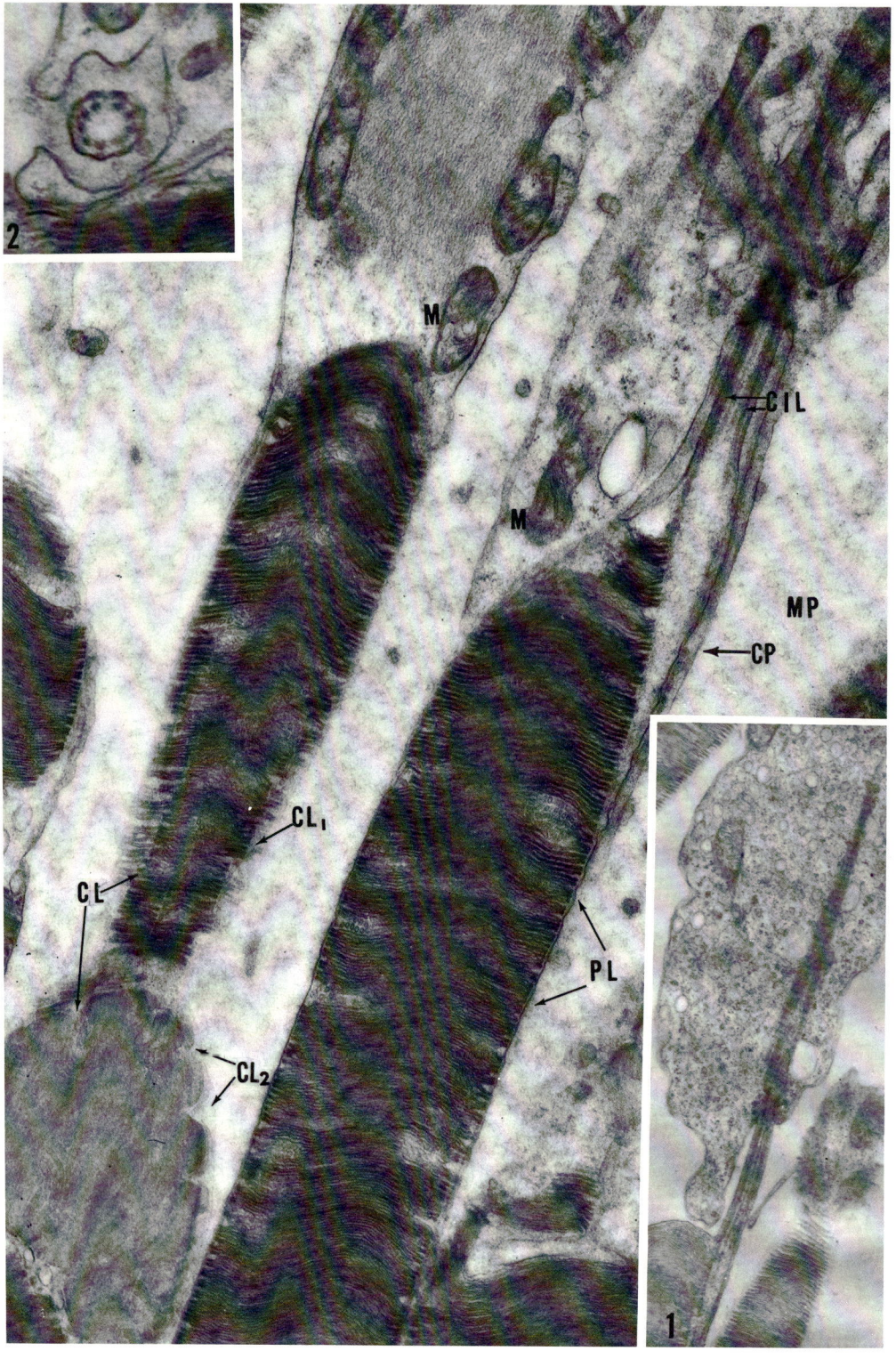

FIG. 6–26. (*Opposite page.*) Rod outer segments. Double membranes of discs are loosely attached to each other. Discs are free from one another and from surface plasma membranes, and possess clefts (CL_2) which lie in register forming longitudinal furrows (CL, CL_1). A rod connecting cilium (*CIL*) is present issuing from calyx of inner segment. A calyceal process (*CP*) extends from inner segment along one side of outer segment. *M,* mitochondria; *MP,* intercellular mucopolysaccharides; *PL,* plasma *membrane.* **Inset 1,** from rabbit retina, shows more clearly cross striations of ciliary rootlets. **Inset 2** shows a cilium cross-sectioned through its shaft. Ciliary filaments are of the 9 + 0 arrangement. **Main figure,** human nasal macula, ×25,300; **inset 1,** ×10,800; **inset 2,** ×32,000. (**Main figure** from Fine, B. S. In McPherson, A. (ed.). *New and Controversial Aspects of Retinal Detachment.* New York, Hoeber, 1968.)

FIG. 6–27. A. Junction of inner segment (*IS*) and outer segment (*OS*) of photoreceptor in macula. Specimen was initially fixed in glutaraldehyde, accounting for often "coagulated" appearance (*free arrows*) of interreceptor mucoid. Section was treated with TSC *only.* Dark (glycogen) particles are widely distributed within inner segment between the mitochondria. Note apparent inclusion of glycogen particles within mitochondrion at *M.* Arrow M_1 indicates a neighboring C-shaped mitochondrion partially enclosing cluster of similar particles. Longitudinal sectioning of latter mitochondrion would give *appearance* of intramitochondrial glycogen as shown at *M.* ×15,000. **B.** Inner segments of cone (*C*) and rod (*R*) cells near the terminal bars (*TB*) of the external limiting membrane. The cone inner segment contains TSC positive particles (glycogen), while those of the rods do not. *N,* nucleus of cell; *MC,* lucent cytoplasm of Müller cell. (Glut.-Dalton's, TSC only, ×15,000.)

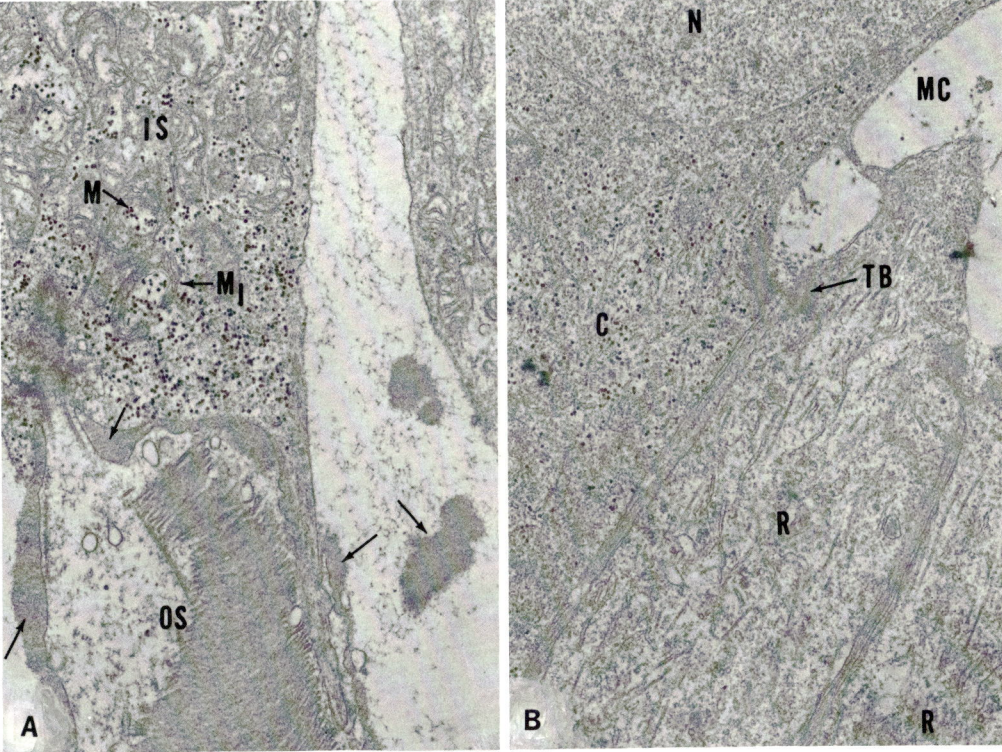

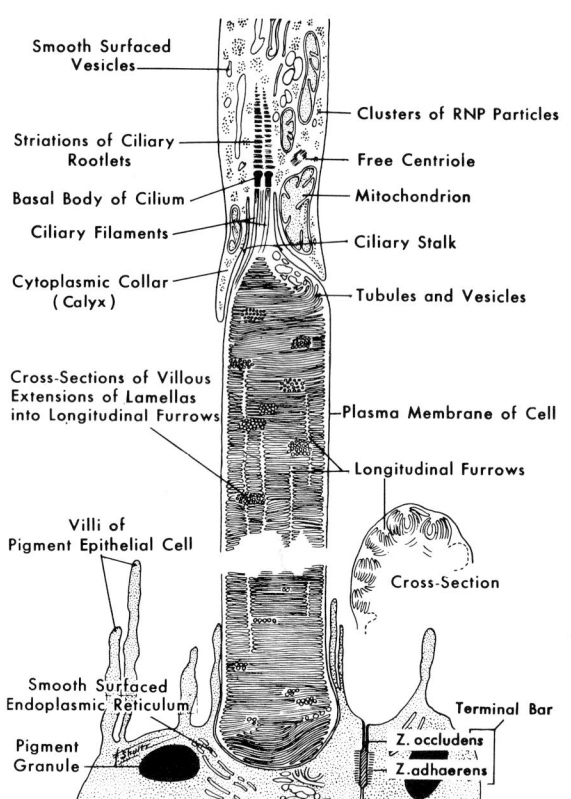

FIG. 6–28. Schematic drawing of rod outer segment and its relation to pigment epithelial cell and its villi. (From Fine, B. S. In McPherson, A. (ed.). *New and Controversial Aspects of Retinal Detachment.* New York, Hoeber, 1968.)

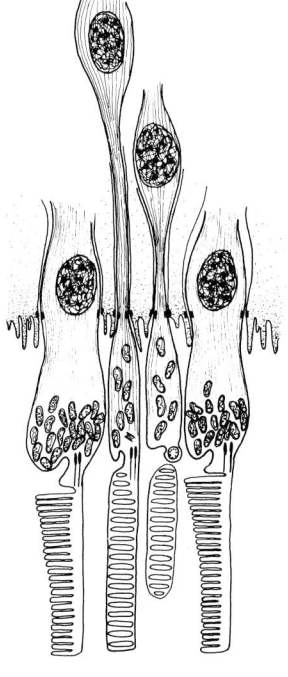

FIG. 6–29. Schematic drawing showing that alignment of rods and cones beyond external limiting membrane is maintained by intervening projections of the Müller cells. Difference in rod and cone cell diameters as well as arrangement of their lamellated outer segments are also illustrated. (From Ts'o, M.O.M., et al. *Amer J Ophthal* 69:350, 1970.)

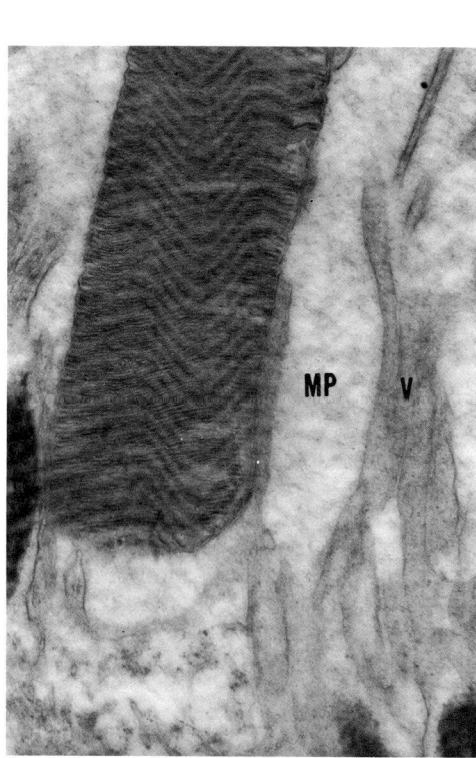

FIG. 6–30. Rod cell impinging on pigment cell and enveloped by apical villi (*V*) of latter cell. *MP*, interreceptor mucoid. ×20,000.

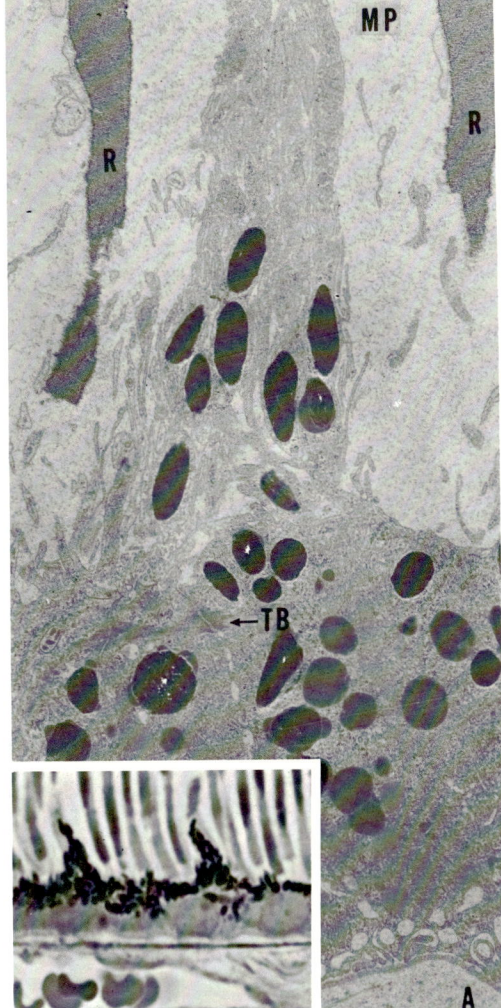
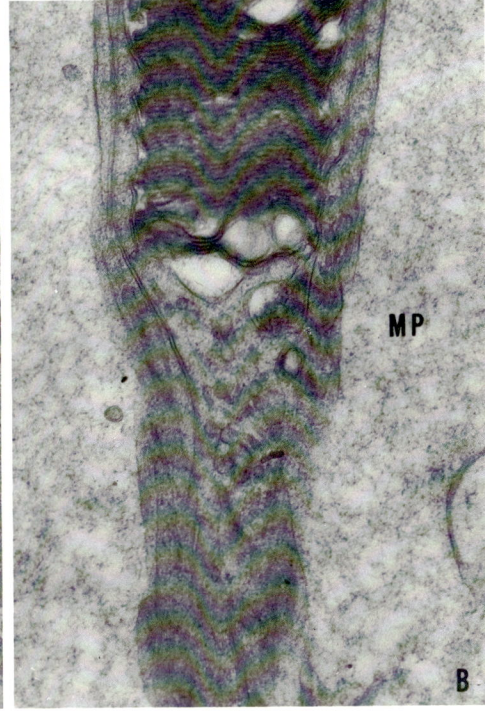

Electron microscopic examination shows that these synaptic expansions are filled with myriad small (200 to 300 Å) vesicles called *synaptic vesicles*.[15,40] These vesicles are believed to contain acetylcholine,[17] which is released at the cell surface upon appropriate stimulation.

Many TSC-positive particles are also present within the cone pedicles, suggesting a small content of glycogen.

The many infoldings at the innermost cell surface of the synaptic expansions of both rod and cone cells are produced by the "plugged-in" blind endings of adjacent bipolar cell dendrites (Fig. 6-39) as well as by both dendrites and axons of such lateral interconnect-

FIG. 6–31. A. Extrafoveal cone outer segment falls short of the usual plane of pigment epithelium which sends a mass of pigment bearing villi inwards (**inset,** PD, ×820, AFIP neg. no. 71–5949) to envelope the outer segment end. MP, interreceptor mucoid; R, rod outer segments; TB, terminal bars. ×8,000. **B.** Outer end of cone enveloped by pigment epithelial villi some distance inward from the pigment epithelial cell layer, MP, interreceptor mucoid. ×20,700.

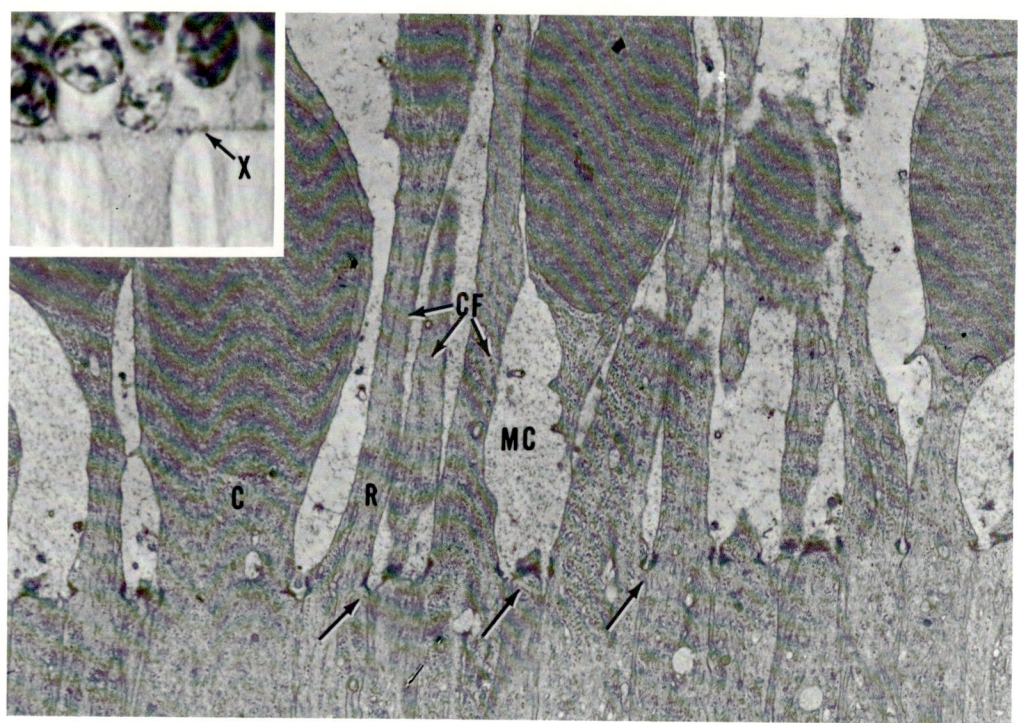

FIG. 6–32. External limiting membrane. Viewed by light microscopy (**inset**), membrane (X) appears as discontinuous series of small dots. Electron microscopy shows series of terminal bars (*free arrows*) which attach Müller cells (MC) to photoreceptor cells (R, C). Nuclei of photoreceptor cells make up outer nuclear layer. CF, connecting fibers. **Main figure,** ×4,500; **inset,** ×1,500 (AFIP Neg. 61–1958).

FIG. 6–33. External limiting membrane sectioned more obliquely to show zonula-adherens-like nature of its structure. A zonula-occludens-like portion is not present. IS, photoreceptor inner segment; MC, Müller cell; V, Müller cell villi. ×23,500.

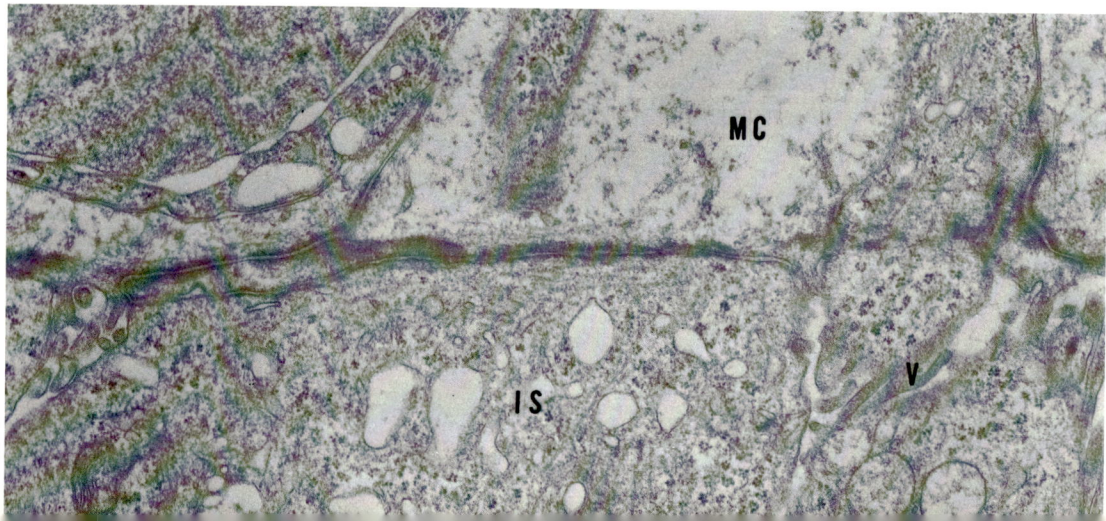

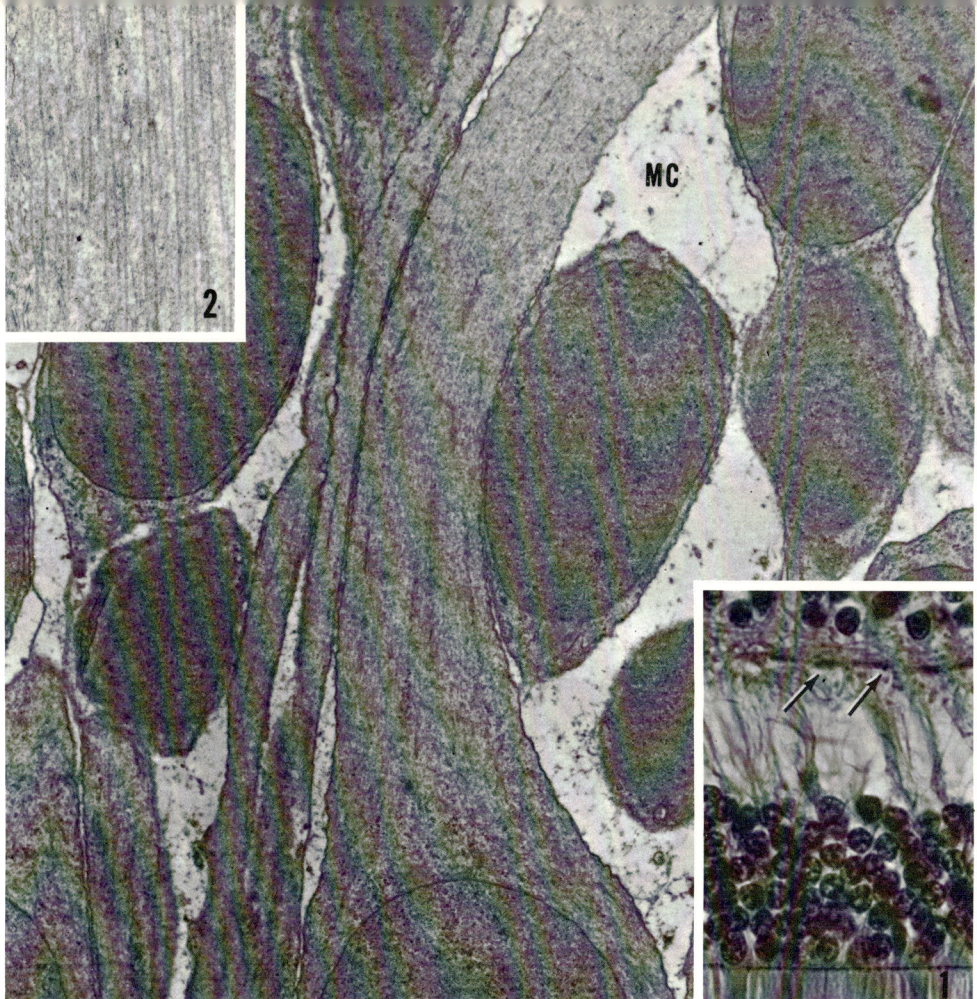

ing neurons as the *horizontal cells*. Dense attachment foci (*synaptic densities*)* are present along this surface. Frequently, within the synaptic expansion and the cleavage formed by two adjacent impinging neurites are seen short dense strips, the synaptic *lamellas*[25,38,40,43] or *ribbons*,[59]† around which a peculiar *"halo"* of synaptic vesicles is arranged. The lamellas are multiple in the cone foot,‡ but

---

* Although morphologists generally tend to equate the site of synaptic density with the physiologic event of transmission of the nerve impulse, such equation remains uncertain. Originally, Sherrington[57] defined a synapse as a *postulated* anatomic site (which he had inadequate means at the time to observe directly) for the physiologic event of transmission of the nerve impulse.

† Although the structure is ribbon-like in three dimensions (hence the name "synaptic ribbon"), it is recognizable in most sections as a cross-section of the ribbon (hence the name "synaptic lamella").

FIG. 6–34. Inset 1. Light micrograph showing outer plexiform layer. Inner third is plexiform and outer two-thirds is fibrous (Henle fibers), union being marked by plane of photoreceptor synapses (*arrows*). **Main figure.** Electron micrograph showing a number of photoreceptor cell bodies, their nuclei, and their axonal extensions (Henle fibers) surrounded by lucent cytoplasm of Müller cells (*MC*). **Inset 2.** Electron micrograph showing cytoplasmic microtubules ("neurotubules") present within axoplasm. **Main figure,** ×6,800; **inset 1,** H&E, ×600; **inset 2,** glutaraldehyde-fixation, ×24,000. (Modified from Fine, B. S., and Zimmerman, L. E. *Invest Ophthal* 1:304, 1962.)

---

‡ Up to 25 lamellas, as determined by serial reconstruction of a single cone synaptic pedicle.[43]

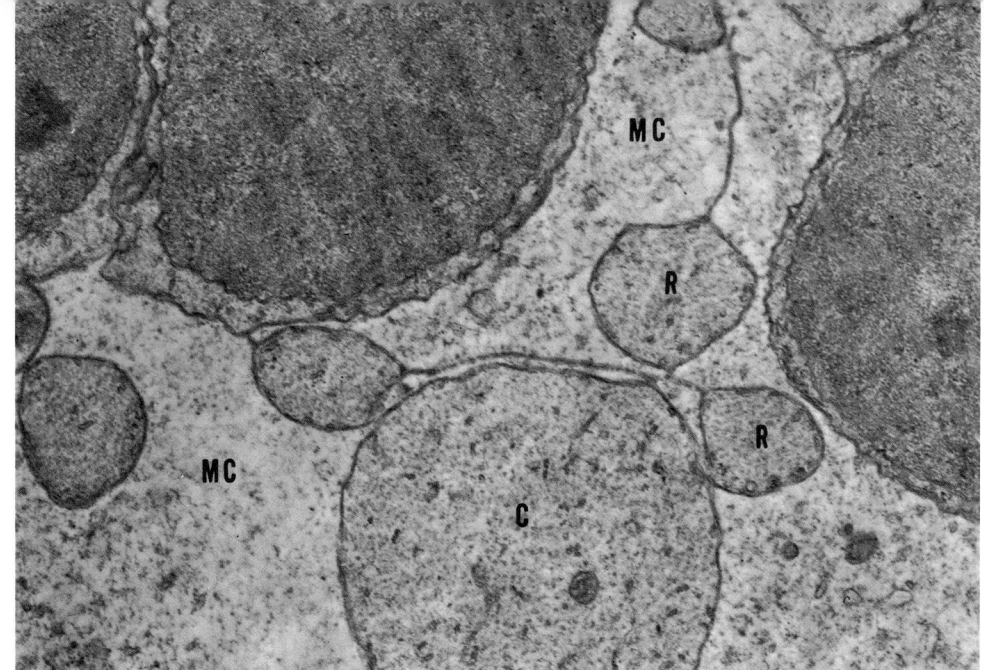

FIG. 6–35. Section through outer nuclear layer parallel to external limiting membrane. Cross-sections of large-diameter cone axons (C), small-diameter rod axons (R), and cell bodies are enveloped by lucent Müller cell cytoplasm (MC). Nasal retina, ×12,000.

FIG. 6–36. Henle fiber layer consisting of packed photoreceptor axons (C, R) surrounded by Müller cell cytoplasm (MC). ×16,000.

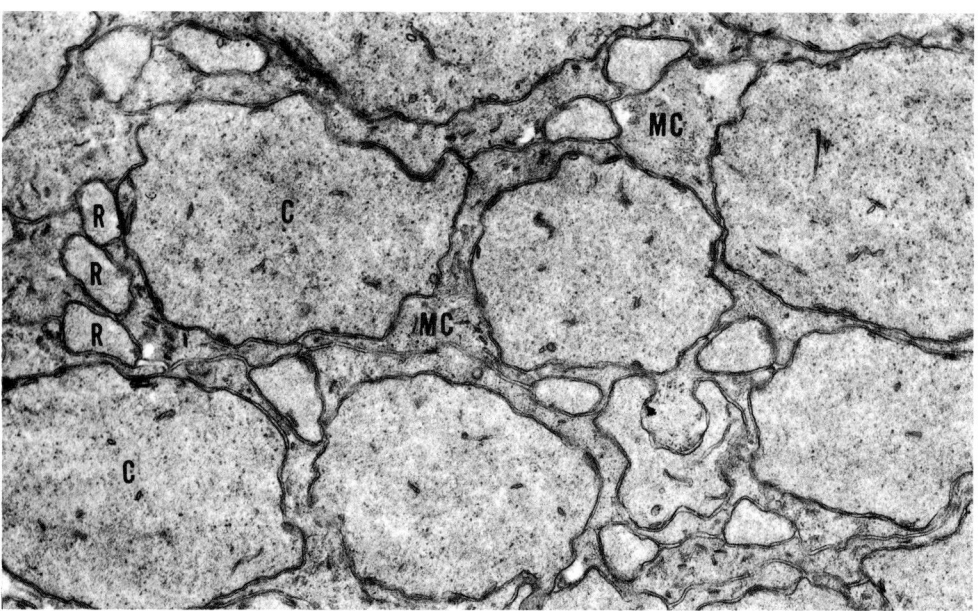

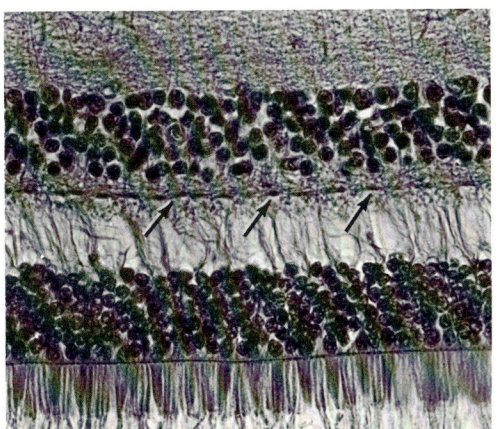

FIG. 6–37. Light micrograph showing plane of photoreceptor synapses (*arrows*) in zone of middle limiting membrane. H&E, ×305.

Occasionally, lateral projections of the photoreceptor expansions may be seen abutting against an adjacent expansion. These abutments are sometimes interpreted to function as lateral communicating channels.[11,13]

Inner Nuclear Layer

The inner nuclear layer (INL) contains the nuclei of the bipolar, the Müller, the horizontal, and the amacrine cells.

generally single within a rod spherule; occasionally, however, two lamellas are present in a spherule. In both varieties of synaptic expansions the lamellas are separated from the surface plasma membrane by a dense band of material known as the *arciform density* (Figs. 6–39 and 6–40).[38]

Mitochondria, although few in number, are also present within the synaptic expansions.

These arrangements within the photoreceptor synaptic expansions are repeated "in miniature" for the smaller synaptic expansions of the bipolar cells in the *inner plexiform layer* (Figs. 6–46 and 6–47).[21,25,33,40,52]

Similar arrangements are present in all synaptic expansions of the central nervous system except for the lack of synaptic lamellas and arciform densities. The latter structures are peculiar to the retina, the hair cells of the cochlea of the inner ear,[63] the photoreceptor-like cells in the pineal gland of amphibia,[32] and specialized sensory cells of certain fish.[3] Although their peculiar localization is clear, their significance is not.

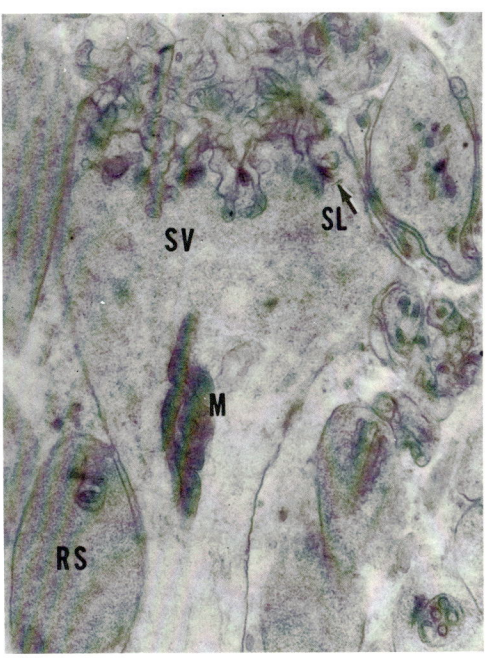

FIG. 6–38. Cone (foot) synaptic expansion. Basilar plasma membranes are inpocketed by bipolar dendrites. Mitochondria (*M*), region of synaptic vesicles (*SV*), and synaptic lamellas (*SL*) are present within expansion. *RS*, adjacent rod spherule synaptic expansions located along a plane external to that of cone feet. ×6,000.

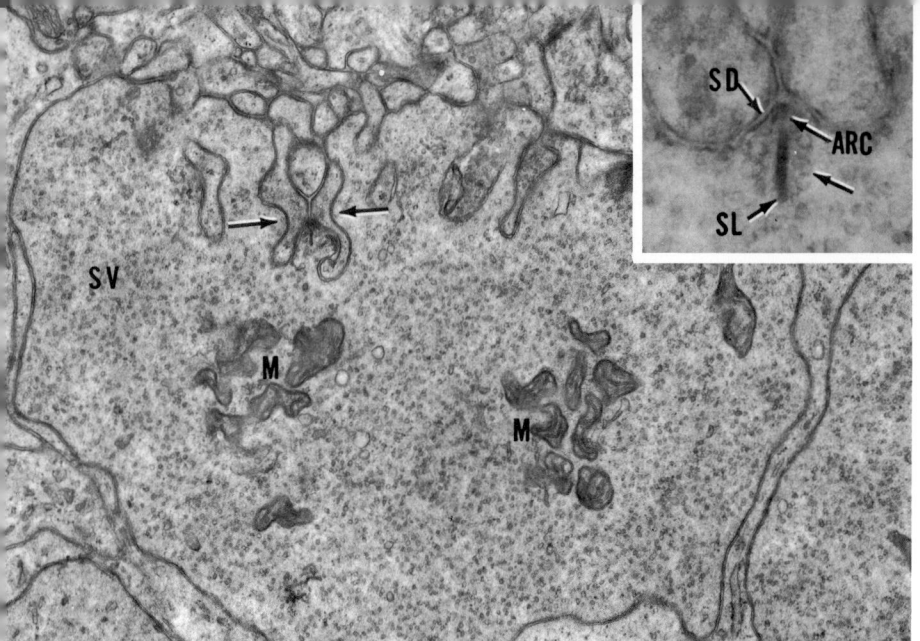

FIG. 6–39. Cone foot synaptic expansion. Synaptic vesicles (*SV*) and mitochondria (*M*) are evident. Impinging two dendrites (*arrows*) of adjacent bipolar cells envelop a portion of cone foot containing a synaptic lamella. **Inset** shows arrangement of synaptic lamella (*SL*), special halo of synaptic vesicles (*free arrow*), arciform density (*ARC*), and density of adjacent membranes (*synaptic densities, SD*). **Main figure,** ×16,000; **inset,** ×66,000.

Bipolar Cells. Bipolar cells possess typical dendrites which make up most of the narrow, truly plexiform portion of the so-called outer plexiform layer (Fig. 6–41). The dendritic stalks (Fig. 6–42) are filled with a considerable number of poorly oriented mitochondria, segments of agranular reticulum, and, especially with glutaraldehyde-fixation, large numbers of neurotubule-like structures or microtubules.

The many branches of the dendritic stalk are interwoven with other dendrites and with the axons of laterally communicating cells such as the *horizontal* cells. This basketweave arrangement of dendrites and axons forms a layer the strength of which is augmented by many desmosome-like attachment plates (Fig. 6–41). The zone of desmosome-like attachments of the photoreceptor synaptic expansions (*synaptic densities*) can be seen by light microscopy as a series of dashes in the innermost part of the outer plexiform layer (inset, Fig. 6–41) and is termed the *middle limiting membrane* (MLM).[29] This membrane has considerable strength and restraining ability much like that of the external limiting membrane. Its strength is explained by three morphologic observations: (1) the interweaving of the neurites, (2) the desmosome-like attachment plaques (or maculae adherentes), and (3) the synaptic densities (which possess the adhering property of desmosomes).

In pathologic tissues such a "membrane" may act as a temporary internal barrier to exudates which may accumulate between the photoreceptor axons ("Henle fiber layer" or "outer plexiform layer"). Since this membrane also demarcates the outermost capillary reach of the normal retinal vasculature, it may serve to direct neovascularization inward at least in the early stages of this pathologic form of endothelial proliferation.

The middle limiting membrane is observed easily in properly oriented conventional H&E-stained sections (Figs. 6–34, 6–37, and 6–41) and resembles the external limiting membrane (terminal bars) more closely than the internal limiting membrane (basement membrane). These relations

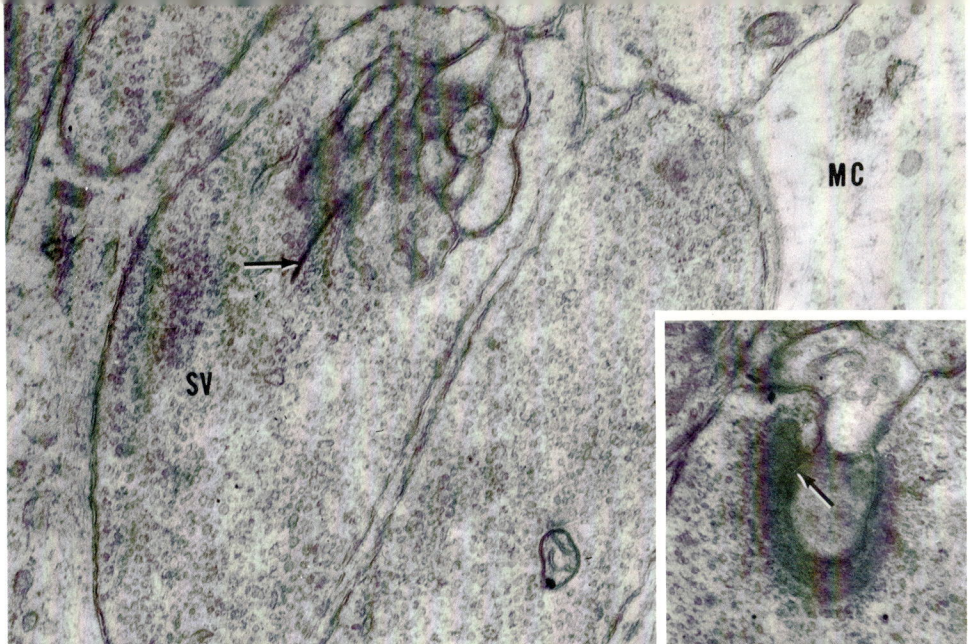

and differences are augmented by special staining techniques such as iron hematoxylin, which accentuates terminal bars and desmosomes, but not basement membranes.

The bipolar cell perikaryon (Figs. 6–43, 44) is generally scanty, and the axonal extension is quickly lost in the complexity of the inner plexiform layer. The axons, however, are easily recognized by the axial arrangement of their cytoplasmic components (like photoreceptor axons), particularly when their highly oriented microtubules are accentuated by initial glutaraldehyde fixation.

Accessory Neurons. Many forms of bipolar cells have been described based on their varying morphology as determined by metal impregnation or coating techniques.[6,49] Their various functional significances, however, are generally unknown. The nuclei of the Müller (glial) cells may lie at any level of the bipolar layer. Two special cells which often are considered to act as laterally communicating neurons are also present in this zone.[6] One, the *horizontal cell*, lies just anterior to the middle limiting membrane; the other, the *amacrine cell*, an analogous cell, lies near the innermost bipolar cell bodies (Fig. 6–18).

The horizontal cell is generally identified in the *human* retina by the presence of a

FIG. 6–40. Rod spherule synaptic expansion with its synaptic lamella (*arrow*) in cleavage plane formed by impinging dendrites of adjacent bipolar cells. *SV*, synaptic vesicles; *MC*, Müller cell. **Inset** shows almost complete longitudinal section of single lamella, illustrating its three-dimensional ribbon-like appearance. Special vesicle halo and adjacent arciform density (*arrow*) are also recognizable. **Main figure,** ×22,860; **inset,** ×32,800. (**Main figure** from Fine, B. S. *J Neuropath Exp Neurol* 22:255, 1963. **Inset** from Fine, B. S. In McPherson, A. (ed.) *New and Controversial Aspects of Retinal Detachment.* New York, Hoeber, 1968.)

peculiar body in its cytoplasm known in light microscopy as a Kolmer crystalloid.[34] By electron microscopy (Figs. 6–44 and 6–45),[41,72] this structure is seen to be composed of a superimposed stack of special tubules, large in diameter, and studded with ribonucleoprotein particles (identified by specific enzyme digestion and negative for PAS by the thiosemicarbazide method) on their surface (Fig. 6–45). The structure only superficially resembles a stack of granular endoplasmic retic-

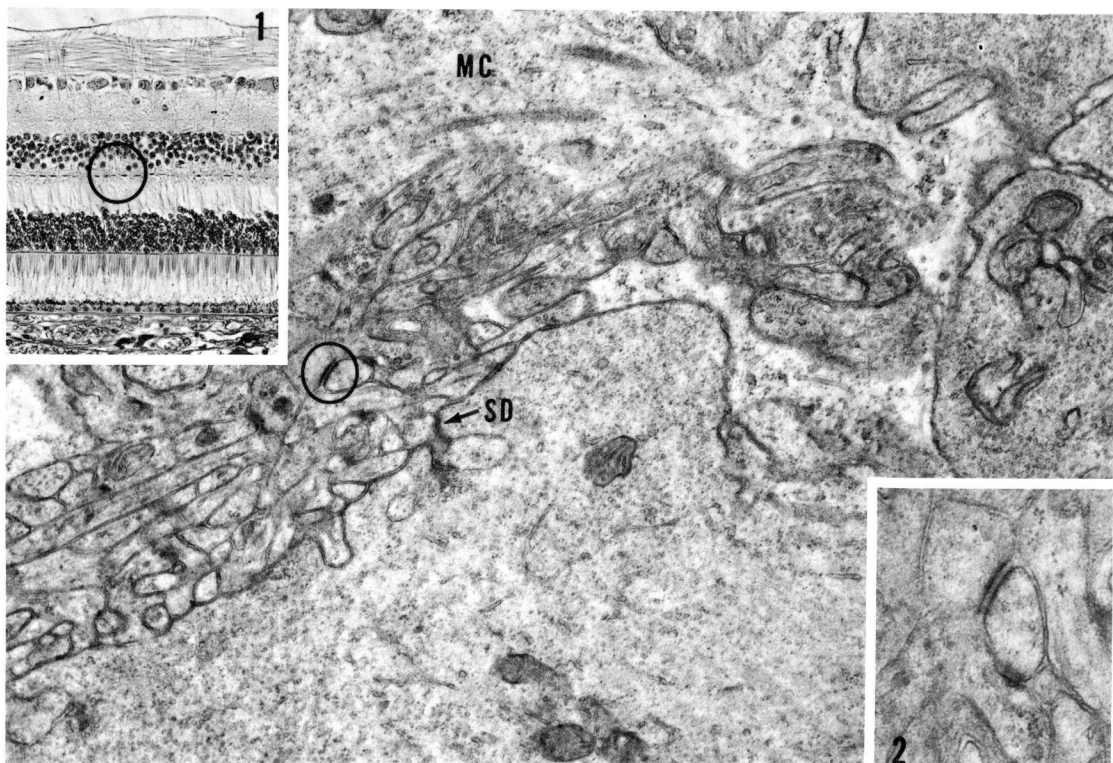

FIG. 6–41. Inset 1. Light micrograph with circle indicating approximate region illustrated in electron micrograph. **Main figure.** Electron micrograph showing cone foot synaptic expansion indented by bipolar dendrites. Desmosome-like densities (*circle*) are present attaching adjacent neurites to one another within plexiform zone. **Inset 2.** Enlargement of circled area. *MC,* Müller cell; *SD,* synaptic density. **Main figure,** ×18,000; **inset 1,** H&E, ×130 (AFIP Neg. 61-1960); **inset 2,** ×32,400. (**Main figure** from Fine, B. S. In McPherson, A. (ed.). *New and Controversial Aspects of Retinal Detachment.* New York, Hoeber, 1968.)

ulum; its function is entirely unknown. The analogous amacrine cell lacks such a characteristic structure but is generally identified by its lobulated nucleus[20] and its position along the innermost boundary of the bipolar cell layer.

Pathologic accumulations that may occur within the inner nuclear layer generally are localized to discrete pockets by the anteroposterior orientation of the entire bipolar cells much as they are similarly localized by the Henle fibers or the outer plexiform layer.

Inner Plexiform Layer

The inner plexiform layer (IPL) is composed of the axons of the bipolar and amacrine cells and their synapses, and the dendrites of the ganglion cells. The bipolar cell synaptic expansions are most easily identified as miniature photoreceptor synaptic expansions with mitochondria, typical synaptic vesicles, and a typical but much smaller synaptic lamella lying generally within the cleavage of two dendrites inpouching from a ganglion cell (Fig. 6–46). The appearance of these two inpouching dendrites is sometimes known as a *dyad** configuration,[21] but, if considered together with the associated synaptic expan-

THE RETINA   75

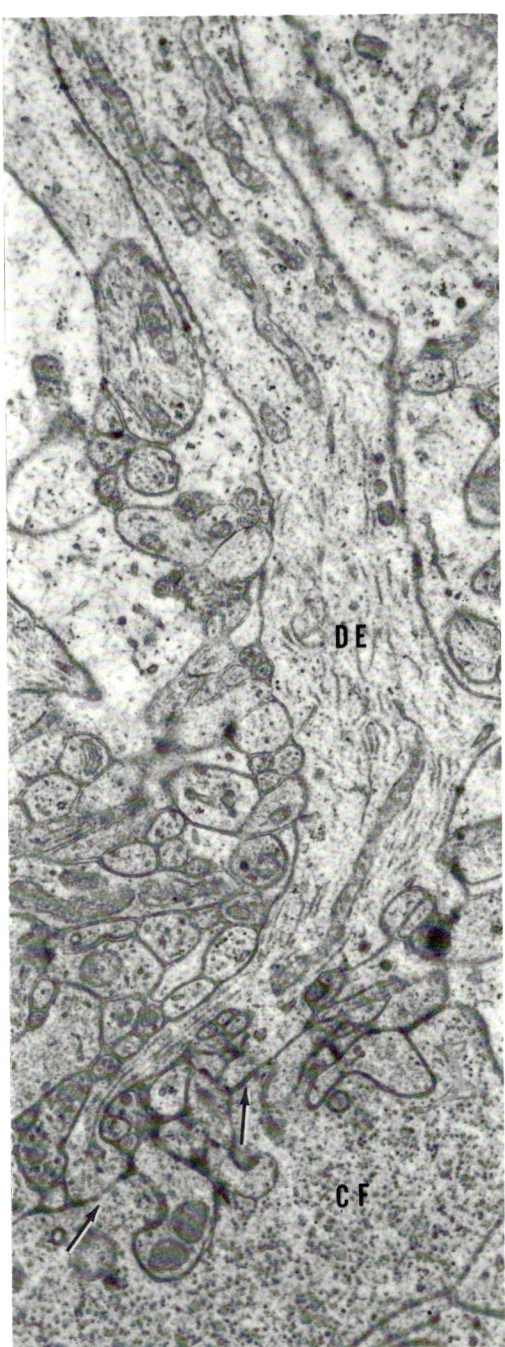

FIG. 6–42. Branches of single bipolar dendrite (*DE*) impinging (*arrows*) on cone foot synaptic expansion (*CF*). ×14,000.

FIG. 6–43. Light micrograph with *circle* indicating approximate region illustrated in Figure 6–44. Note that sectioning *across* Henle fibers produces an outer plexiform layer (*OPL*) that only *superficially* resembles the inner plexiform layer (*IPL*). H&E, ×300.

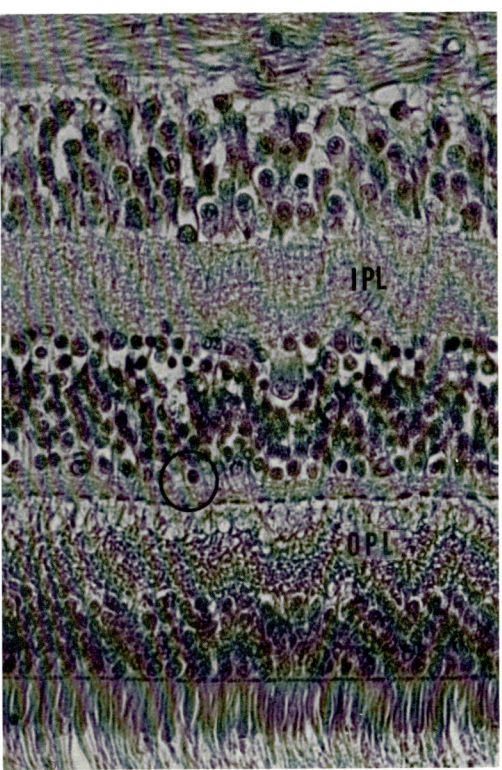

sion and its lamella, it might be called a *triad** configuration (Fig. 6–47).

Myriad synapses occur within the inner plexiform layer. In sections of bird retinas this layer is clearly subdivided into three or more zones, each of which is presumably a different level for synapse. In sections of human retinas only a very slight suggestion of such layering can be seen.

* These terms are useful only in proper context because the same terms, *dyad* and *triad*, are currently also used for specialized regions of agranular endoplasmic reticulum ("sarcoplasm") of striated muscle cells.[53]

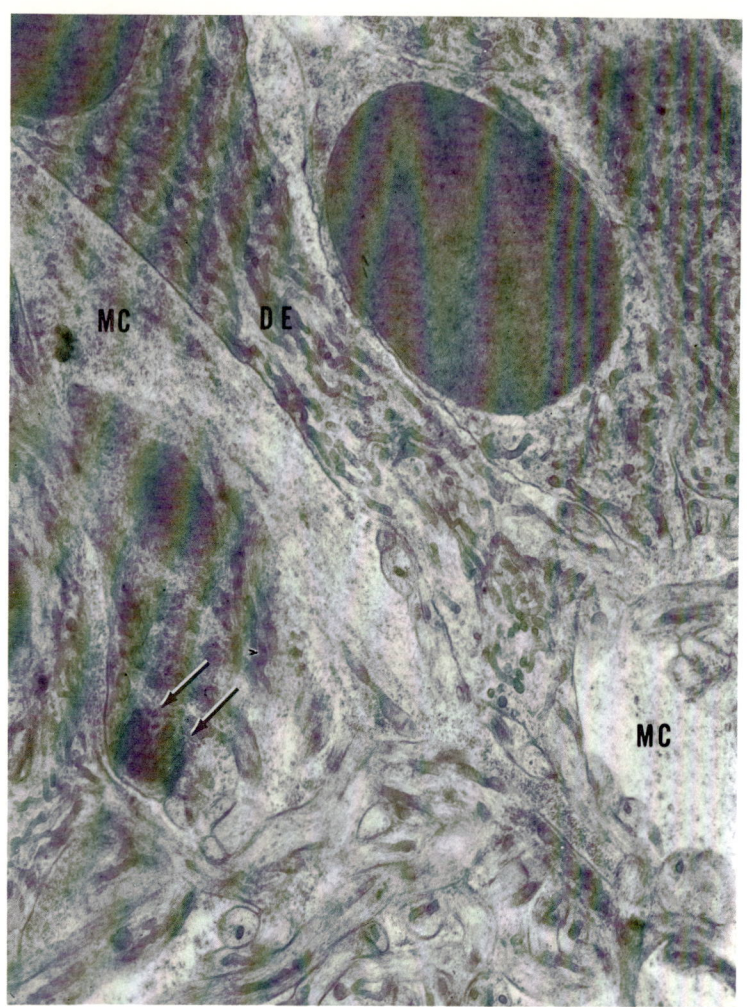

FIG. 6–44. Outer boundary of bipolar cell layer. Stout mitochondria-filled dendrites of a bipolar cell (*DE*) are prominent. A cluster of large tubules (Kolmer crystalloid) is present (*arrows*) in adjacent cell lying just internal to plane of middle limiting membrane. *MC*, Müller cell. ×7,500.

Ganglion Cell Layer

The large ganglion cell bodies form a single ganglion cell layer (GCL) throughout most of the retina (Fig. 6–48). The dendrites of the ganglion cells branch and resemble closely the dendrites of bipolar cells (Fig. 6–49). The cytoplasmic organelles of the ganglion cells differ little from those of other cells (mitochondria, Golgi complex, filaments or tubules) except for the presence of large aggregates of granular endoplasmic reticulum (*Nissl* substance or bodies) and many small bodies of varying density (Figs. 6–50 and 6–51).[4,24] With increasing age, the small bodies increase in density.

Dissolution of the basophilic areas of Nissl substance is an early sign of either physiologic or pathologic disturbance to the ganglion cell.

Nerve Fiber Layer

The axons of the ganglion cells are aggregated into nerve fiber bundles which pass parallel to the retinal surface through the arcades formed by the stout columns and footplates of the Müller (glial) cells (Figs. 6–52 and 6–53) to compose the nerve fiber layer (NFL) of the retina. The axons in tissues initially fixed in osmium tetroxide (Fig. 6–53) look different from those in tissues fixed initially in glutaraldehyde (Fig. 6–54).[26,35] With the former, the axoplasm contains

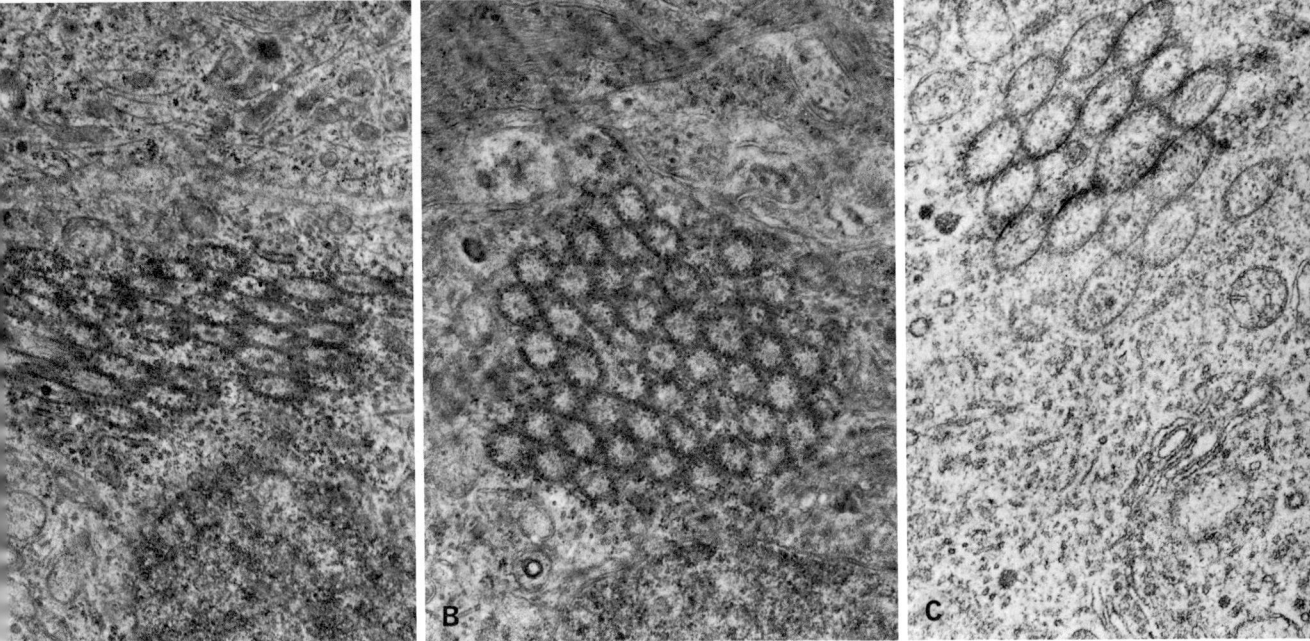

FIG. 6–45. Kolmer crystalloid. **A.** Oblique section showing ribosome-studded superimposed large tubules. **B.** Cross-section. **C.** Oblique section following TSC treatment. Granules are unaffected by TSC treatment. **A,** ×14,800; **B,** ×24,000; **C,** ×25,000.

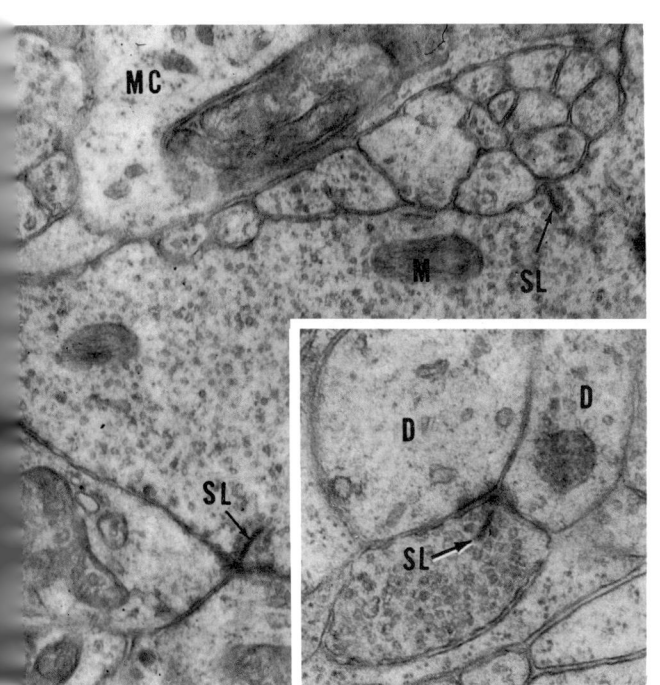

FIG. 6–46. Bipolar synaptic expansion within inner plexiform layer. Miniature synaptic lamellas (*SL*) as well as synaptic densities and vesicles are present. **Inset** shows triad arrangement of bipolar synapse with its lamella (*SL*) and the two adjacent dendrites (*D*). *MC,* Müller cell; *M,* mitochondrion. **Main figure,** ×21,800; **inset,** ×35,200. (From Fine, B. S. *J Neuropath Exp Neurol* 22:255, 1963.)

OCULAR HISTOLOGY

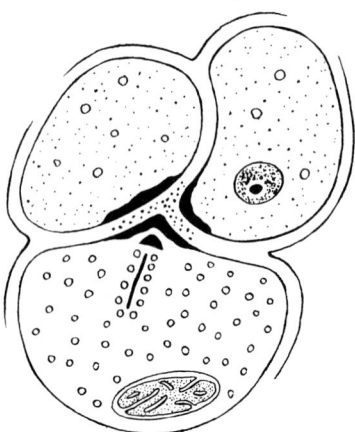

FIG. 6–47. Schematic drawing of bipolar triad. Synaptic expansion contains one or more mitochondria, synaptic vesicles, a synaptic lamella with halo of vesicles, and an arciform density. Dendrites are attached to synaptic expansion by cell plasma membrane densities (synaptic densities) with an intercellular cement.

myriad neurofilaments (Figs. 6–53 and 6–54) embedded in a very lucent ground substance. Mitochondria and some segments of agranular reticulum are present and longitudinally oriented, but the small particles (i.e., glycogenlike) of the photoreceptor axons are lacking. With initial glutaraldehyde fixation the neurofilaments are entirely replaced by *neurotubules*, and the ground substance is more prominent and more dense (Fig. 6–54). In the human retina the axons are unmyelinated because during embryonic development myelination ceases abruptly when it reaches the lamina cribrosa of the optic nerve head (Fig. 12–3) after proceeding from the chiasm down the optic nerve toward the retina.

Occasionally the process of myelination continues into the retina ("skipping" the lamina cribrosa) and may be observed by the ophthalmologist as a white patch with a serrated or "feathered" edge radiating from the nerve head or even as an isolated patch of myelinated nerves widely separated from the nerve head. Such anatomic variations are commonly brought to the attention of the student as points in differential diagnosis to prevent confusion with such pathologic changes as exudates, cottonwool spots, and retinal tumors. Since these white superficial patches are quite opaque, they may produce scotomas which can be clearly delineated on tangent screen (central field) examination. Some animals (e.g., the rabbit) normally possess a band of myelinated retinal nerve fibers (Fig. 6–55).

Internal Limiting Membrane

In the human retina the *internal limiting membrane* (ILM) is a thick (0.5 μ or more) basement membrane (Fig. 6–56).[23] Its inner or vitreal surface is extremely flat, whereas its outer surface follows closely the uneven surface of the Müller cell basal plasma membranes. Frequently the inner surface of the Müller cell footplates is deeply pocketed and the basement membrane fills this pocket (Fig.

FIG. 6–48. Light micrograph of a single layer of ganglion cells. Note abundant cytoplasm, basophilic Nissl bodies (*arrow*), and a dendrite (*DE*) of one of the ganglion cells extending into plexiform layer. H&E, ×520. (From Fine, B. S. *Arch Ophthal (Chicago)* 69:83, 1963.)

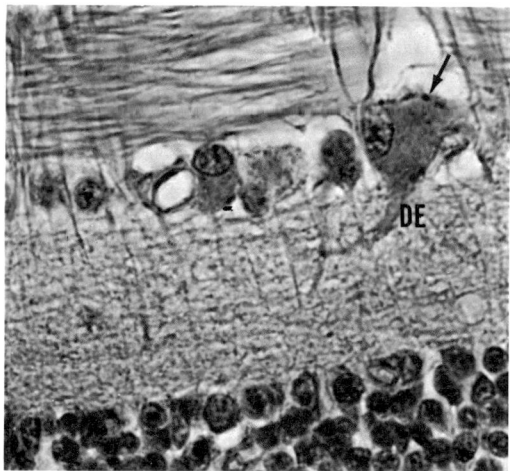

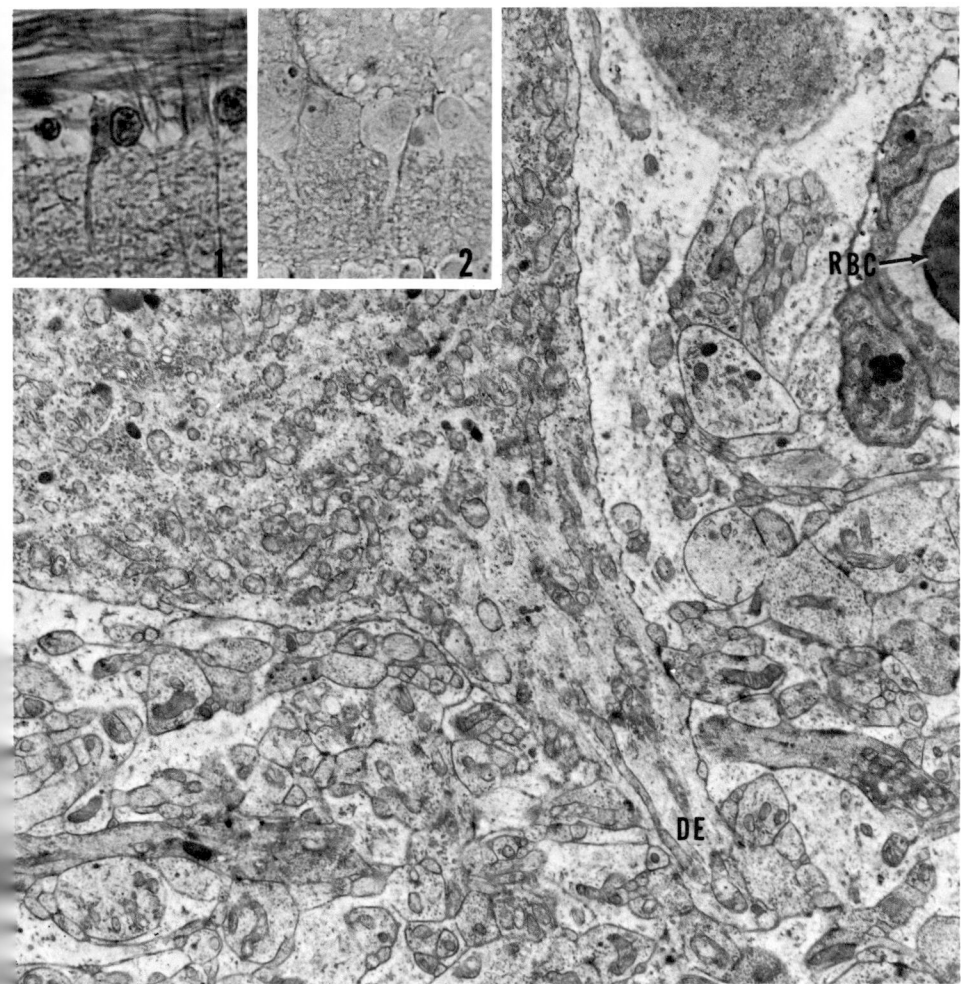

FIG. 6–49. Portion of a ganglion cell dendrite (*DE*) emerging from adjacent inner plexiform layer. Small blood vessel (*RBC*) is present at right. Insets show comparable regions by light microscopy: **inset 1** by conventional 8- to 10-μ section; **inset 2** by plastic-embedded 1.5-μ section. Main figure, ×7,700.

6–57). The inner plane of the basement membrane therefore is maintained. These basement-membrane-filled pockets are called *basement membrane facets* (Fig. 6–57).[27]

Reflections from the inner surface of the glass-like internal limiting membrane produce the retinal "sheen" that may be observed with the ophthalmoscope. Similarly, it is likely that the *basement membrane facets* (Fig. 6–58) are the anatomic reason for the normal glistening spots or "Gunn's dots" so frequently observed in this sheen.

Such anatomic observations may be of some value to the clinician in differentiating ophthalmoscopically a true, otherwise uncomplicated subvitreal hemorrhage from a sub-internal limiting membrane (i.e., superficial intraretinal) hemorrhage, for no sheen will be seen on the surface of the uncomplicated subvitreal hemorrhage, whereas it will be present over the intraretinal hemorrhage. A submembranous intraretinal hemorrhage should also be easily distinguishable from one into the deeper layer of nerve fibers (ganglion cell axons) by means of the characteristic "flame shape" of the latter. The internal limiting membranes of retinas of some animals, such as rabbit

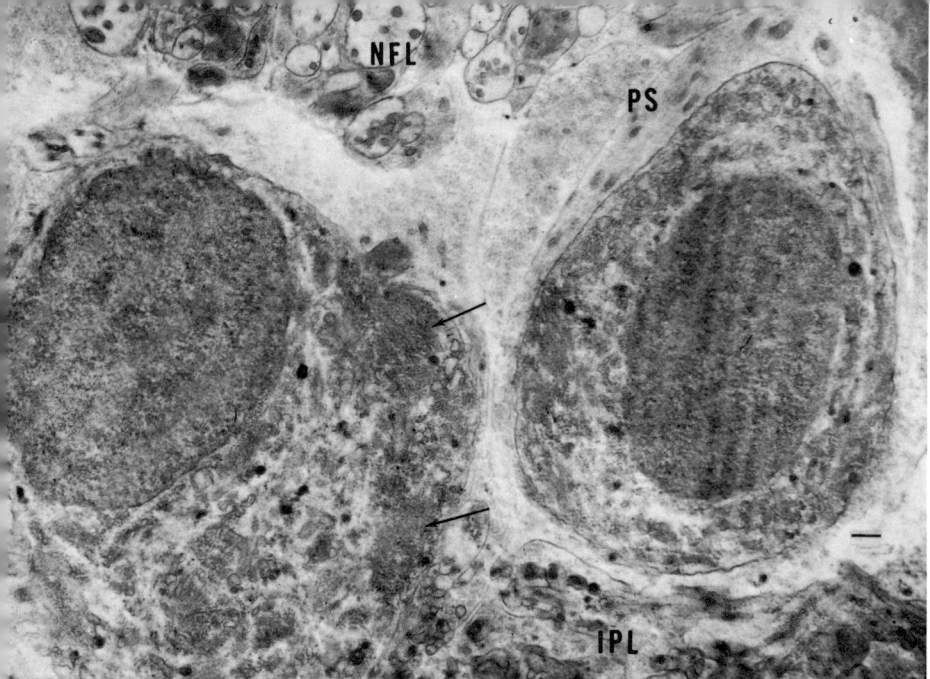

FIG. 6–50. Two ganglion cells from nasal retina lying between nerve fiber layer (*NFL*) and inner plexiform layer (*IPL*). Patches of Nissl substance (rough endoplasmic reticulum, *arrows*) are present, as are many dense cytoplasmic bodies. *PS*, nonspecific periganglion satellite cells. ×4,500. (From Fine, B. S. *Arch Ophthal* (Chicago) 69:83, 1963.)

FIG. 6–51. Higher magnification of portion of ganglion cell. Basophilic Nissl granules or substance are formed of clusters of granular or rough endoplasmic reticulum (*RER*). Many dense bodies (*D*) are present throughout cytoplasm. *M*, mitochondria; *MC*, Müller cell; *NUC*, nucleus of ganglion cell. ×24,000.

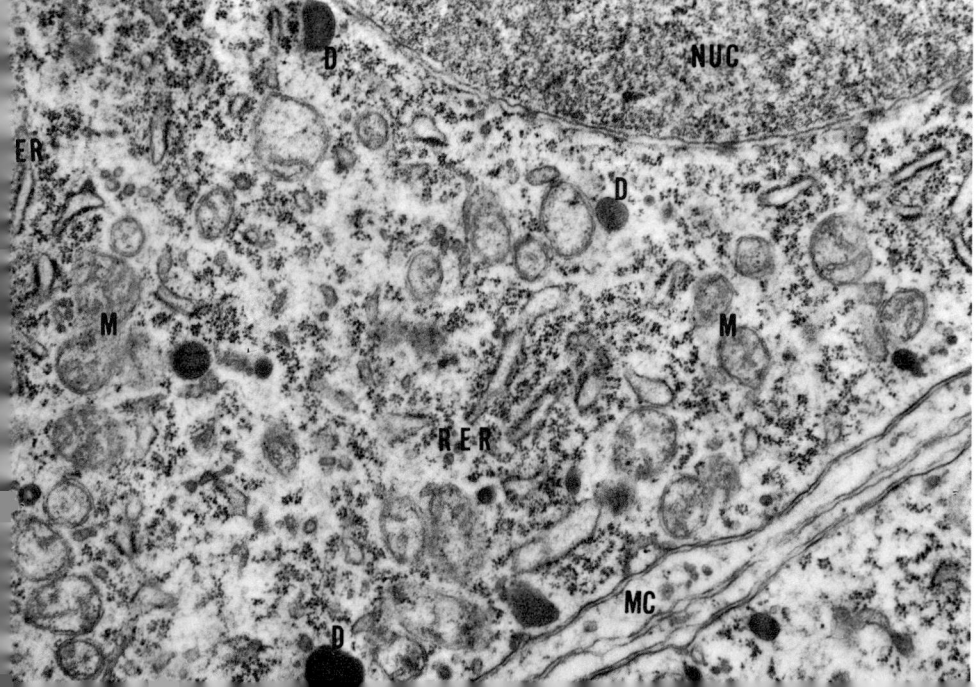

FIG. 6–52. Nerve fiber layer (*NFL*) of retina passing through arcades formed by stout columns and footplates (*arrows*) of Müller cells. *GC*, ganglion cell; *ILM*, internal limiting membrane. 1.5-μ section, PD, ×350. (AFIP Neg. 69–7570.)

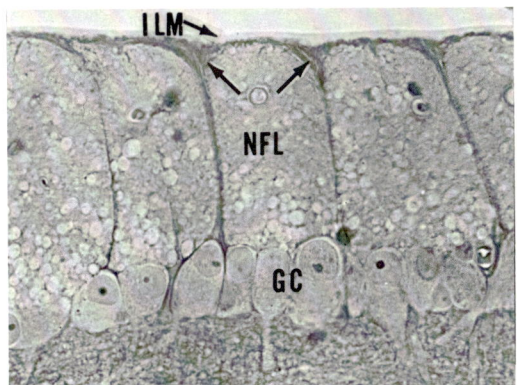

FIG. 6–53. Bundle of nerve fibers cut in cross-section. Their lucency is due to initial osmium fixation (filaments plus lucent ground substance). Adjacent Müller cell cytoplasm (*MC*) is dense, filament-filled in the columns, but relatively filament-free between axons. M_1, Müller cell mitochondrion; *M*, cross-sectioned axonal mitochondria. **Inset.** Light micrograph showing approximate region (*circle*) of electron micrograph. *Arrow* points to anterior surface of internal limiting membrane. **Main figure,** ×14,000; **inset,** 1.5-μ section, PAS-hematoxylin, ×1,500 (AFIP Neg. 61–2158). (**Main figure** from Fine, B. S., and Zimmerman, L. E. *Invest Ophthal* 1:304, 1962. **Inset** from Fine, B. S. *Arch Ophthal (Chicago)* 66:847, 1961.)

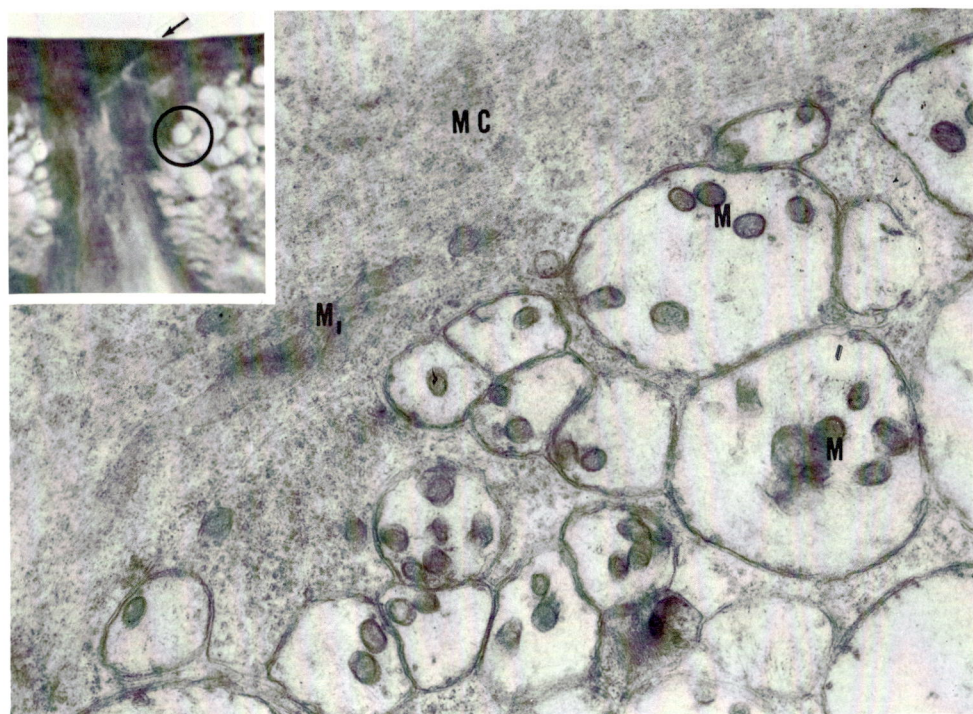

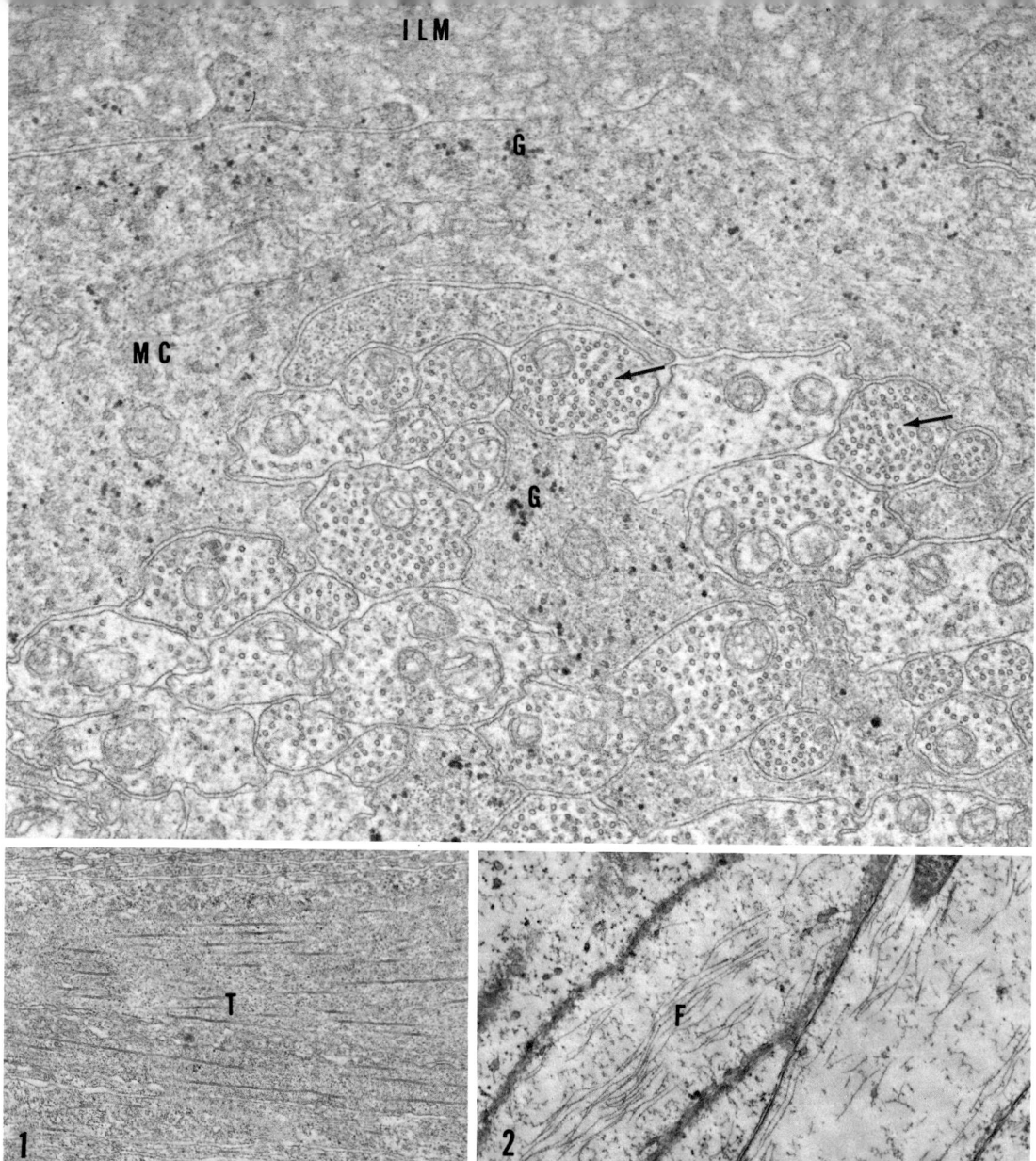

FIG. 6–54. Ganglion cell axons in tissue fixed initially in glutaraldehyde. Microtubules (*arrows*) within axons are seen here mainly in cross-section. Müller cell (*MC*) cytoplasm contains glycogen particles (*G*) and envelops many of the axons. Many axons are not separated by Müller cell cytoplasm. **Inset 1** shows microtubules (*T*) in longitudinal section. **Inset 2** shows filaments (*F*) within axoplasm in tissue fixed initially in osmium tetroxide. *ILM*, internal limiting membrane. All ×16,800.

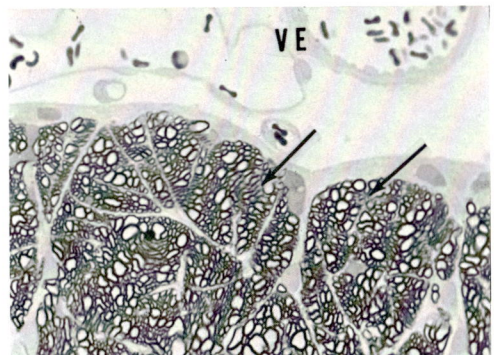

FIG. 6–55. Rabbit retina showing band of myelinated nerve fibers (*arrows*) normally present in the species. Retinal vasculature (*VE*) in this species is preretinal. 1.5-μ section, PD, ×395. (AFIP Neg. 66–6700).

FIG. 6–56. Internal limiting membrane of retina (*ILM*), a thick basement membrane closely applied to basal plasma membranes of Müller cells (*arrows*). Inner surface of this basement membrane is smooth. *FIL*, Müller cell cytoplasmic filaments; *MC*, Müller cell. ×11,700.

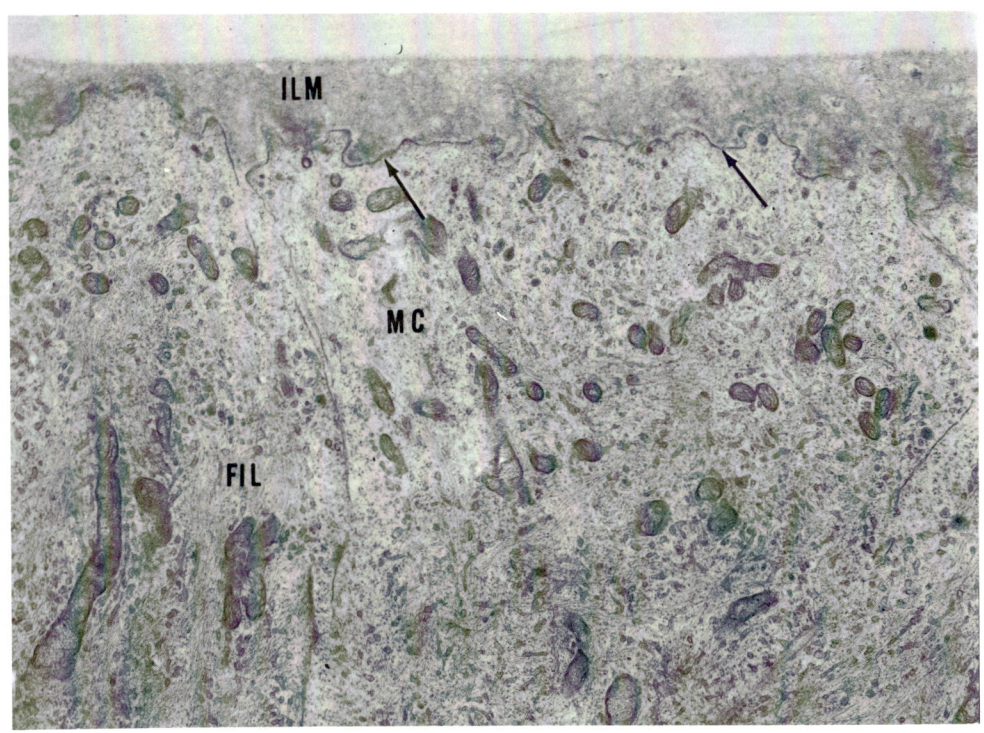

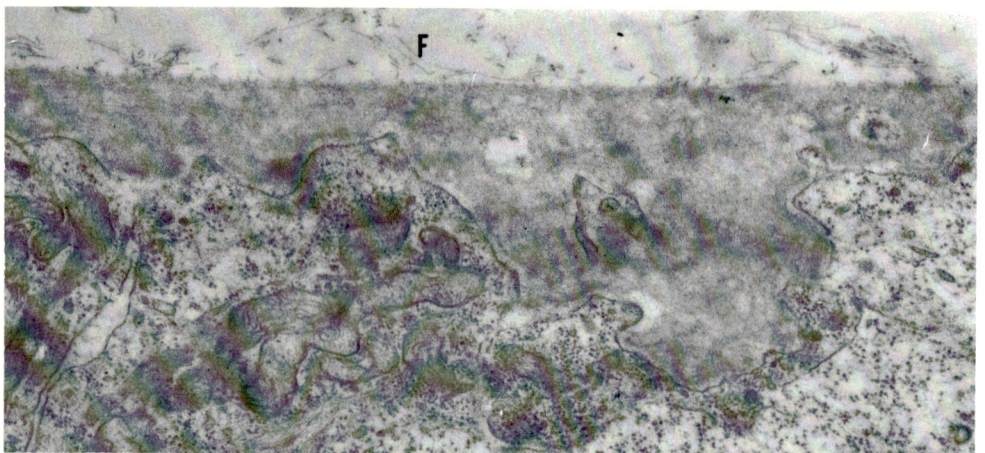

FIG. 6–57. Deep pocket filled with internal limiting membrane, a basement membrane facet. Vitreous filaments (F) are present, attached to smooth inner surface of membrane. ×22,500. (From Fine, B. S. In McPherson, A. (ed.). *New and Controversial Aspects of Retinal Detachment.* New York, Hoeber, 1968.)

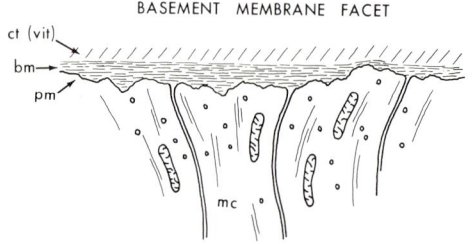

FIG. 6–58. Schematic drawing of basement membrane facet. *bm,* basement membrane; *ct (vit),* vitreous filaments; *MC,* Müller cell; *pm,* plasma membrane.

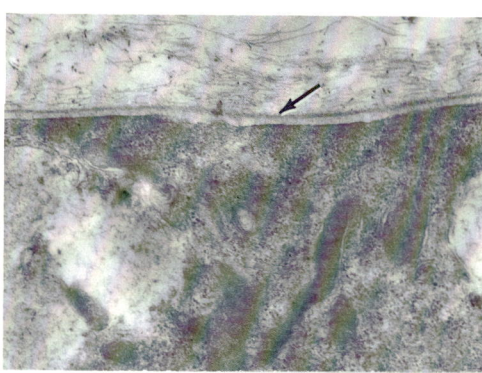

FIG. 6–60. Thin basement membrane (internal limiting membrane, *arrow*) of frog retina. ×13,250.

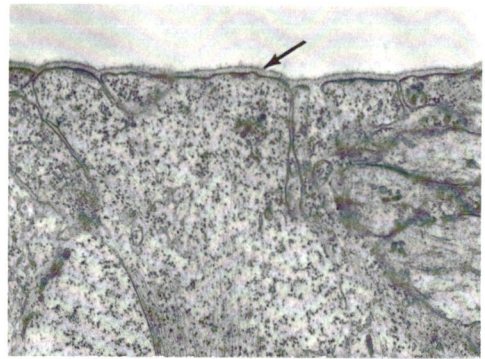

FIG. 6–59. Thin basement membrane (internal limiting membrane, *arrow*) of rabbit retina. ×16,000.

FIG. 6–61. (*Opposite page.*) Intermediate zone of Müller cell lying just apical to its nucleus. Golgi complex (*GC*) is present in this region. Lucent cytoplasm (*MC*) of Müller cell permeates plexiform layer as well as all levels occupied by photoreceptor cells internal to external limiting membrane. *N,* nuclei of bipolar cells; *PS,* photoreceptor synaptic expansions. ×15,500. (From Fine, B. S., and Zimmerman, L. E. *Invest Ophthal* 1:304, 1962.)

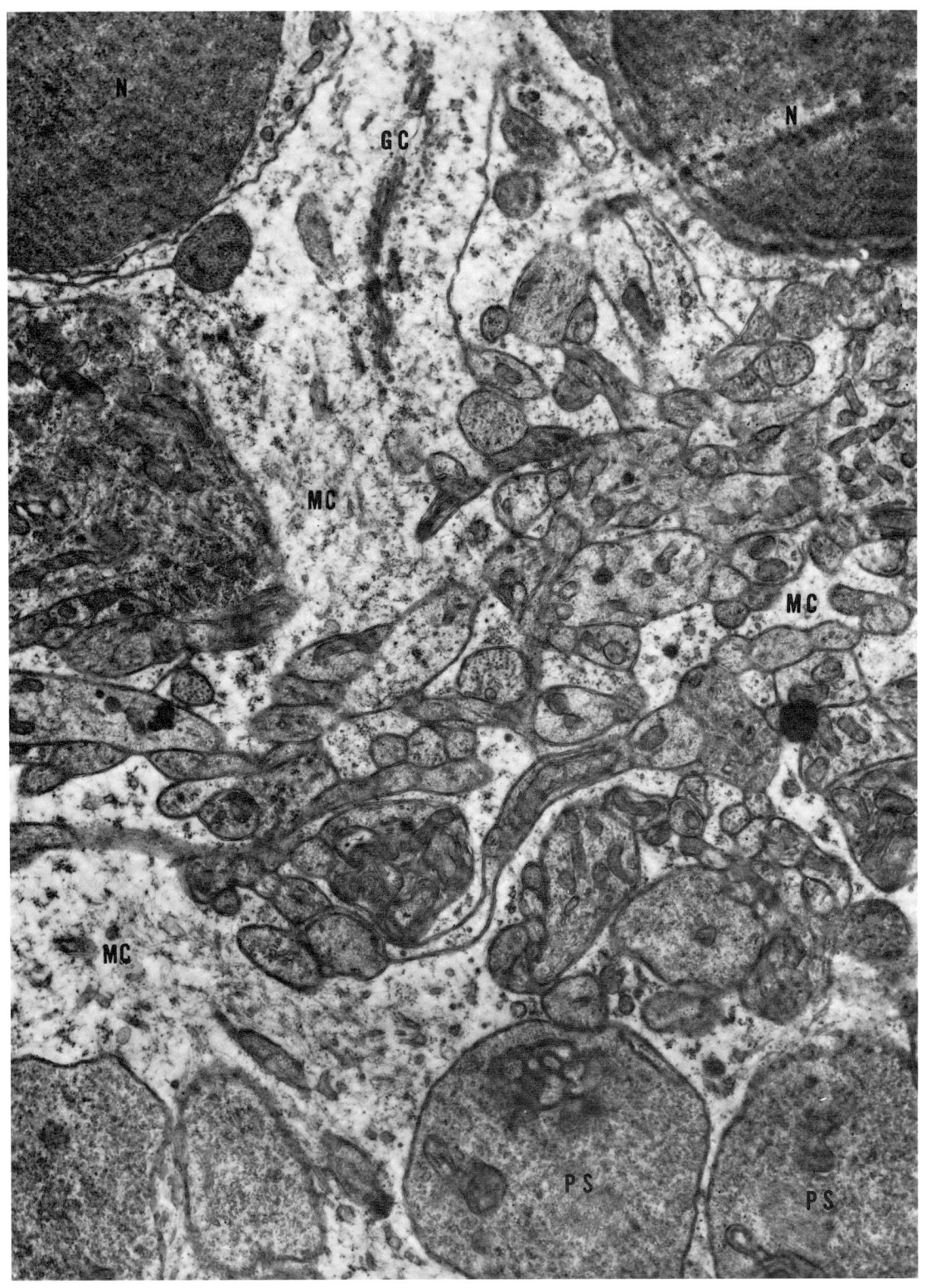

(Fig. 6–59) or frog (Fig. 6–60), are exceedingly thin and therefore produce no ophthalmoscopic sheen.

THE GLIAL SYSTEM

The Müller Cell

This, the largest glial cell in the retina, makes up the largest volume of glial tissue in the retina. The cell cannot be demonstrated in its entirety by conventional staining techniques, but requires special silver stains such as those of Golgi.[6] With that metal-impregnation method the cell can be seen to have at least two distinct parts: a stout inner trunk with rather small lateral branches and a delicate honeycomb-like outer part which forms delicate sheaths of cytoplasm around the cell bodies of the photoreceptor cells in the outer nuclear layer (Figs. 6–32, 6–34, and 6–35). Delicate villi, once known as the "fiber baskets of Schultze," can be seen protruding outward through the plane of the external limiting membrane (Fig. 6–33) to terminate between the inner segments of the photoreceptors.

By electron microscopic examination, the cytoplasm of the Müller cell is seen to be subdivided into three parts: the inner filamentous cytoplasm, the outer lucent cytoplasm, and the intermediate portion or transitional zone.

1. *The inner, dense cytoplasm.* The inner part of the Müller cell, especially its footplates or expansions, separates the nerve fiber bundles from the internal limiting membrane (Fig. 6–54) except in a few areas where smaller glia intervene. The stout portion of the cell contains many filaments aligned along its axis. From the stout filamentous columns, delicate, almost filament-free processes, containing glycogen particles (Figs. 6–53 and 6–54), pass between many of the axons in each nerve fiber bundle. The processes occupy much of the retinal space except for the typical 150 to 200 A intercellular spaces. The intercellular spaces, although appearing small, are probably highly functional as a pathway for ionic and nutritional exchanges.[46,62]

2. *The outer, lucent cytoplasm.* This easily disrupted part of the cell contains clusters of mitochondria near the external limiting membrane and some filaments (i.e., microtubules in tissue fixed initially with glutaraldehyde).

3. *The intermediate portion or transition zone.* This zone of the Müller cell lies apical to its nucleus (Fig. 6–61) and contains the poorly developed Golgi complex as well as a few segments of granular endoplasmic reticulum. Individual particles of glycogen are widely dispersed within the lucent cytoplasm.

Presumably, the intracellular filaments are partly the reason why, in formalin-fixed retinas (Fig. 6–16), the stout inner columns of the Müller cells are seen easily and thus called Müller fibers. The inner segments of the cells also appear more "fibrous" when the nerve fiber bundles are cross-sectioned (Fig. 6–52). The outer segments or cytoplasmic leaflets lack the large numbers of filaments and dense ground substance, so they are almost impossible to detect in conventional H&E-stained sections except perhaps by their absence or "negative" staining. Small hemorrhages can track easily along the axons within the nerve fiber layers because the delicate Müller cell cytoplasmic filament-free processes offer little resistance. The distribution of blood in the nerve fiber layer produces the "feathered-edge" hemorrhage observed by the clinician, an observation which identifies for him the retinal layer involved.

Accessory Glia

The smaller glia, fibrous and protoplasmic astrocytes and oligodendrocyte-like cells, normally are found only in the nerve fiber, ganglion cell, and inner plexiform layers. As their names imply, they are roughly identified by the concentration of their cytoplasmic filaments. The small astrocytes (Fig. 6–62) probably are responsible for much of the retinal gliosis in pathologic conditions. They may also undergo a form of degeneration which produces small intracytoplasmic bodies known

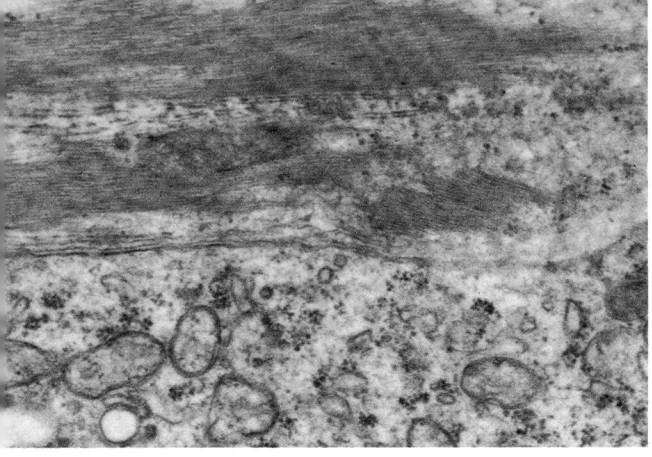

FIG. 6–62. Fibrous astrocytes, recognizable by their cytoplasmic content of filaments. Macula, ×24,000.

as *corpora amylacea*[27,51] found in the inner retinal layers (Fig. 6–63). The properties of the oligodendrocyte-like cells in the retina* are not clear because in the retina the cells do not produce myelin as they do in the central nervous system. Other cells are found which lack some of the clear-cut criteria for astrocytes, but because of their close association with ganglion cells they have been called, in a nonspecific way, periganglion cells (like perineuronal cells of the central nervous system).

Microglia, the central nervous system counterpart of the histiocytes or macrophages found elsewhere, are difficult to identify in normal tissue. In fact, they generally are identified only when they have become macrophagic (have actually ingested ma-

---

* The presence of true oligodendroglia in the retina has long been disputed by those who seek to identify them by metal-impregnation techniques.[49] Some authors say that the Müller cells function in the retina as the oligodendroglia. By current electron microscopic cytologic criteria, however, a few cells are present that possess a number of the characteristics of oligodendroglia. Some of these latter cells may also represent "resting" microglia. They may also be another source for preretinal gliosis (i.e., in addition to fibrous astrocytes).

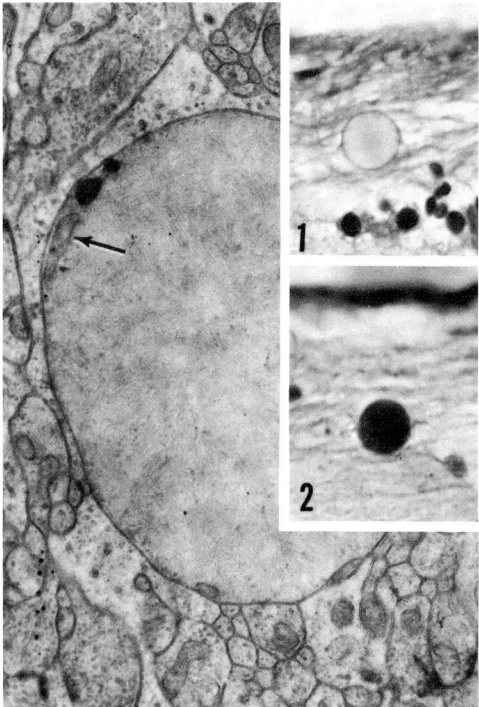

FIG. 6–63. Corpora amylacea formed from fibrous astrocytes. **Inset 1** shows single body in nerve fiber layer. **Inset 2** shows similar body stained with PAS. Electron micrograph (**main figure**) shows that a single body lies within a cell membrane and contains a few dense bodies as well as a mitochondrion (*arrow*). **Main figure,** × 12,000 (From Fine, B. S.[27]). **inset 1,** H&E, ×395 (AFIP Neg. 70–3670); **inset 2,** PAS, ×530 (AFIP Neg. 70–3846).

terial). Many of the macrophagic cells observed in pathologic tissues probably originate from the blood stream.

Various "nourishing" and "structural" functions frequently are attributed to these glial cells. Some evidence for their nutritional function is their high glycogen content[36] (Müller cells) (Fig. 6–54) and their uptake of necrotic materials (Müller cells and newly arrived macrophages). Evidence for their structural function is the aggregation of cytoplasmic filaments (inner ends of Müller cells, fibrous astrocytes).

THE VASCULAR SYSTEM

The human retina is vascularized in its inner layers outward to the level of the middle limiting membrane.* The choriocapillaris of the choroid mostly is responsible for nourishing the photoreceptor and enveloping Müller cell cytoplasm that lies external to the middle limiting membrane.

The retina has no real arteries or veins except possibly just adjacent to the optic disc. The largest vessels are arterioles and venules; the smallest are capillaries.

The capillaries are easily identified by their single endothelial lining and basement membrane (Fig. 6–64).[27,42] Outside the endothelial basement membrane is an interrupted layer of cells. These cells, pericytes (periendothelial or "mural"[37] cells), are also surrounded by their own basement membrane material which fuses with that of the capillary endothelium, giving an appearance of a cell "embedded in basement membrane." With aging, some of the basement membrane degenerates, producing dense bodies and a vacuolated appearance.

The observation of preferential loss of pericytes in relation to endothelial cells from the retinal capillaries is considered to be highly characteristic of diabetes mellitus.[71]

To the pericyte basement membrane are attached glial cells (generally Müller cells) or, in the larger vessels, collagenous connective tissue.

Terminal bars near the lumen attach the capillary endothelial cells to one another (Fig. 6–65). Near the attachments, the endothelial cells frequently project into the lumen in the form of villi or flaps which may, on occasion, be so marked as to produce bizarre configurations.

The endothelial cells possess micropinocytotic vesicles, as do similar cells of the cerebral capillaries.[27] The blood-brain barrier is considered to be within the capillary endothelial cell since the basement membrane and adjacent extracellular spaces are demonstrably permeable. It is highly likely that the blood-retina barrier is similar.

An arteriole is easily differentiated from a venule by its thicker wall, which takes a darker stain in flat preparations, and by its more circular cross-section. The arteriole lacks an internal elastic lamina and a continuous layer of smooth muscle cells. Smooth muscle cells are typical of those found elsewhere in the body and are separated by their own basement membranes (Fig. 6–66). The outermost muscle cells are generally separated by a layer of collagen fibrils or filaments ("adventitia") from the adjacent parenchymal cells.

The major retinal vessels (Fig. 6–67) arise as branches of their corresponding central retinal vessel in the optic nerve head. The central retinal artery lies on the nasal side of the central retinal vein. As they enter the retina, the artery and the vein divide into four main branches: the upper and lower nasal and the upper and lower temporal. The arterioles are slightly smaller in diameter than the corresponding venules. Generally mirroring the pattern of contiguous nerve fibers, these major vessels lie near the internal limiting membrane or not far beneath it.

---

* The zone where, in the terminology of the older anatomists,[54] the "neuroepithelial layers meet the cerebral layers," i.e., where the cerebral layers are vascularized.

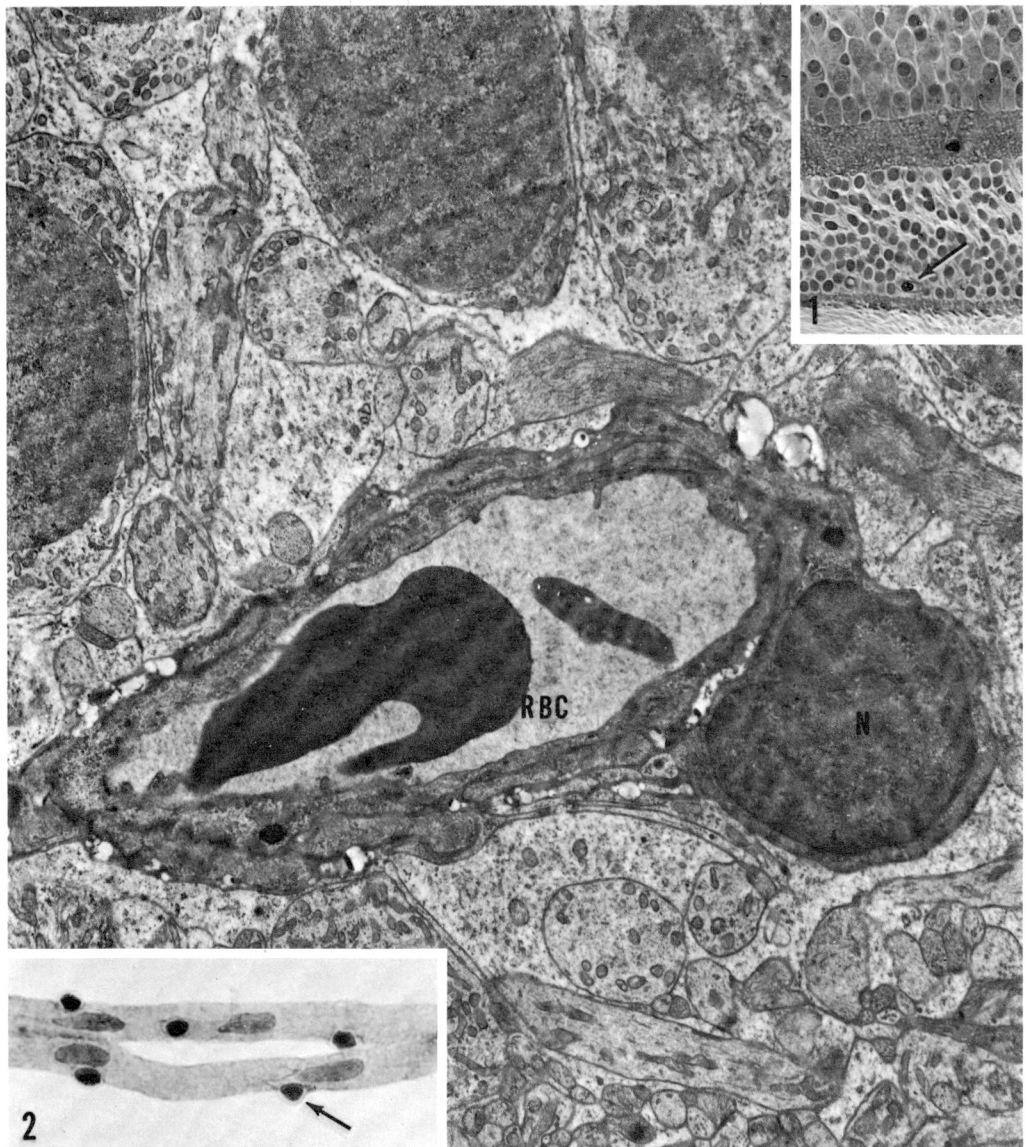

FIG. 6–64. Retinal capillaries. Inset 1. Light micrograph showing outermost retinal capillaries (*arrow*) lying just anterior to middle limiting membrane. **Main figure.** Electron micrograph of a similar capillary. Dense nucleus (*N*) of an adjacent pericyte is clearly seen. Pericyte is separated from both endothelium and adjacent parenchymal cells by its basement membrane. *RBC*, a red blood cell in lumen. **Inset 2.** Light micrograph showing segments of retinal capillaries following retinal digestion with trypsin. Large axial pale-staining nuclei of endothelial cells are easily distinguished from denser, rounded nuclei of pericytes. A thin "shell" of PAS-positive basement membrane can be seen around pericyte present in profile (*arrow*). **Main figure,** ×7,200; **inset 1,** 1.5-μ section, PD, ×210 (AFIP Neg. 63–6895); **inset 2,** PAS, ×440 (AFIP Neg. 64–7004–B). (From Fine, B. S. In McPherson, A. (ed.). *New and Controversial Aspects of Retinal Detachment.* New York, Hoeber, 1968.)

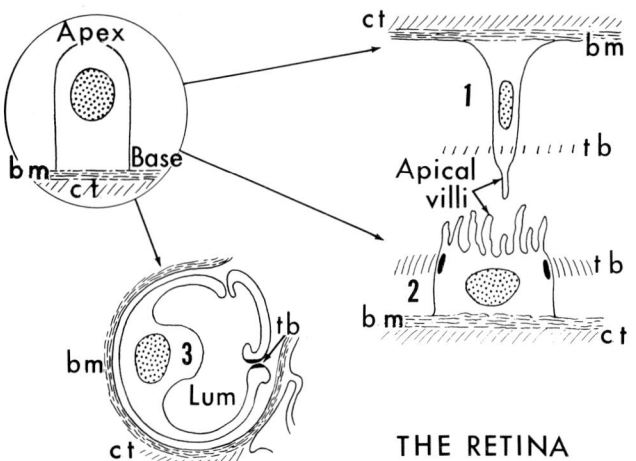

FIG. 6–65. Schematic drawing showing arrangement of retinal cells to form (1) Müller cells, (2) pigment epithelial cells, and (3) vascular endothelium. All these cells possess apical attachments and basement membranes (*bm*), *ct,* connective tissue; *tb,* terminal bars. (From Fine, B. S. In McPherson, A. (ed.). *New and Controversial Aspects of Retinal Detachment.* New York, Hoeber, 1968.)

THE RETINA

FIG. 6–66. Arteriole, with lumen (*LU*), endothelial lining (*EN*) with apical attachments (*TB*), and wall of smooth muscle cells (*SM*), all surrounded by layers of basement membrane. Parenchymal cells (*P*) are separated from outermost smooth muscle cells by basement membrane (*free arrow*) and layer of connective tissue (*CT*). **Inset.** Light micrograph showing comparable retinal arteriole near internal limiting membrane. **Main figure,** ×16,000; **inset,** H&E, ×300.

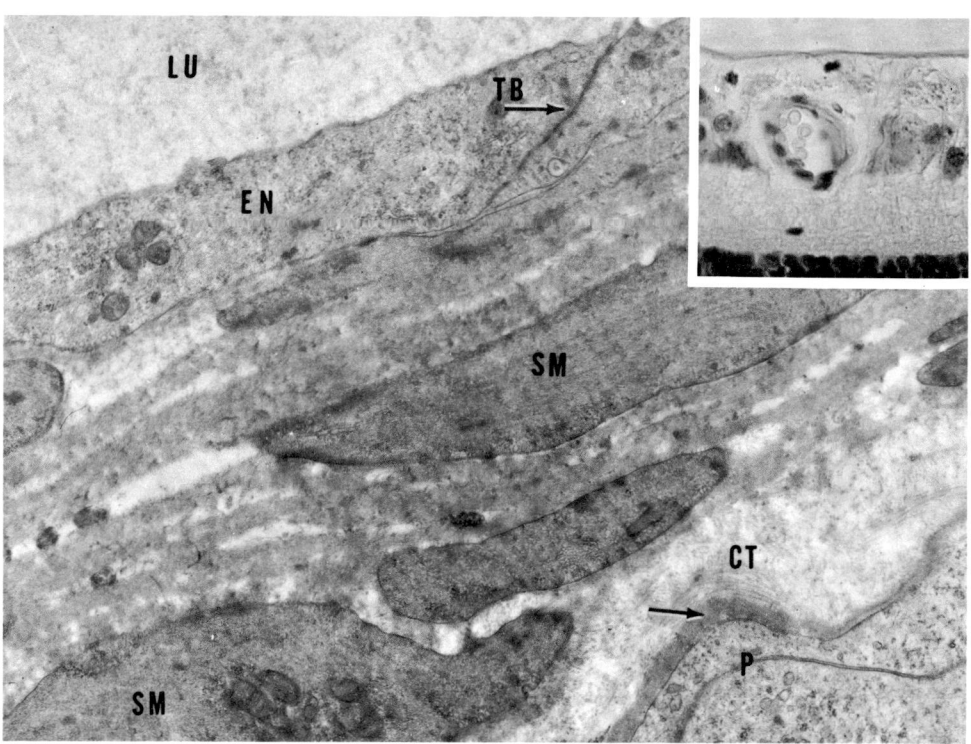

Where the vessels approach the surface, the internal limiting membrane may be extremely attenuated (Fig. 6–68),* thereby causing a "weakness" in this basement membrane analogous to that in the regions of the optic disc and fovea (Fig. 6–84).

The arteriole generally overlies (lies internal to) the corresponding venule (Fig. 6–73). Because both vessel walls normally are transparent, the clinician sees one column of lighter colored blood (oxygenated in the arteriole) overlying a column of slightly darker blood (deoxygenated in the venule). With aging or with disease processes which accelerate aging (e.g., diabetes, hyperten-

* By both light and electron microscopy the internal limiting membrane becomes attenuated wherever the cytoplasmic footplates of the Müller cells become attenuated (Figs. 6–68 and 6–69). A thin internal limiting membrane (Figs. 6–70 through 6–72) is therefore sporadically present along the distribution of the major retinal vessels. The foci of thin basement membrane may indicate sites of "weakness" through which early changes of preretinal disease processes (e.g., preretinal gliosis, early neovascularization) may gain access to the vitreous compartment. (Compare with the basement membrane lining the optic cup of the optic nerve head, Fig. 12–4).

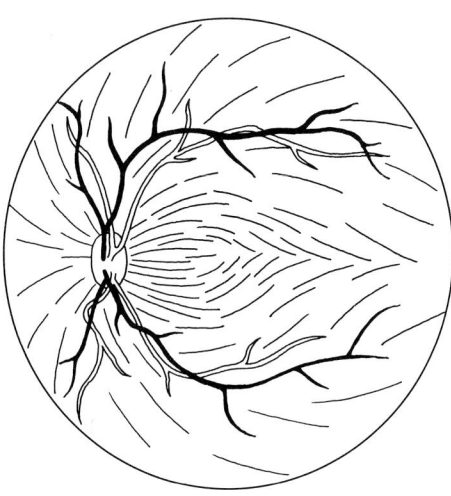

FIG. 6–67. Topography of major retinal vessels and their relation to adjacent nerve fiber layer.

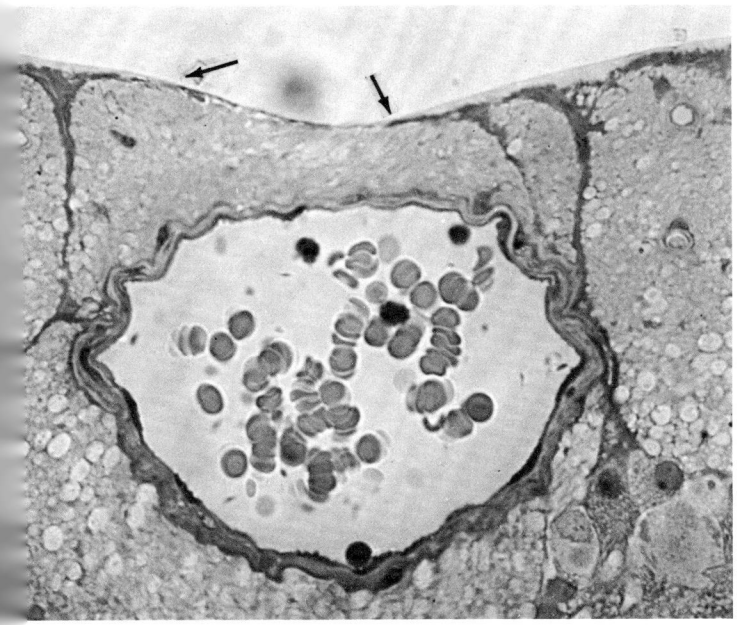

FIG. 6–68. Large vessel near retinal surface. Internal limiting membrane is attenuated (arrows), mirroring attenuation of adjacent Müller cell cytoplasm. 1.5-μ section, PD, ×485. (AFIP Neg. 69-7567.)

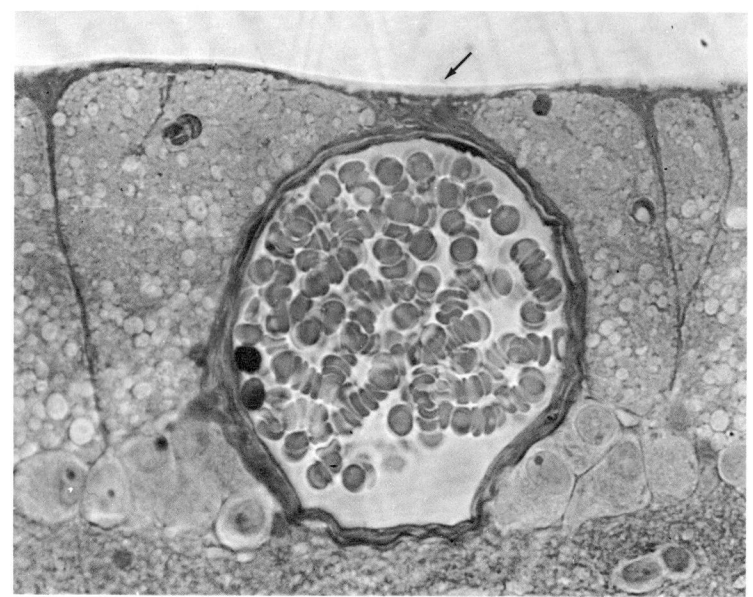

FIG. 6–69. Large vessel near retinal surface. Internal limiting membrane (*arrow*) remains thick, mirroring thick underlying layers of Müller cell cytoplasm. 1.5-µ section, PD, ×485. (AFIP Neg. 69–7566.)

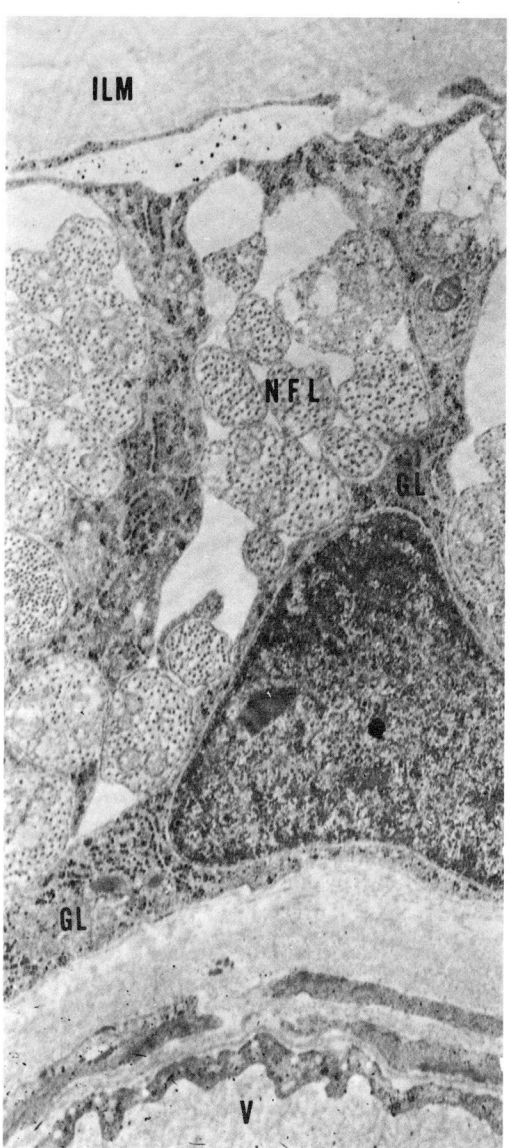

FIG. 6–70. Thin internal limiting membrane (*ILM*) overlying thin cytoplasmic leaflets of Müller and superficial glial cells (*GL*). The latter cells are perivascular astrocytes, as determined by their cytoplasmic content of filaments and glycogen. Cells are also associated with a basement membrane. Edge of large retinal vessel (*V*) is present just external to small glial cell. *NFL*, nerve fiber layer. ×16,500.

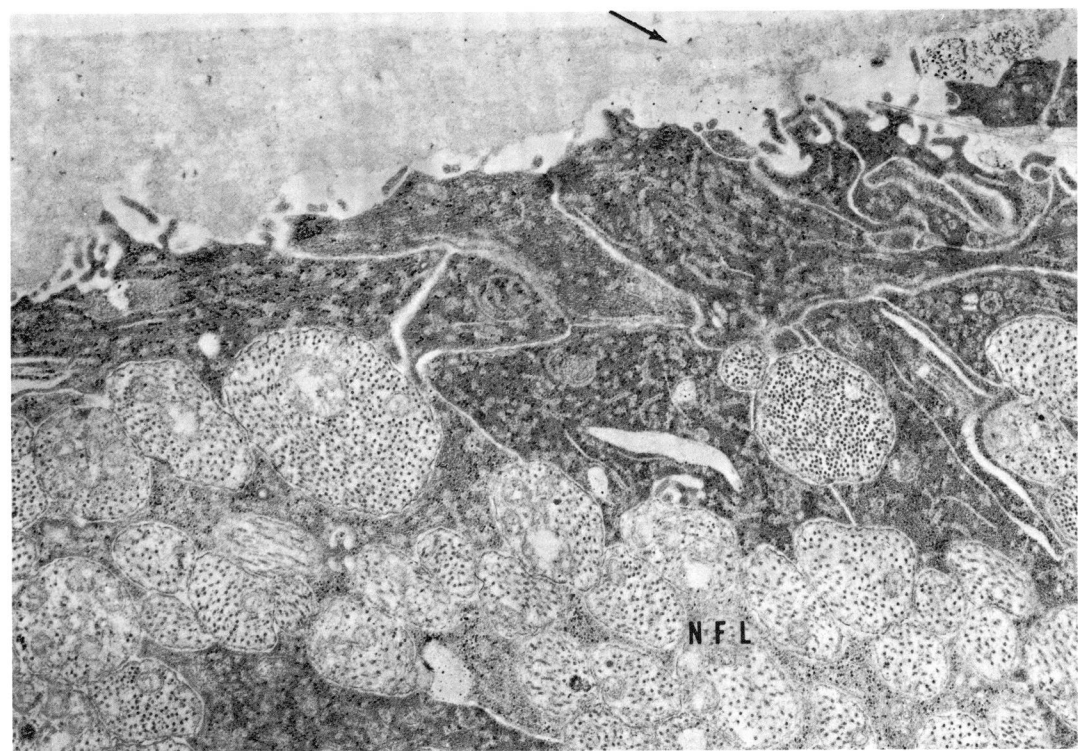

FIG. 6–71. Internal limiting membrane peripheral to that in Figure 6–70, overlying large retinal blood vessel. Transition of a thick to a thin internal limiting membrane can be seen at *arrow*. Dense cytoplasm underlying membrane is that of Müller cell. Axons of nerve fiber layer (*NFL*) are clearly seen, mostly cut in cross-section. ×16,500.

sion), the arterial wall may thicken (arteriolar sclerosis) and hide from view a small segment of the underlying venular blood column ("arteriolar-venous nicking").

In approximately 25 per cent of the population a retinal vessel arises directly from the choroidal vasculature at the temporal edge of the optic disc (the *cilioretinal artery*) to nourish much of the macula and the papillomacular bundle.

Central retinal artery occlusion occurring in a person having a cilioretinal artery may have little effect on his central visual acuity. Conversely, embolization of such a cilioretinal vessel (a rare event) may severely damage central acuity, leaving the peripheral vision intact. An accompanying vein is rare.

A relatively wide capillary-free zone lies adjacent to the arterioles (Figs. 6–74 and 6–75). A similar but much less prominent capillary-free zone accompanies the venules.

The retinal vasculature terminates in delicate arcades in the periphery ∼ 1 mm short of the *ora serrata*, and the arterioles are therefore sometimes known as "end arteries."

TOPOGRAPHIC VARIATIONS

Macula

Temporal to the optic disc is an area of posterior fundus of larger size than the optic disc which appears clinically to be avascular. This is the clinical *macula*. A tiny glistening reflex of light arises from the center of this area. The reflex arises from an anatomic depression in the macula, the *fovea*. On gross examination an oval zone of yellow coloration can often be seen around the fovea. This yellow zone is the *macula lutea* (yellow spot).

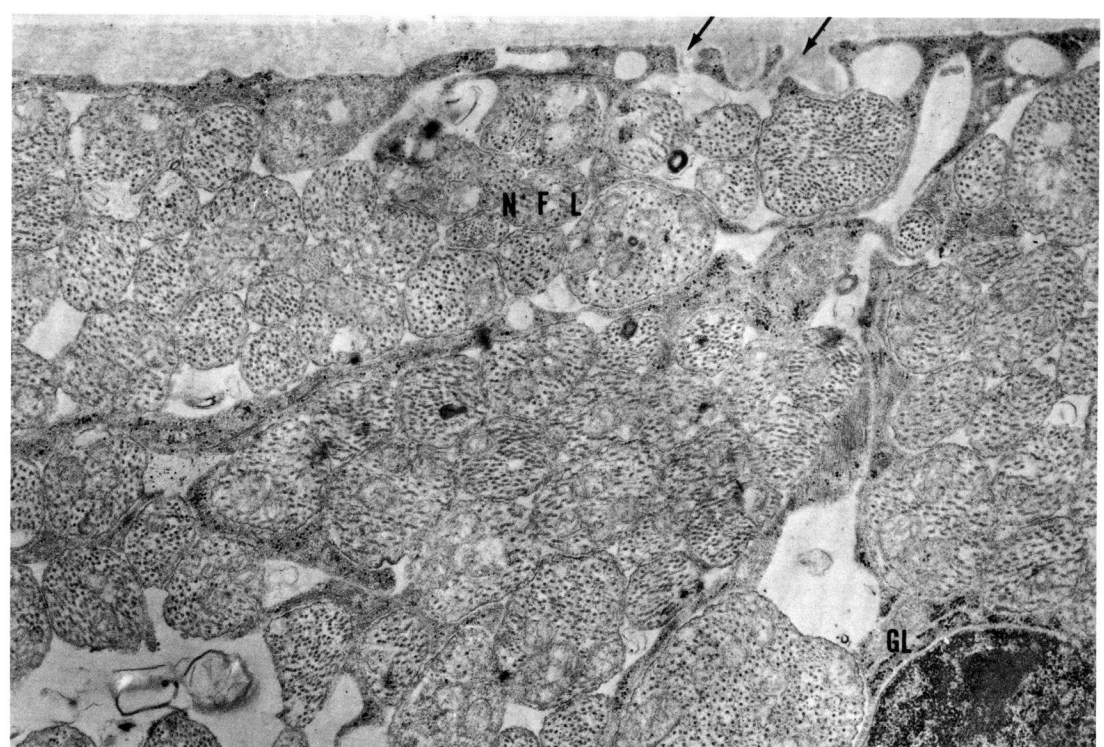

FIG. 6–72. Internal limiting membrane, thinnest where underlying glial cell cytoplasm is minimal. These regions of membrane thinning suggest regions of weakness. Note "loose" spaces occupied by basement membrane (*arrows*) along retinal surface in close relation to its vasculature. *GL,* perivascular astrocyte; *NFL,* nerve fiber layer. ×16,000.

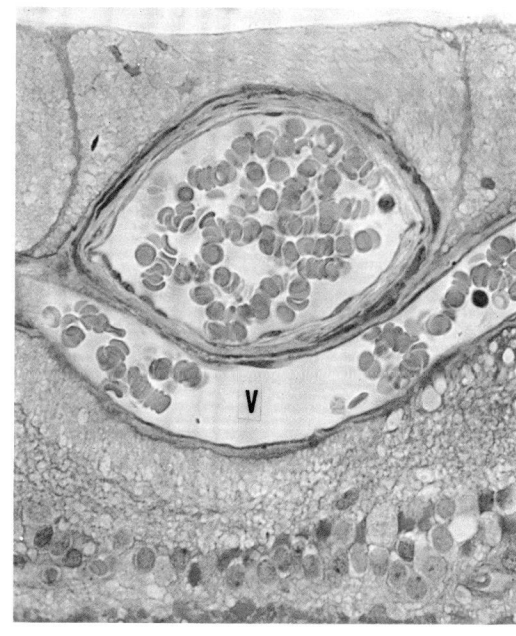

FIG. 6–73. An arteriovenous crossing. Arteriole generally crosses internal to venule (*V*). Adventitia of the two vessels is here in continuity. 1.5-μ section, PD, ×350. (AFIP Neg. 69–7564.)

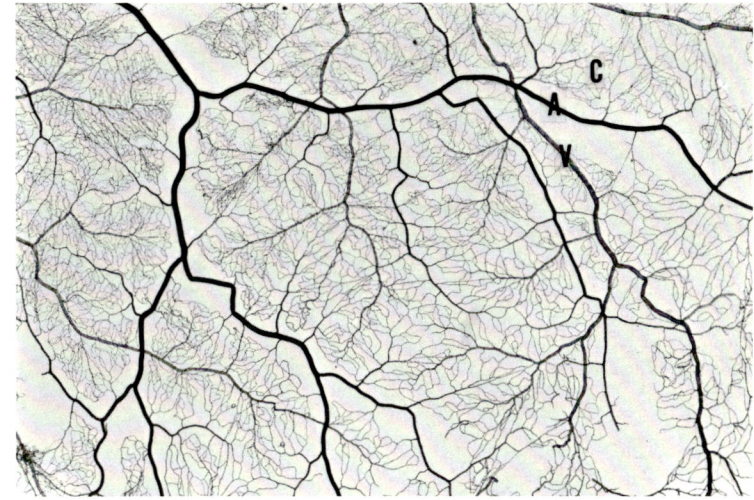

FIG. 6–74. Trypsin-digested specimen of retina showing distribution of vasculature. *A,* arterioles; *C,* capillaries; *V,* venules. PAS, ×8. (AFIP Neg. 64–7007.)

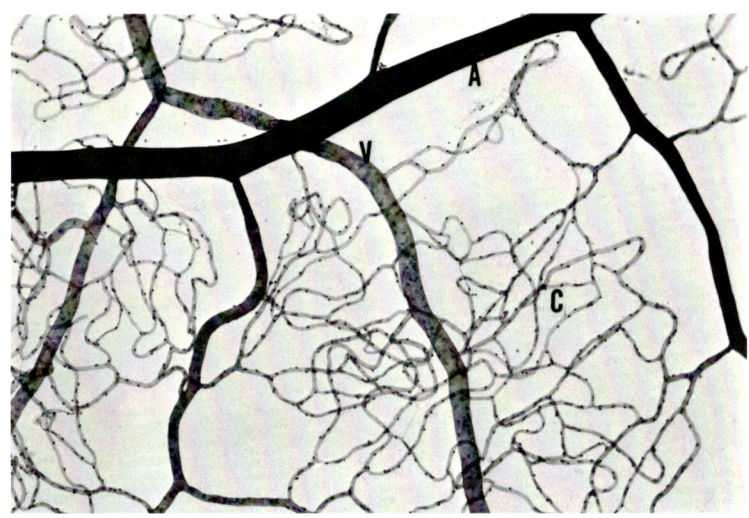

FIG. 6–75. Trypsin-digested specimen of retina. Large dark-staining vessel has a thick wall, indicating it is an arteriole (*A*). Relatively capillary-free zone is present alongside arterioles. Cellular large vessel represents a small venule (*V*). *C,* capillaries. PAS, ×50. (AFIP Neg. 64–7002–B.)

Although the clinical macula is a large, apparently avascular retinal area surrounding the fovea, histologically the retinal vasculature (Fig. 6–76 and 6–79*A*) approaches very near to the *foveal* slopes and is lacking only in the foveal region itself.

The *histologic macula* is identified by at least three criteria: The ganglion cell layer is more than one cell layer thick; the Henle fibers of the outer plexiform layer are obliquely arranged; and there is a large concentration of cones among the photoreceptors. The ganglion cells in the macular area are somewhat smaller than elsewhere in the retina, and may reach five to seven layers in thickness at the edge of the fovea centralis (Figs. 6–77 and 6–78). Two layers of ganglion cells may frequently be present at the temporal edge of the optic disc. The cytoplasmic organelles and inclusions of these ganglion cells are identical to those elsewhere in the retina.

The dense cytoplasmic inclusions closely resemble the dense or yellow-orange colored cytoplasmic inclusions common to neurons in the central nervous system.

FIG. 6–76. Trypsin-digested specimen of retina showing vasculature in region between optic disc on left and avascular foveal region (F) on right. Vasculature terminates at foveal declivity as series of capillary arcades. PAS, ×22. (From Scheie, H. G., and Albert, D. M. *Adler's Textbook of Ophthalmology,* Philadelphia, Saunders, 1969.)

FIG. 6–77. Section through center of rhesus monkey fovea. Ganglion cell layer is five to seven cell layers thick at periphery. At foveal floor all layers except internal limiting membrane and entire length of photoreceptor cells and their associated pigment epithelial cells are absent. 1.5-μ section, PD, ×90. (AFIP Neg. 66–8920.)

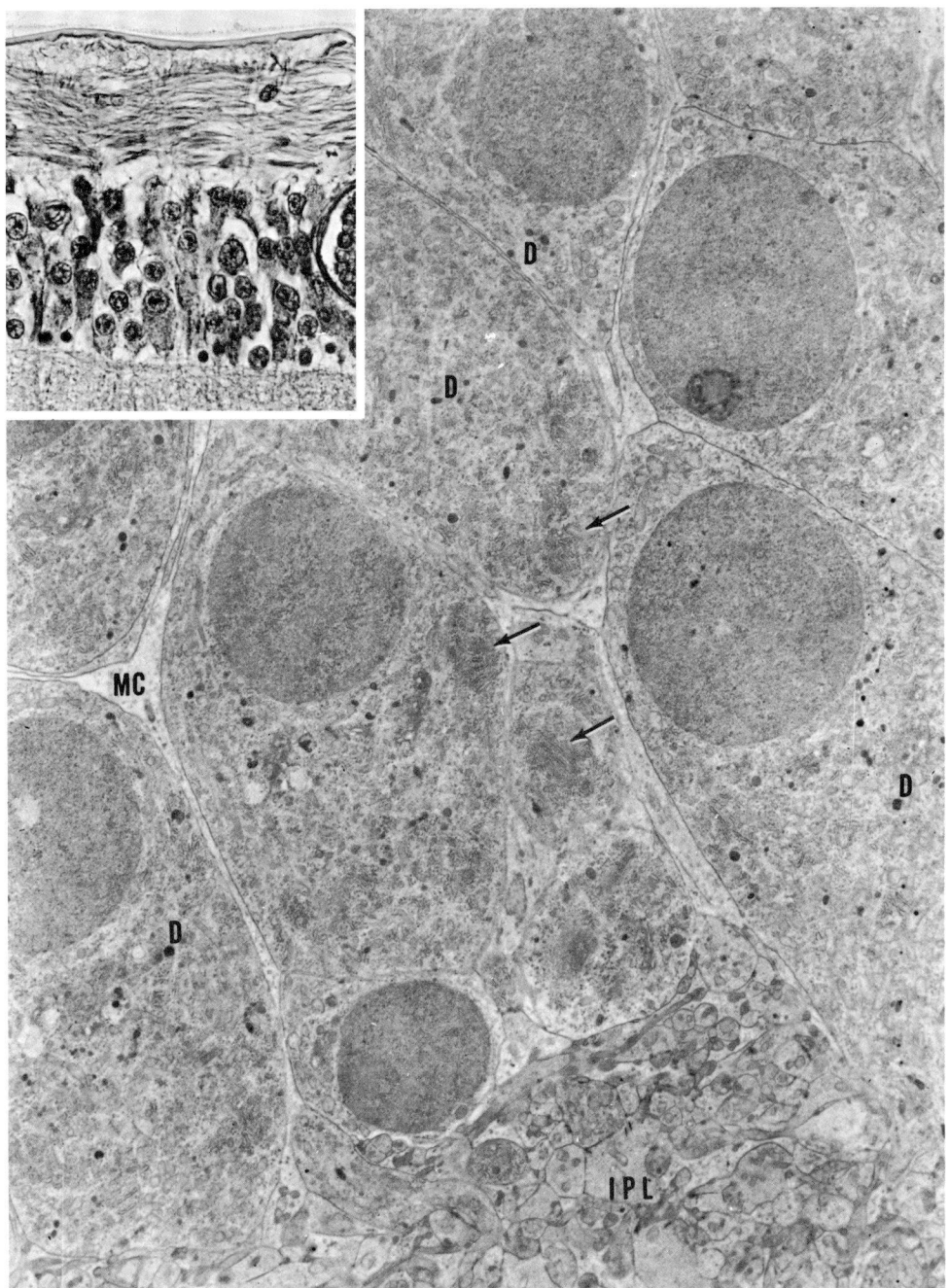

FIG. 6–78. Inset. Light micrograph showing multiple ganglion cell layers present in macular region at some distance from fovea. Note thick internal limiting membrane and nerve fiber layer. **Main figure.** Electron micrograph showing sampling of multiple ganglion cell layers. Except for being slightly smaller in size, ganglion cells appear identical to others throughout retina. *Arrows* indicate granular endoplasmic reticulum (Nissl body or substance). *D*, dense (? lysosome) bodies; *IPL*, inner plexiform layer; *MC*, Müller cell cytoplasmic leaflets. **Main figure,** ×3,700; **inset,** H&E, ×350 (AFIP Neg. 61–1959). (From Fine, B. S. *Arch Ophthal (Chicago)* 69:83, 1963.)

Superimposition of these presumably yellow bodies may account for some of the yellow color of the macula lutea.[24] Gross examination of the free edge of the retina cut through near the fovea reveals that the yellow color may be present in many retinal layers.

The Henle fibers (outer plexiform layer) in the macula are obliquely arranged and longer than elsewhere in the retina. They represent the axonal extensions of the photoreceptors. As elsewhere, they are enveloped by the lucent cytoplasm of the Müller cell.

The oblique arrangement of Henle's fibers in the macula allows greater ease of anteroposterior separation of the retinal layers in the macula compared with only slight lateral separation in the other parts of the retina. Accumulations of pathologic materials (hemorrhages, exudates) therefore follow these natural cleavage planes and appear to the ophthalmoscopist either as radiating lines ("macular stars") in the macula or as more localized spherical accumulations elsewhere (focal hemorrhages or exudates). A tendency of the Müller (glial) cell to disintegrate in this region during preparation of retinal tissue for study often accentuates these cleavage planes. The lucent areas of most light microscopic preparations (i.e., between the presumed eosinophilic Henle fibers, Fig. 6–16) probably represent the interaxonal, broken-down or separated glial cytoplasm.

The tripartite arrangement of the Müller cell cytoplasm is not nearly so pronounced in the macula. Here the Müller cell cytoplasm is more uniformly lucent.

The anatomic observation that the Müller cell cytoplasm is more uniformly lucent in the macula may be related to the well-known observation that both *edema* and *cystoid spacing* can occur as easily in *both* outer and inner layers of the macula as they can in outer layers of the retina elsewhere.

Fovea

The fovea centralis is a small depression lying approximately in the center of the macula. The fovea lacks several of the inner retinal layers (Figs. 6–77 and 6–79), namely, the nerve fiber layer, the ganglion cell layer, the inner plexiform layer, the bipolar cell layer, and the middle limiting membrane. The retinal vasculature is also absent (Figs. 6–76 and 6–79A).

The lack of vasculature in the fovea is a necessary specialization to permit light to pass unobstructed into this region of highest visual acuity.[66] The thinning of the retinal layers in the fovea has often been considered a form of specialization to minimize obstruction to incoming light. Because all retinal layers are exceedingly transparent to visible light, the partition in the foveal region is more likely due to the need of the cells to remain within the sphere of nutritional influence provided by their blood supply. The blood vessels are the only significantly opaque structures normally lying between the photoreceptors and the incoming light.

The pigment epithelial cells of the macula and foveal region are more columnar than elsewhere and generally contain larger numbers of lipofuscin granules. These granules increase in number with age or with mild to moderate traumatic insults.

The ratio of cones to rods increases toward the fovea until the population becomes exclusively cones. The foveal cones resemble the rods superficially (Figs. 6–80 and 6–81). The foveal cones possess all the cytologic characteristics of cones elsewhere (Fig. 6–24) except for their shape. Their outer segments are cylindrical and elongated like rods but possess lamellas that are typical of cones. The lamellas are of the "tight" or closely apposed type, with direct connections to the surface plasma membrane. Longitudinal furrows are absent. The calyces differ slightly in being elongated into villous-like extensions around the inner portion of the outer segment (Figs. 6–80 and 6–81). Their synaptic expansions are all typical broad cone "feet" containing the characteristic multiple synaptic lamellas. Their relation to the pigment epithelium here resembles that of *rods* elsewhere (i.e., they reach to the surface of the pigment cell and are enveloped by extremely delicate apical villi of the pigment epithelial cells) (Fig. 6–82).

Although the fovea has generally been defined in the past as twice the distance from the center to the first recognizable rods, such a determination is difficult to make on histologic sections (Fig. 6–77). On such sections the fovea can be defined on the basis of a measurement made from the last ganglion cell on one side to a similar cell on the opposite side, a similar measurement made in the plane of the bipolar cells, or identification of a zone in which discrete photoreceptors cannot be detected at 400× to 500× magnification (Fig. 6–83). In the last case, such a zone represents roughly the region of foveal cone outer segments.

As previously mentioned, the internal limiting membrane of the human retina is a thick basement membrane. In the fovea, beginning at the edge of the declivity or descent, the basement membrane becomes thin and continues as a thin basement membrane along the declivity or *clivus* as well as along the foveal floor (Fig. 6–84).[68] A few overlapping, attenuated cytoplasmic processes from adjacent Müller cells lie beneath the thin base-

FIG. 6–79. The fovea. **A.** Exact center of rhesus monkey fovea. Only internal limiting membrane (see Fig. 6–84), a few glial cells and processes, part of the photoreceptor cell length and pigment epithelium remain of the thickness of the same layers elsewhere in retina. Photoreceptor nuclei (*ONL*) appear piled up, there is a slight anterior bowing of external limiting membrane ("external fovea"), and photoreceptor outer and inner segments are delicate cylinders closely packed together. Choriocapillaris here is at its greatest concentration of aperture versus area. Retinal capillary closest to fovea is seen at *arrow*. **B** and **C.** Two sections through floor of rhesus fovea showing approximate center (**B**), indicated by presence of a few cross-sectioned photoreceptor axons (Henle fibers, *arrow*), and exact center (**C**), where no cross-sectioned photoreceptor axons are present (*arrow*). All 1.5-μ section, PD. **A,** ×220 (AFIP Neg. 66–8921); **B,** ×395 (AFIP Neg. 66–8923); **C,** ×395 (AFIP Neg. 66–8922).

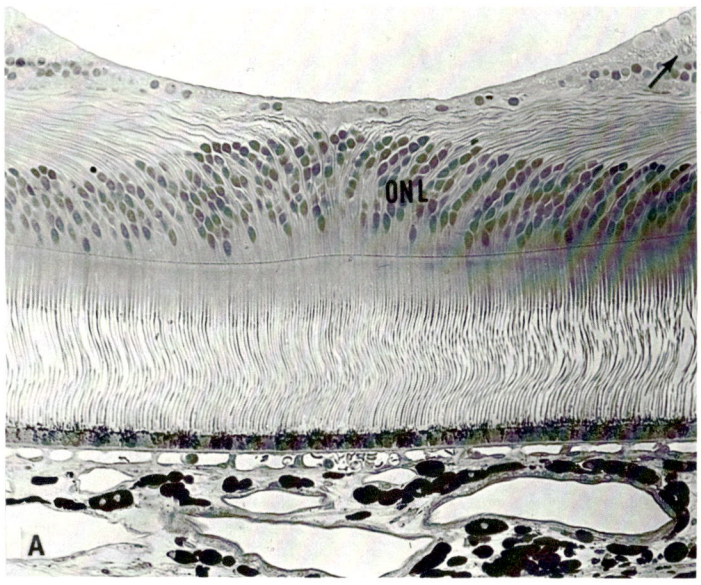
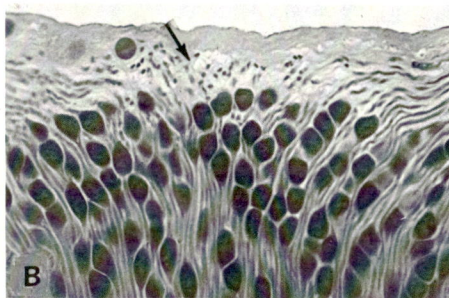
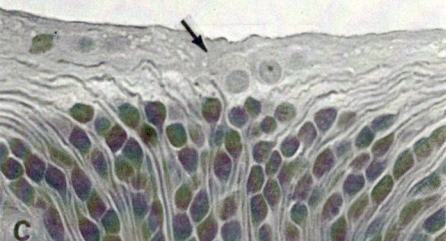

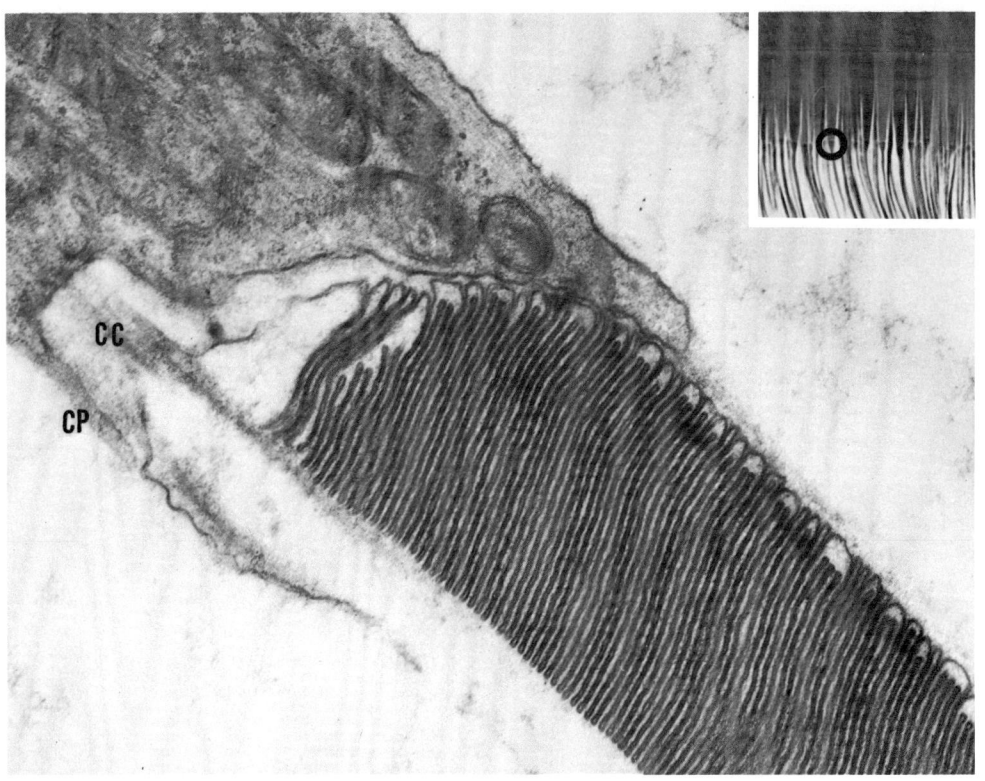

FIG. 6–80. Junction between outer and inner segments of parafoveal cone from rhesus monkey (circled area in **inset**). Connecting cilium (*CC*) is only slightly sectioned and calyceal processes (*CP*) are better appreciated in cross-sectional views (Fig. 6–81*B* and *C*). **Main figure,** ×29,000; **inset,** 1.5-μ section, PD, ×300.

FIG. 6–81. (*Opposite page.*) Parafoveal receptors from rhesus monkey. **A.** Light micrograph. Circles *1, 2,* and *3* indicate approximately parts of photoreceptors illustrated in **B, C,** and **D.** **B.** Oblique section of photoreceptor near its junction of inner and outer parts. Note long calyceal processes (*CP*) that extend for some distance along outer segment. **C.** Cross-sectioned cone is characterized by its "fingerprint-like" appearance. Outer segment is ringed by long calyceal processes (*CP*). **D.** Distal to termination of calyceal processes, long villi of pigment epithelial cells (*V*) envelop photoreceptor outer segments. Fingerprint-like outer segment cross-sections are cones; cross-sectioned outer segments with scalloped periphery are rods. Mixture of rods and cones indicates a perifoveal location for this section. *MP,* interreceptor mucopolysaccharides. **A,** 1.5-μ section, PD, ×300 (AFIP Neg. 66–8679); **B,** ×29,000; **C,** ×17,400; **D,** ×17,000.

ment membrane. A few small glia are also present, as indicated by the few nuclei in the region (compare with retinal vessels and optic nerve head).

Although the retinal sheen is caused by the reflection of light from the thick retinal basement membrane, the light reflex from the fovea results from the geometry of the foveal pit and is presumably unrelated to its thin basement membrane. The presence of this thin basement membrane overlying a number of small glia in the foveal region may represent another anatomic basis for

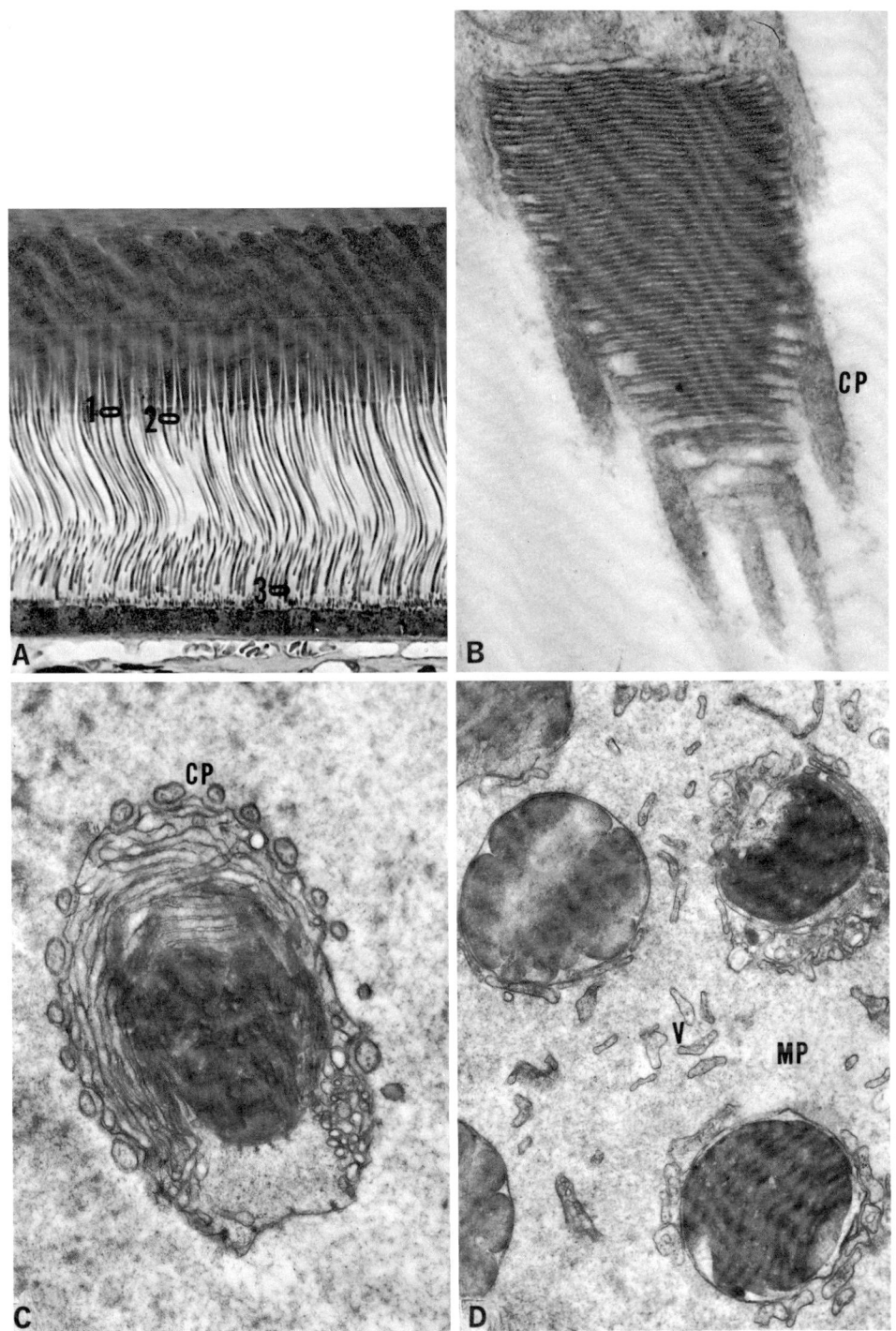

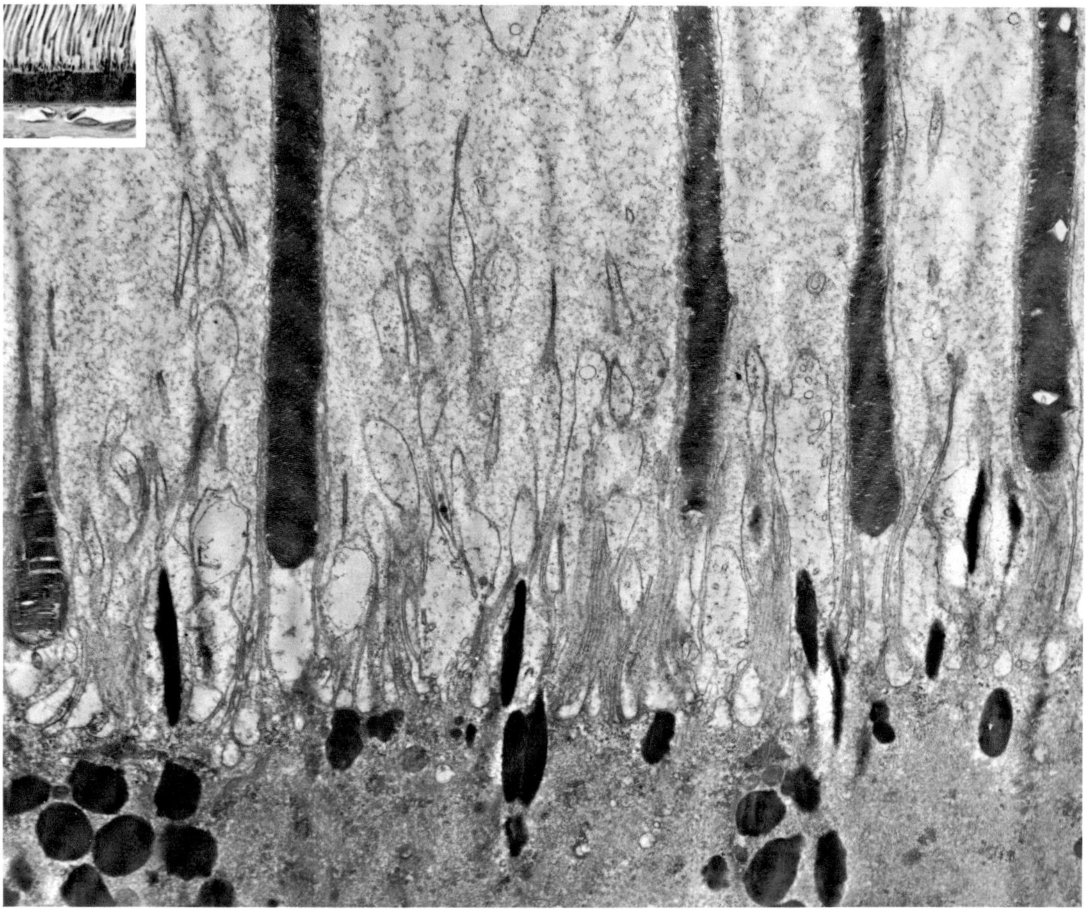

FIG. 6–82. Junction of foveal cones with pigment epithelium. **Inset.** Light micrograph of junction region in rhesus monkey. **Main figure.** Electron micrograph of similar region. Cylindrical cones make contact with many delicate villous processes of pigment epithelial cells. **Main figure,** ×7,200; **inset,** 1.5-μ section, PD, ×300 (AFIP Neg. 66–8679).

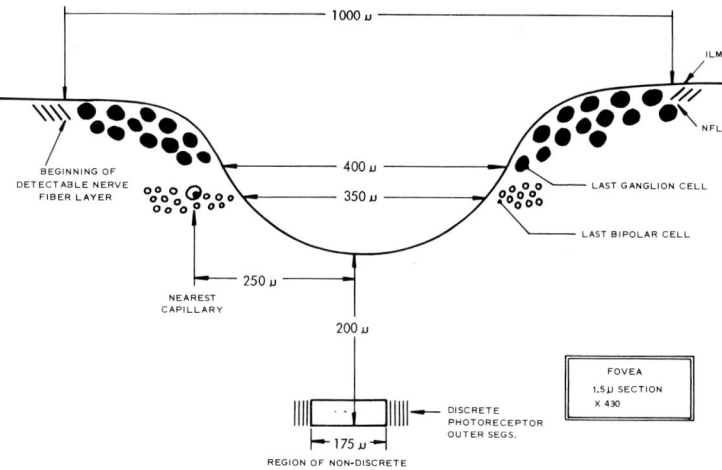

FIG. 6–83. Schematic representation of histologic section through fovea of rhesus monkey. Various measurements are indicated, giving a variety of criteria for measurement of fovea centralis from such a preparation.

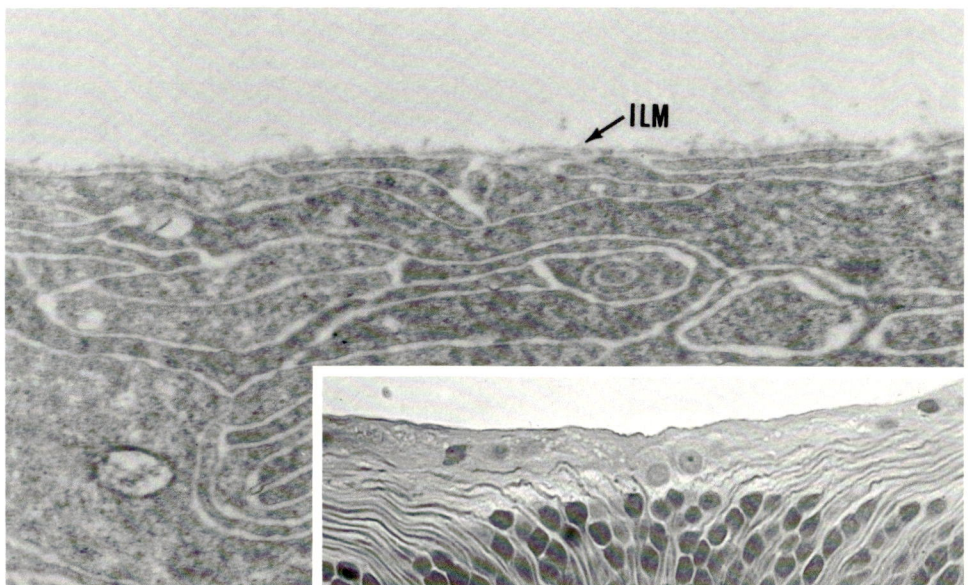

FIG. 6–84. Foveal floor. **Inset.** Light micrograph of section through center of foveal floor of rhesus monkey (see Fig. 6–79). **Main figure.** Electron micrograph of adjacent section showing presence of a thin basement membrane (*ILM*) covering cytoplasmic processes that make up underlying foveal floor. **Main figure,** ×19,500; **inset,** 1.5-μ section, PD, ×395 (AFIP Neg. 66–8922).

some of the mysterious proliferations that may occur clinically in this region, proliferations that distort or wrinkle the macula to such a degree that visual acuity may be severely impaired. Traction by the adjacent vitreous body may be the stimulus which initiates the migrations and proliferations by the small glial cells in the regions of the fovea and macula.

The Subretinal Space

The subretinal space* ends blindly in two circumferential cul-de-sacs. The anterior and larger circumferential cul-de-sac occurs at the ora serrata; the smaller, posterior or circum-

---
* Although this space lies embryonically *between* the two layers of neuroepithelium, it is convenient to continue to use the well established clinical term of subretinal.

papillary cul-de-sac encircles the intraocular portion (i.e., retinal layer) of the optic nerve head.

Ora Serrata and Anterior Subretinal Cul-de-Sac. The peripheral neural retina thins abruptly anteriorly and becomes continuous with the nonpigmented epithelium of the ciliary body. Grossly (Fig. 6–85), the transition occurs circumferentially in a serrated fashion, the serrations, or teeth, being most prominent on the nasal side of the globe. Deep to the neural retina (Fig. 6–86), within the anterior cul-de-sac of the subretinal space, the photoreceptor outer and inner segments (mainly cones) become malformed and disappear as the external limiting membrane unites with the fenestrated membrane (terminal bars) of the pigment epithelium to form the tight seal that continues anteriorly between the two layers of ciliary epithelium. This seal produces the large circumferential anterior cul-de-sac of the subretinal space.

Within the peripheral neural retina the cells become separated by an accumulating watery material as the retina ages (Fig. 6–87). The *cystoid degeneration* of the peripheral retina is more prominent on the temporal side of the retina and is accentuated by aging or by many disease processes. The separation of

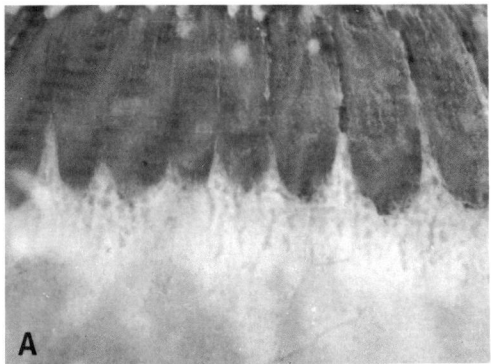

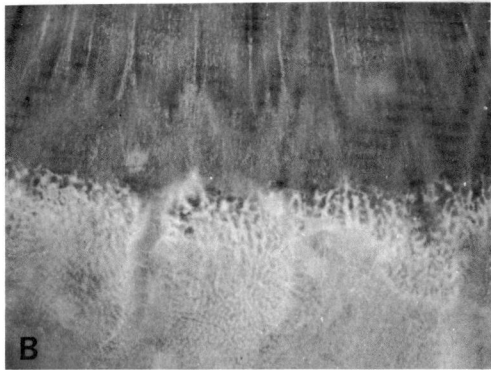

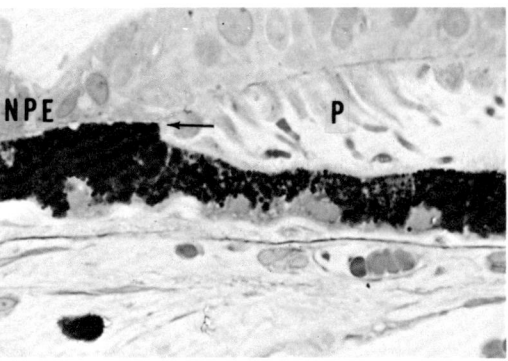

FIG. 6–86. Ora serrata and anterior subretinal cul-de-sac. Sensory retina is continuous with nonpigmented ciliary epithelium (*NPE*). External limiting membrane unites with fenestrated membrane of pigment epithelium (*arrow*). Photoreceptors (*P*), mostly short malformed cones, disappear. 1.5-µ section, PD, ×485. (AFIP Neg. 70–3848.)

FIG. 6–85. Gross appearance of ora serrata. Tooth-like configuration is much more prominent on nasal side of eye (**A**) than on temporal (**B**). Cystoid changes within peripheral retina are also more pronounced on temporal side.

cells within the neural retina occasionally traverses the ora serrata to appear within the nonpigmented epithelium of the ciliary body. When exaggerated, the latter fluid-filled pockets are called *cysts of the pars plana*.

Posterior Subretinal Cul-de-Sac. Posteriorly the subretinal space ends blindly as a smaller circumferential cul-de-sac around the retinal layer of the intraocular portion of the optic nerve head (Fig. 6–88). The photoreceptor

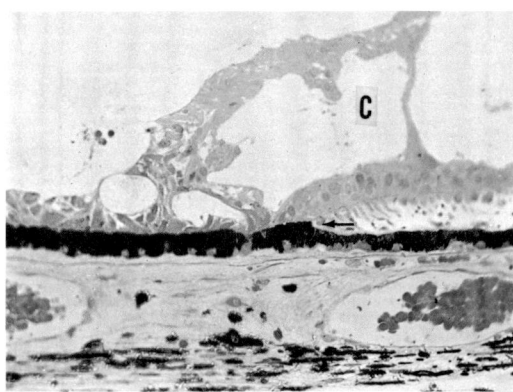

FIG. 6–87. Cystoid degeneration (*C*) in peripheral retina. Union of external limiting membrane to pigment epithelium is seen at *arrow*. PD, ×165.

inner and outer segments cease, and the external limiting membrane of the neural retina continues via terminal-bar-like attachments

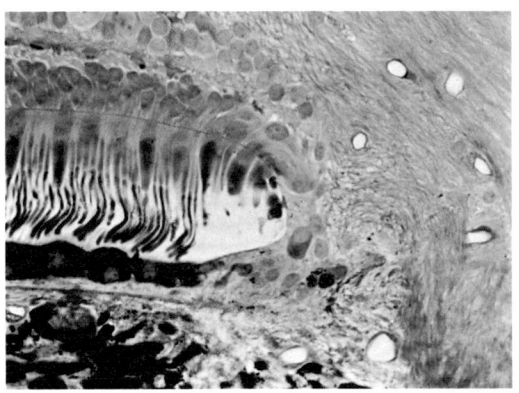

FIG. 6–88. Posterior or circumpapillary subretinal cul-de-sac of rhesus monkey. Photoreceptors terminate as a few malformed cones. External limiting membrane continues with comparable attachment girdles near free surface of pigment epithelial cells. ×300. (AFIP Neg. 66–8674.)

between a number of large glial (?Müller) cells to become continuous with the comparable attachment girdles (terminal bars) near the apexes of the pigment epithelial cells thereby forming a seal. This seal produces the smaller, posterior or *circumpapillary cul-de-sac* of the subretinal space.

Recent experiments by Zauberman and Berman[76] point up the importance of the concept of an anatomically "sealed" subretinal space. From these experiments Zauberman[75] postulated that the forces which maintain apposition of the neural retina to the pigment epithelium in vivo may be in part related to a *negative pressure*, an explanation which we believe is the most useful concept to date and one which is well supported by the anatomic evidence. The presence of the watery interreceptor mucoid would, in all probability, serve to improve this adhesion.

REFERENCES

1. Angelucci, A. Histologische Untersuchungen über das retinale Pigmentepithel der Wirbelthiere. *Arch Physiol (Leipzig)*, 1878, p. 353.

2. Arey, L. B. Retina, Choroid and Sclera. In Cowdry, E. V. (ed.). *Special Cytology*, ed. 2, vol. 3. New York, Haeber, 1932.

3. Barets, A., and Szabo, T. Appareil synaptique des cellules sensorielles de l'ampoule de Lorenzini chez la torpille, Torpedo marmorata. *J Micr* 1:47, 1962.

4. Beams, H. W., Tahmisian, T. N., Anderson, E., and Devine, R. Studies on the fine structure of ultracentrifuged spinal ganglion cells. *J Biophys Biochem Cytol* 8:793, 1960.

5. Breathnach, A. S., and Wyllie, L. M. A. Ultrastructure of retinal pigment epithelium of the human fetus. *J Ultrastruct Res* 16:584, 1966.

6. Cajal, S. R. *Studies on Vertebrate Neurogenesis*. Tr. by L. Guth. Springfield, Ill., Thomas, 1960.

7. Cohen, A. I. The ultrastructure of the rods of the mouse retina. *Amer J Anat* 107:23, 1960.

8. Cohen, A. I. The fine structure of the extrafoveal receptors of the rhesus monkey. *Exp Eye Res* 1:128, 1961.

9. Cohen, A. I. A possible cytological basis for the "R" membrane in the vertebrate eye. *Nature* 205:1222, 1965.

10. Cohen, A. I. New details of the ultrastructure of the outer segments and ciliary connectives of the rods of human and macaque retinas. *Anat Rec* 152:63, 1965.

11. Cohen, A. I. Some electron microscopic observations on interreceptor contacts in the human and macaque retina. *J Anat* 99:595, 1965.

12. Cohen, A. I. New evidence supporting the linkage to extracellular space of outer segment saccules of frog cones but not rods. *J Cell Biol* 37:424, 1968.

13. Cohen, A. I. Rods and Cones and the Problem of Visual Excitation. In Straatsma, B. R., Hall, M. O., Allen, R. A., and Crescitelli, F. (eds.). *The Retina: Morphology, Function and Clinical Characteristics*. UCLA Forum in Medical Sciences No. 8, Berkeley, University of California Press, 1969.

14. DeRobertis, E. Electron microscopic observations on the submicroscopic organization of the retinal rods. *J Biophys Biochem Cytol* 2:319, 1956.

15. DeRobertis, E. Submicroscopic morphology of the synapse. *Int Rev Cytol* 8:61, 1959.

16. DeRobertis, E. Some observations on the ultrastructure and morphogenesis of photoreceptors. *J Gen Physiol 43* (Suppl): 1, 1960.
17. DeRobertis, E., DeIraldi, A. P., Rodriquez, G., and Gomez, C. J. On the isolation of nerve endings and synaptic vesicles. *J Biophys Biochem Cytol 9*:229, 1961.
18. DeRobertis, E., and Lasansky, A. Submicroscopic organization of retinal cones of the rabbit. *J Biophys Biochem Cytol 4*:743, 1958.
19. DeRobertis, E., and Lasansky, A. Ultrastructure and Chemical Organization of Photoreceptors. In Smelser, G. K. (ed.). *The Structure of the Eye.* New York, Academic Press, 1961, p. 29.
20. Dowling, J. E., and Boycott, B. B. Neural Connections of the Primate Retina. In Rohen, J. W. (ed.). *The Structure of the Eye.* Stuttgart, Schattauer, 1965, p. 55.
21. Dowling, J. E., and Boycott, B. B. Organization of the primate retina: Electron microscopy. *Proc Roy Soc (Biol) 166*:80, 1966.
22. Fernandez-Moran, H. Fine structure of the light receptors in the compound eyes of insects. *Exp Cell Res 5* (Suppl):586, 1958.
23. Fine, B. S. Limiting membranes of the sensory retina and pigment epithelium: An electron microscopic study. *Arch Ophthal (Chicago) 66*:847, 1961.
24. Fine, B. S. Ganglion cells in the human retina, with particular reference to the macula lutea: An electron microscopic study. *Arch Ophthal (Chicago) 69*:83, 1963.
25. Fine, B. S. Synaptic lamellas in human retina: An electron microscopic study. *J Neuropath Exp Neurol 22*:255, 1963.
26. Fine, B. S. Observations on the Axoplasm of Neural Elements in the Human Retina. In Titlbach, M. (ed.). *Proceedings, Third European Regional Conference on Electron Microscopy.* Prague, Czechoslovak Academy of Sciences, 1964, p. 319.
27. Fine, B. S. Retinal Structure: Light- and Electron-Microscopic Observations. In McPherson, A. (ed.). *New and Controversial Aspects of Retinal Detachment.* New York, Hoeber, 1968, p. 16.
28. Fine, B. S., and Geeraets, W. J. Membranes and Ground Substance in Photic Injury to the Retina. In Uyeda, R. (ed.). *Proceedings, Sixth International Congress for Electron Microscopy,* Kyoto, Japan, 1966. Tokyo, Maruzen, 1966, p. 503.
29. Fine, B. S., and Zimmerman, L. E. Müller's cells and the "middle limiting membrane" of the human retina. *Invest Ophthal 1*:304, 1962.
30. Fine, B. S., and Zimmerman, L. E. Observations on the rod and cone layer of the human retina. *Invest Ophthal 5*:446, 1963.
31. Fitzpatrick, T. B., Miyamoto, M., and Ishikawa, K. The evolution of concepts of melanin biology. *Arch Derm 96*:305, 1967.
32. Kelley, D. E. Ultrastructure and development of amphibian pineal organs. *Progr Brain Res 10*:270, 1965.
33. Kidd, M. Electron microscopy of the inner plexiform layer of the retina in the cat and the pigeon. *J Anat 96*:2, 179, 1962.
34. Kolmer, W. Uber Kristalloïden in Nervenzellen der Menschlichen Netzhaut. *Anat Anz 51*:314, 1918.
35. Kuwabara, T. Microtubules in the Retina. In Rohen, J. W. (ed.). *The Structure of the Eye.* Stuttgart, Schattauer, 1965, p. 69.
36. Kuwabara, T., and Cogan, D. G. Retinal glycogen. *Arch Ophthal (Chicago) 66*:680, 1961.
37. Kuwabara, T., and Cogan, D. Retinal vascular patterns. VI. Mural cells of retinal capillaries. *Arch Ophthal (Chicago) 69*:492, 1963.
38. Ladman, A. J. The fine structure of the rod–bipolar cell synapse in the retina of the albino rat. *J Biophys Biochem Cytol 4*:459, 1958.
39. Misotten, M. L. Etude des batonnets de la rétine humaine au microscope électronique. *Ophthalmologica 140*:200, 1960.
40. Misotten, M. L. Etude des synapses de la rétine humaine au microscope électronique. In Houwink, A. L., and Spit, B. J. (eds.). *Proceedings European Regional Conference on Electron Microscopy,* Delft, 1960. Delft, Nederlandse Vereniging voor Electronenmicroscopie.
41. Misotten, M. L. L'ultra-structure des cellules horizontales externes de la rétine humaine. *Bull Soc Belg Ophtal 128*:207, 1961.
42. Misotten, M. L. Etude des capillaires de la rétine et de la choriocapillaire au microscope électronique. *Ophthalmologica 144*:1, 1962.
43. Misotten, M. L. The Synapses in the Human Retina. In Rohen, J. W. (ed.). *The Structure of the Eye.* Stuttgart, Schattauer, 1965, p. 17.
44. Moyer, F. H. Genetic effects on melanosome fine structure and ontogeny in normal and malignant cells. *Ann NY Acad Sci 100*:584, 1963.
45. Mund, M. L., Rodrigues, M. M., and Fine, B. S. Light and electron microscopic observations on the developing pigmented layers of the human eye. In press.
46. Nicholls, J. G., and Kuffler, S. W. Extracellular space as a pathway for exchange between blood

and neurons in the central nervous system of the leech: Ionic composition of glial cells and neurons. *J Neurophysiol* 27:645, 1964.
47. Nilsson, S. E. G. Receptor cell outer segment development and ultrastructure of the disc membranes in the retina of the tadpole. *J Ultrastruct Res* 11:581, 1964.
48. Parsons, H. *The Pathology of the Eye:* vol. 2, Histology, part I, p. 445. London, Hodder, 1905.
49. Polyak, S. L. *The Retina: The Anatomy and Histology of the Retina in Man, Ape, and Monkey.* Chicago, University of Chicago Press, 1941.
50. Porter, K. R., and Yamada, E. Studies on the endoplasmic reticulum. V. Its form and differentiation in pigment epithelial cells of the frog retina. *J Biophys Biochem Cytol* 8:181, 1960.
51. Ramsey, H. J. Ultrastructure of corpora amylacea. *J Neuropath Exp Neurol* 24:25, 1965.
52. Raviola, G., and Raviola, E. Light and electron microscopic observations on the inner plexiform layer of the rabbit retina. *Amer J Anat* 120:403, 1967.
53. Rosenbluth, J. Ultrastructure of dyads in muscle fibers of *Ascaris lumbricoides. J Cell Biol* 42:817, 1969.
54. Salzmann, M. *The Anatomy and Histology of the Human Eyeball in the Normal State: Its Development and Senescence.* Tr. by E. V. L. Brown. Chicago, University of Chicago Press, 1912.
55. Scheie, H. G., and Albert, D. M. *Adler's Textbook of Ophthalmology.* Philadelphia, Saunders, 1969.
56. Seiji, M. Formation of mammalian melanin. *Jap J Dermat (Tokyo)* series B 73:4, 1963.
57. Sherrington, C. *The Integrative Action of the Nervous System.* New Haven, Yale University Press, 1961, p. 17.
58. Sjöstrand, F. S. The ultrastructure of the outer segments of rods and cones of the eye as revealed by the electron microscope. *J Cell Comp Physiol* 42:15, 1953.
59. Sjöstrand, F. S. Ultrastructure of retinal rod synapses of the guinea pig eye as revealed by three dimensional reconstruction from serial sections. *J Ultrastruct Res* 2:122, 1958.
60. Sjöstrand, F. S. The ultrastructure of the retinal receptors of the vertebrate eye. *Ergebn Biol* 21:128, 1959.
61. Sjöstrand, F. S. Electron Microscopy of the Retina. In Smelser, G. K. (ed.). *The Structure of the Eye.* New York, Academic Press, 1961, p. 1.
62. Smelser, G. K., Ishikawa, T., and Pei, Y. F. Electron Microscopic Studies of Intraretinal Spaces: Diffusion of Particulate Materials. In Rohen, J. W. (ed.). *The Structure of the Eye.* Stuttgart, Schattauer, 1965, p. 109.
63. Smith, C., and Sjöstrand, F. S. A synaptic structure in the hair cells of the guinea pig cochlea. *J Ultrastruct Res* 5:184, 1961.
64. Spitznas, M., and Hogan, M. J. Outer segments of photoreceptors and the retinal pigment epithelium: Interrelationship in the human eye. *Arch Ophthal (Chicago)* 84:810, 1970.
65. Ts'o, M. O. M., Fine, B. S., and Zimmerman, L. E. The nature of retinoblastoma. II. Photoreceptor differentiation: An electron microscopic study. *Amer J Ophthal* 69:350, 1970.
66. Weale, R. A. Fovea in human retina: Theory for existence. *Nature* 212:255, 1966.
67. Yamada, E. Observations on the fine structure of photoreceptive elements in the vertebrate eye. *J Electron Micr (Chiba)* 9:1, 1960.
68. Yamada, E. Some structural features of the fovea centralis in the human retina. *Arch Ophthal (Chicago)* 82:151, 1969.
69. Yamada, E., Tokuyasu, K., and Iwaki, S. The fine structure of retina studied with electron microscope. II. Pigment epithelium and capillary of the choriocapillary layer. *J Electron Micr (Chiba)* 6:42, 1958.
70. Yamada, E., Tokuyasu, K., and Iwaki, S. The fine structure of retina studied with electron microscope. III. Human retina. *J Kurume Med Ass* 21:1979, 1959.
71. Yanoff, M. Diabetic retinopathy. *New Eng J Med* 274:1344, 1966.
72. Yoshida, M. The fine structure of the so-called crystalloid body of the human retina as observed with the electron microscope. *Jap J Electron Micr* 14:285, 1965.
73. Young, R. W., and Bok, D. Participation of the retinal pigment epithelium in the rod outer segment renewal process. *J Cell Biol* 42:392, 1969.
74. Young, R. W., and Bok, D. Autoradiographic studies on the metabolism of the retinal pigment epithelium. *Invest Ophthal* 9:524, 1970.
75. Zauberman, H. Personal communication, 1970.
76. Zauberman, H., and Berman, E. R. Measurement of adhesive forces between the sensory retina and the pigment epithelium. *Exp Eye Res* 8:276, 1969.
77. Zimmerman, L. E., and Eastham, A. B. Acid mucopolysaccharide in the retinal pigment epithelium and visual cell layer of developing mouse eye. *Amer J Ophthal* 47:488, 1959.

chapter 7

The Vitreous Body

Embryology
Components
   Fibrous component
   Mucinous component
Boundaries
Attachments
   Posterior attachments
   Lateral attachments
   Anterior attachments
Zones
Modifications

The vitreous body is an example of a tissue composed predominantly of extracellular materials. It completely fills the large posterior (vitreous) compartment of the eye (Fig. 7–1). The vitreous body thus accounts for most of the volume of the eye (Fig. 7–2); it is a very delicate, transparent connective tissue—probably the most delicate of all the connective tissues in the body.

EMBRYOLOGY

The vitreous body develops in three stages. In the first stage the cavity of the optic cup is filled with a mass of ectodermal and mesodermal filaments from which the primary vitreous forms. The ectodermal components originate mostly from the cells of the retina, and the mesodermal elements, including the hyaloid vessels, enter into the vitreous space through the fetal fissure inferiorly. The primary vitreous is thus vascularized by the hyaloid vascular system.

In the second stage, which occurs after the embryo reaches a size of 13 mm, new vitreous (secondary vitreous) is formed by the cells of the inner layer of the optic cup when the hyaloid vascular channels in the primary vitreous stop growing and begin to atrophy. Eventually the secondary vitreous surrounds the primary vitreous, which then becomes limited to a cone-shaped region running through the vitreous compartment (see Fig. 7–7). Because the primary and secondary vitreous bodies differ in optical density, a line of demarcation (the wall of Cloquet's canal) can be detected at their interface. The hyaloid artery courses within Cloquet's canal

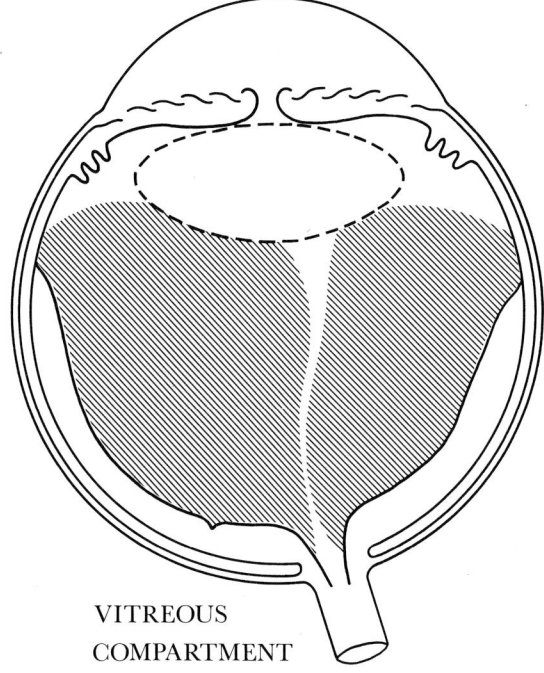

FIG. 7–1. Schematic drawing of vitreous body (*shaded area*) occupying larger or posterior compartment of the eye (vitreous compartment).

to supply the vascular network surrounding the lens.

In the third stage, the tertiary vitreous or zonular fibers develop. This begins at about the 65-mm stage when the primitive nonpigmented ciliary epithelium is in contact with the lens capsule (see Fig. 8–4). As the eye grows anteriorly, the epithelium is displaced posteriorly, "spinning out" the vitreous-like collagen of the zonular fibers until, by the 110-mm stage, they are well visualized.

THE VITREOUS BODY

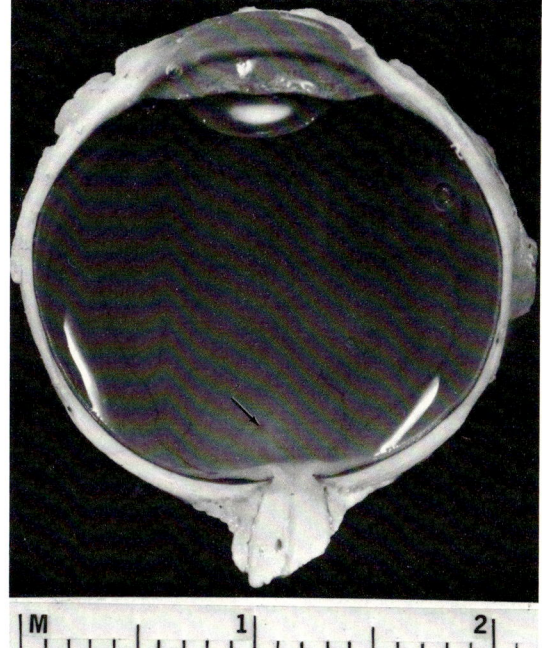

FIG. 7–2. Gross appearance of vitreous body. This delicate, transparent tissue occupies most of the volume of eye, extending from back of lens anteriorly to retinal surface posteriorly. Major temporal vessels of retina can be seen below, encircling slightly edematous macula, characteristic of fixed opened eyes. Fovea centralis (*arrow*) lies in center of depression in macula. (AFIP Neg. 66–257–4.)

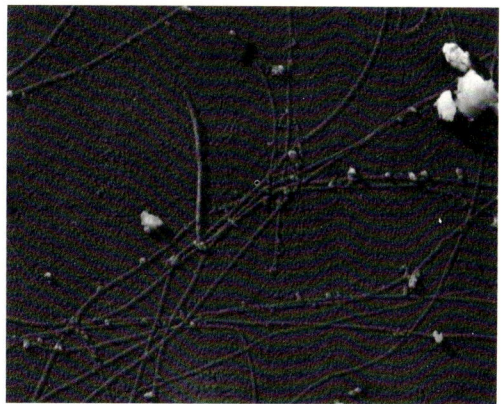

FIG. 7–3. Filaments isolated by mechanical separation from rhesus monkey vitreous body. White, irregular-shaped particles represent debris. U-shad, ×15,000.

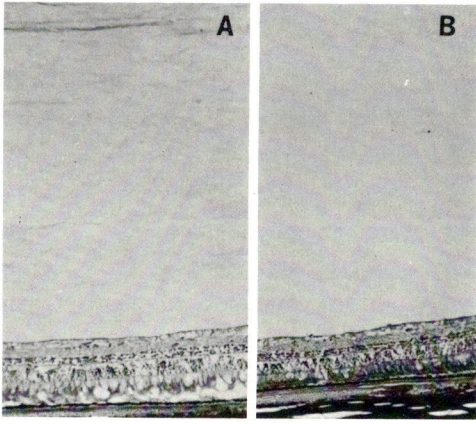

FIG. 7–4. Chemical composition of vitreous body. **A.** In untreated tissue, colloidal iron stains acid mucopolysaccharides.[12] **B.** No staining occurs when tissue is first treated with hyaluronidase. Vitreous body contains hyaluronic acid.[7] Colloidal iron stain with and without hyaluronidase, ×70. (AFIP Neg. 57–1284.)

COMPONENTS

The vitreous body is readily separable into two component parts: the *fibrous* and the *mucinous*. Each part is most easily observed by mechanical separation of the components and subsequent examination by electron microscopy.

Fibrous Component

The fibrous component (Fig. 7–3) is composed of a delicate (∼ 150 to 200 Å diameter) *filament* with an apparent periodicity (∼ 220 Å). In aggregates, these filaments form the

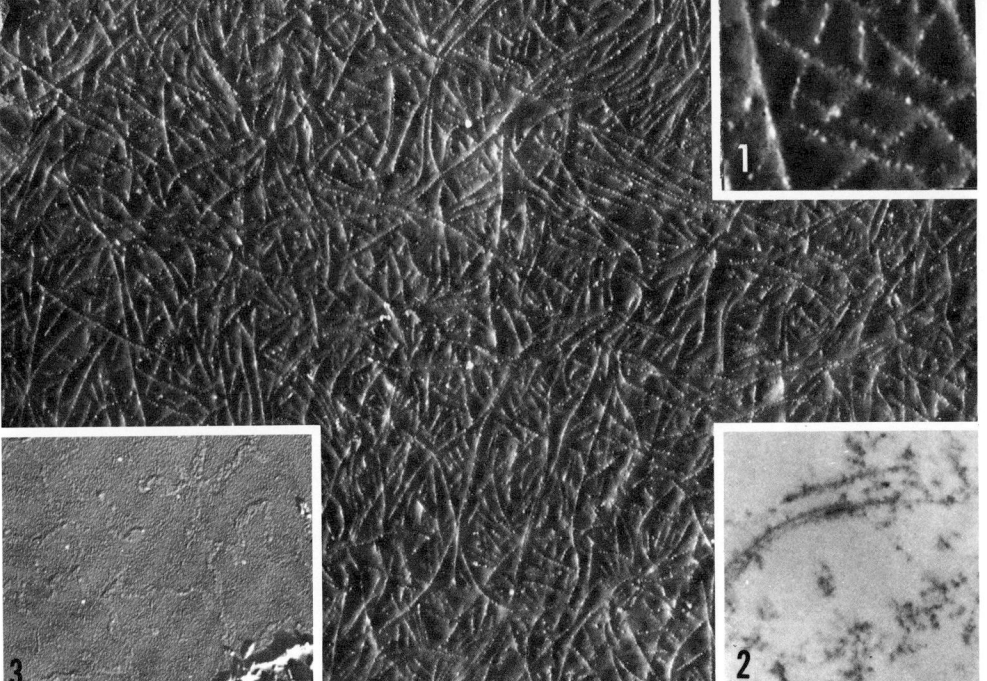

FIG. 7–5. Posterior vitreous body. **Main figure.** Shadow-cast section showing interweaving of filaments of uniform dimension to form sheet-like arrangement. **Inset 1** shows filaments at high magnification, which emphasizes their apparent periodicity. **Inset 2** shows similar filaments in thin section with "fluffy" adherent material representing mucinous component of vitreous, better observed as drying patterns in shadow-cast material (**inset 3**). **Main figure,** U-shad, ×15,800; **inset 1,** ×47,400; **inset 2,** ×33,000; **inset 3,** U-shad, ×20,000. (**Main figure** from Fine, B. S., and Tousimis, A. J. Arch Ophthal (Chicago) 65:95, 1961.)

substructure for the small and large sheet-like arrangements which in total make up the structural framework of the vitreous body (Figs. 7–4, 5). The filaments appear morphologically alike, whether in the anterior or posterior vitreous,[3] and closely resemble, in both diameter and periodicity, the delicate filaments observed commonly in fetal tissues and in young cultures of fibroblasts.[11] By chemical assay, this structural material[9,10] has been shown to contain a significant amount of hydroxyproline. The foregoing criteria identify the filaments as a *collagen*.

The somewhat peculiar delicacy of this structural collagenous framework in the adult eye has stimulated much controversy regarding its true nature. If one considers that, for all practical purposes, the filamentous material is produced in the eye mainly during its growth period,[1] the "fetal" or infantile appearance of the filaments is not surprising. "Formed vitreous" (i.e., that which possesses a modicum of rigidity or a clinically detectable structure) has long been known clinically never to re-form in the mature eye to any significant degree once removed or lost from the eye. When the neuroepithelial cells have matured, they gradually (and perhaps even sequentially) lose their ability to elaborate collagen, but they may be called upon later to synthesize both *mature* and *immature* fibrils and filaments, as well as mucopolysaccharides, in various pathologic processes.

Mucinous Component

The mucinous component, accounting for by far the larger volume of the vitreous body, can be observed morphologically by either light or electron microscopy (Figs. 7–4 through 7–6). In thin-section electron micrographs, this mucinous material can be ob-

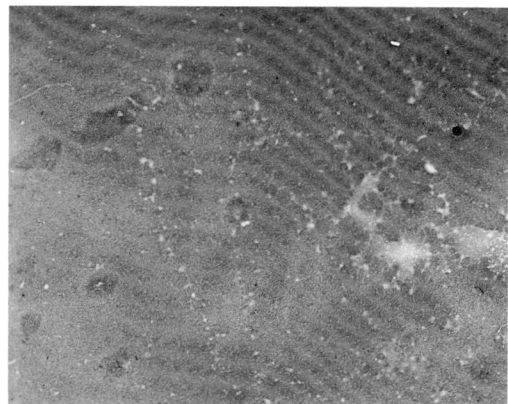

FIG. 7–6. Drying patterns of vitreous supernatant, mechanically separated from fibrous component illustrated in Figure 7–3. Drying patterns represent acid mucopolysaccharides and are similar to patterns observed in shadow-cast tissue sections (*inset 3*, Figure 7–5). U-shad, ×14,000.

served as a rather "fluffy" debris irregularly adherent to the filaments of the vitreous framework (*inset 2*, Fig. 7–5). In shadow-cast preparations of either the mechanically separated components (Fig. 7–6) or the relatively in situ components of tissue sections (*inset 3*, Fig. 7–5), the mucinous material appears as flattened, irregularly distributed macromolecules ("drying patterns") of a nonfibrous material from which the water and various ions have been removed.

BOUNDARIES

The vitreous body is bounded posteriorly by the internal surface of the retina (internal limiting membrane), axially by the surface condensations that extend in tubular fashion from the periphery of the optic disc to the posterior surface of the lens (canal of Cloquet), laterally at the ora serrata by the internal limiting membrane of the retina and of the posterior pars plana, and anteriorly by the posterior surface of the lens (Fig. 7–7).

FIG. 7–7. Schematic outline of major vitreous body attachments and relations: (*1*) attachment of orbiculoanterior zonular fibers, (*2*) attachment of orbiculoposterior zonular fibers, (*3*) attachment of anterior vitreous face to posterior lens capsule, (*4*) anterior extremity of canal of Cloquet, (*5*) anteriormost attachment of vitreous base to mid pars plana (origin of vitreous face), (*6*) region of vitreous "base," (*7*) region of diminishing adherence of vitreous base to retinal surface, (*8*) vitreous-retinal attachment, (*9*) vitreous-retinal attachment at margin of fovea centralis, (*10*) attachment of posterior vitreous around optic disc, (*11*) posterior extremity of canal of Cloquet (area Martegiani), (*12*) cortical vitreous, (*13*) central vitreous. Density of lines indicates approximate relative degrees of strength of attachment.

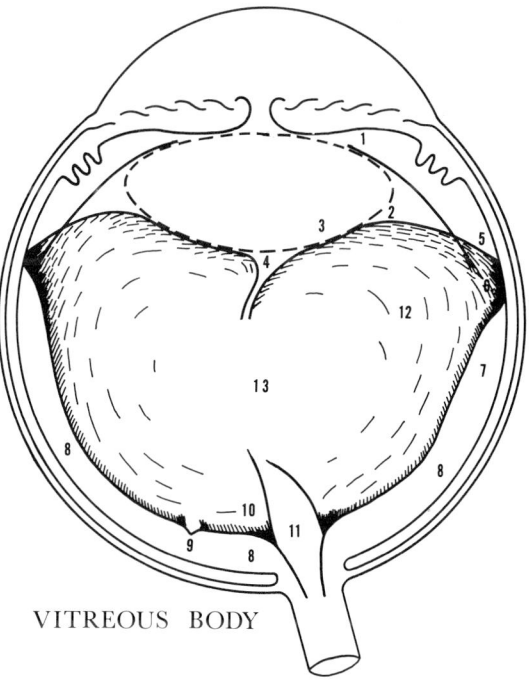

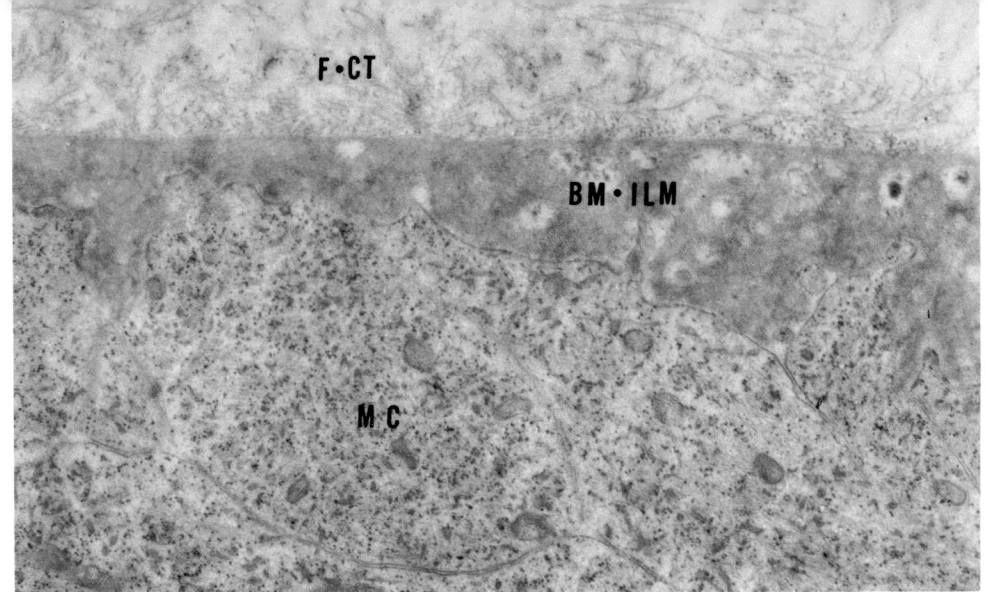

FIG. 7–8. Delicate vitreous filaments (*F·CT*) attached to smooth inner surface of thick retinal basement membrane (*BM·ILM*). *MC*, Müller cell. ×16,000.

FIG. 7–9. Shadow-cast section of tissue similar to that seen in thin section in Figure 7–8. Vitreous filaments stand out clearly, but retinal detail is lost. ×22,500.

ATTACHMENTS

Posterior Attachments

The vitreous framework is rather uniformly and tenuously attached to the thick basement membrane of the retina (Figs. 7–8 and 7–9) except for stronger attachments in the regions of the optic nerve head and the ora serrata, over some regions of the larger retinal vessels, and probably around the fovea centralis.

The delicate attachment along the internal limiting membrane of the retina is easily disrupted. This ease of separation in most regions is accounted for by two electron microscopic observations: (1) the number of filamentous attachments per unit area of basement membrane is rather low in comparison with regions of known stronger attachment, and (2) the internal surface of the retinal basement membrane is smooth, whereas the surface near the ora serrata is multilaminar (Fig. 7–17).

The filaments in the preretinal or *cortical* zone of the vitreous body are arranged in aggregates (Fig. 7–10) which interconnect in a somewhat honeycomb arrangement. These delicate sheet-like aggregations, observed by electron microscopy, lie beyond the resolution of biomicroscopic examination.

The sample micrographs of vitreous structure used for illustrative purposes here are to be understood as just samples in *time* as well as in *location* (Fig. 7–11). The vitreous body framework continually changes its arrangement throughout life,* beginning in infancy as a rather diffuse structure (*diminished relucency*) in which there are only small aggregations in the peripheral cortex (Fig. 7–12). As with collagen elsewhere, the tendency is for larger "sheets" of filaments or fibrils eventually to form. The largest can be observed, especially anteriorly, by clinical methods. Further aggregation of collagen in connective tissue produces collapse of the "sheets" into

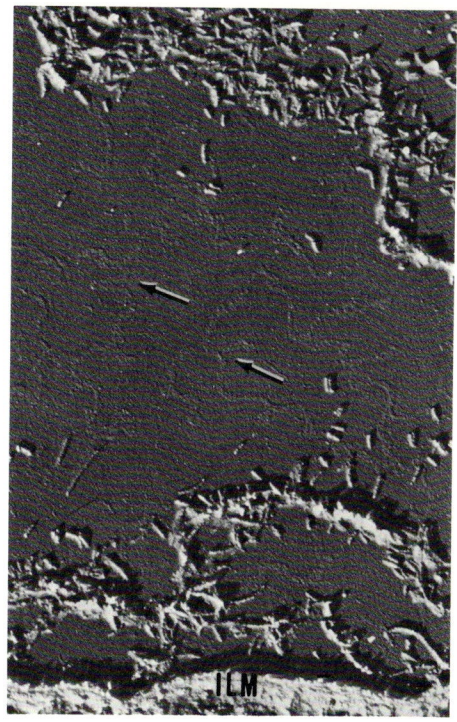

FIG. 7–10. Shadow-cast section showing vitreous filaments forming exceedingly delicate aggregates near posterior retinal surface. Drying patterns (*arrows*) of mucinous part of vitreous body are clearly seen between filamentous aggregates. *ILM*, internal limiting membrane of retina. U-shad, ×13,200.

"bands" or "strands," because aggregation of collagen with aging is associated with shrinkage or contraction. Retraction of the vitreous framework, particularly from its loose attachment to the retinal surface, produces a new boundary—the newly formed "posterior hyaloid," which is detectable both clinically and histologically (Fig. 7–13). Spaces are formed between the hyaloid membrane (condensed vitreous framework, Fig. 7–14) and the internal limiting membrane of the retina (Fig. 7–15). The spaces are occupied by the watery mucinous component of the vitreous body. When the vitreous body contracts (posterior vitreous detachment), if the component is more watery than mucinous, it can pass through the hyaloid

---

* Although the vitreous body may vary considerably *between* individuals, the lamellar structures differ very little between the two eyes of a single individual if there has been no pathologic interference.

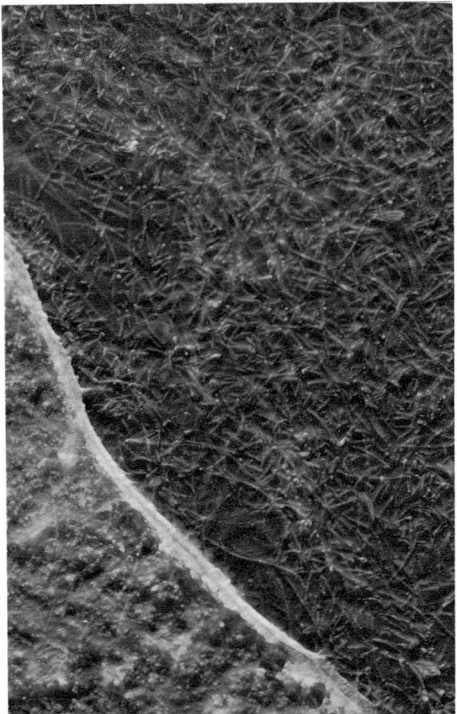

FIG. 7–11. Shadow-cast section of posterior vitreous near retinal surface. Filaments are numerous and in this minute sample do not appear aggregated. U-shad, ×11,000. (From Fine, B. S. In McPherson, A. (ed.). *New and Controversial Aspects of Retinal Detachment.* New York, Hoeber, 1968.)

face of the retinal internal limiting membrane (*inset B*, Fig. 7–15). By electron microscopy (Fig. 7–15) the distinction is easily made.

The vitreous framework has a firm attachment to the basement membrane at the edge of the optic nerve head (or disc). Where the thick basement membrane of the retina thins over the nerve head, the vitreous face turns anteriorly to form the lining walls of the canal of Cloquet. The floor of the canal is called the *area of Martegiani* and is lined by a thin basement membrane covering the astrocytic glia of the nerve head (*central supporting tissue meniscus*, see Fig. 12–4).

Retraction and separation of posterior vitreous body from the region of the nerve head often is accompanied by tearing of the glial tissue from around the optic disc to such a degree that the suspended ring of separated material is sometimes referred to clinically as the vitreous "peephole," a most appropriate term when the ring is complete.

Temporal to the optic disc, in the region of the fovea centralis, the vitreous framework has an attachment to the basement membrane directly. If it is more mucinous, however, a break in the hyaloid membrane is presumed. When the vitreous framework contracts, any vitreoretinal adhesion can cause retinal tears. Both the retinal basement membrane (internal limiting membrane) and the retracted vitreous framework ("posterior hyaloid") stain equally with eosin or with the PAS technique, a situation which may cause some confusion in interpretation. If two smooth layers are present, one in continuity with the retina, the inner one is the posterior hyaloid (*inset A*, Fig. 7–15). Separation of the internal limiting membrane is generally accompanied by tags of Müller cells, which can be observed attached to the outer sur-

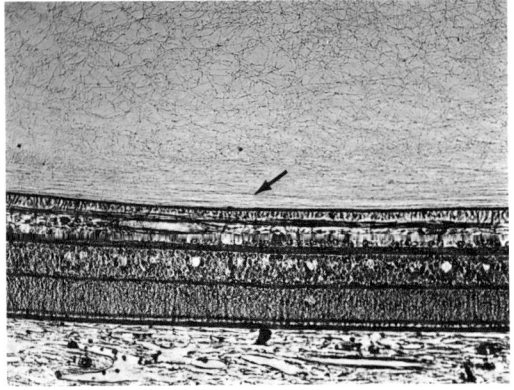

FIG. 7–12. Cortical vitreous of infant eye. Note laminated (i.e., aggregated) appearance near retinal surface (arrow) and irregularly arranged (i.e., less aggregated) appearance toward central region. Wilder reticulum stain, ×115. (AFIP Neg. 60–4411.)

FIG. 7–13. Artifactitiously retracted posterior vitreous. **Inset.** Light micrograph. **Main figure.** Electron micrograph showing retinal surface below and retracted vitreous framework above. Newly formed posterior boundary consists of aggregated vitreous filaments. Drying patterns (*arrows*) of mucinous component are present between retracted vitreous framework and retinal surface. **Main figure,** ×6,900 (From Fine, B. S., and Tousimis, A. J. *Arch Ophthal (Chicago)* 65:95, 1961). **Inset,** PAS, ×575 (AFIP Neg. 60–961). (From Fine, B. S. In McPherson, A. (ed.). *New and Controversial Aspects of Retinal Detachment.* New York, Hoeber, 1968.)

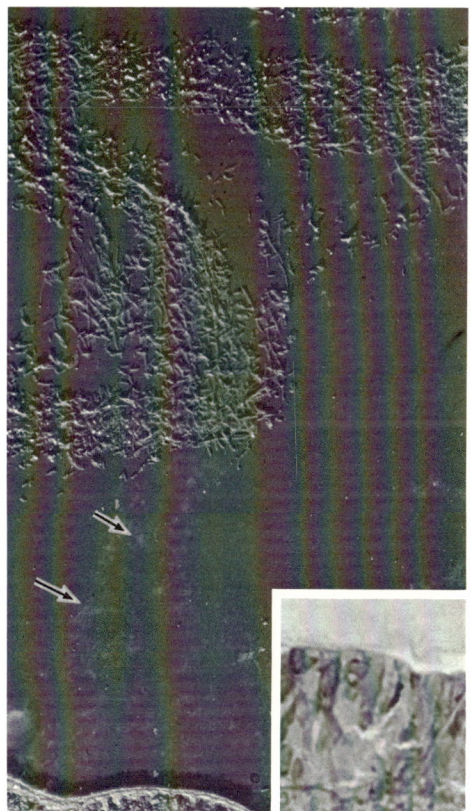

FIG. 7–14. Posterior boundary ("posterior hyaloid") of retracted vitreous framework further removed from retinal surface but adjacent to section shown in Figure 7–13. *Arrow* points to subvitreal space. ×7,900. (From Fine, B. S., and Tousimis, A. J. *Arch Ophthal (Chicago)* 65:95, 1961.)

membrane that is slightly stronger than elsewhere over the posterior retina. Where the foveal declivity begins, the basement membrane abruptly changes from thick to thin. The *slope (clivus)* of the fovea as well as the floor are lined by thin basement membrane (Fig. 6–84), and there is a paucity of vitreous filaments here. The surface anatomic relations here are analogous to those described for the nerve head.

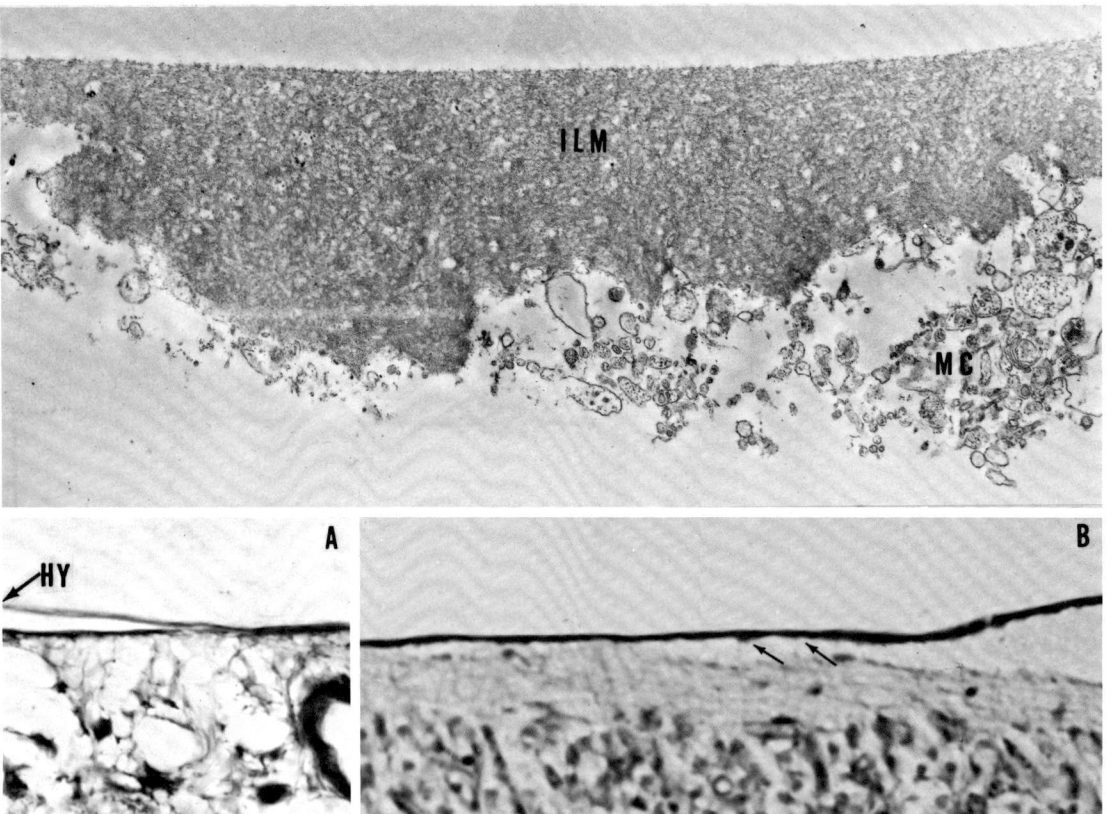

FIG. 7–15. Main figure. Electron micrograph showing detached internal limiting membrane (*ILM*) of retina and a few fragments of adhering Müller cells (*MC*). **Inset A** shows two PAS-positive membranes near inner surface of retina. Posterior hyaloid (*HY*) is smoother and inner of the two. **Inset B** shows detached PAS-positive internal limiting membrane with a few outer irregularities (*arrows*) to which fragments of Müller cells frequently remain adherent. **Main figure,** ×11,000; **inset A,** PAS, ×530 (AFIP Neg. 70–3850); **inset B,** PAS, ×270 (AFIP Neg. 67–4572).

Lateral Attachments

The vitreous framework is condensed into sheets forming a somewhat compressed honeycomb arrangement in the *cortical* zone (Fig. 7–16).[5] The sheets are coarse in contrast to the extremely delicate ones over the posterior retinal surface.

An attachment zone of increasing strength begins at the inner surface of the peripheral retina a few millimeters behind the ora serrata. It extends anteriorly across the ora serrata and along the free surface of the ciliary nonpigmented epithelial cells, and ceases rather abruptly in the *mid pars plana* of the ciliary body. The zone of strong attachment is called the *base* of the vitreous body. Three structural modifications occur in this region that account for this strong attachment (Fig. 7–17): (1) the basement membrane (internal limiting membrane) becomes multilaminar and only a thin layer is applied to the basal plasma membranes of the surface cells; (2) the basilar surfaces of the cells project into the multilaminar basement membrane and are interwoven with it; (3) large aggregates of vitreous framework filaments also are interwoven with this multilaminar basement mem-

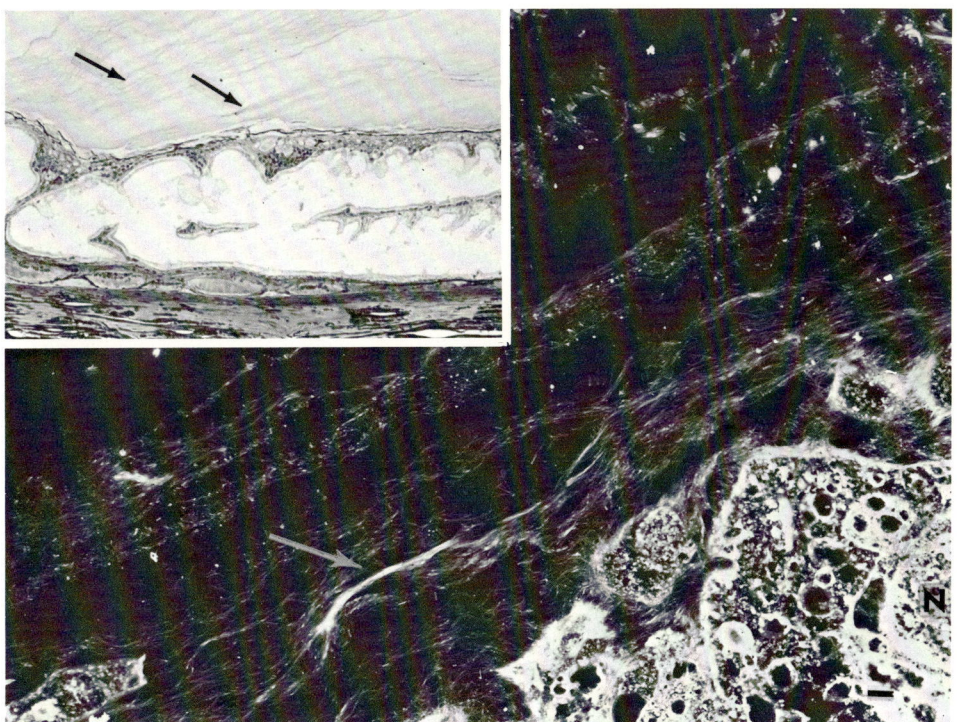

FIG. 7–16. Vitreous framework over region of ora serrata. **Inset.** Light micrograph showing condensation of framework into numerous sheets (*arrows*). **Main figure.** Electron micrograph from similar region showing configuration produced by dense aggregates of vitreous filaments. *Arrow* points to a more than usually dense aggregate which may represent a fragment of zonule. *N*, nucleus of nonpigmented epithelial cell pars plana. **Main figure,** ×3,440; **inset,** Wilder reticulum stain, ×70 (AFIP Neg. 70–1425). (Modified from Fine, B. S., and Tousimis, A. J. *Arch Ophthal (Chicago)* 65:95, 1961.)

brane. This interweaving of the three structures is most pronounced over the region of the posterior pars plana.

Anterior Attachments

Approximately midway along the surface of the pars plana, the now condensed vitreous framework separates from the nonpigmented epithelial cells as a free surface, the *face* of the vitreous body (Fig. 7–18). Histologically, the anteriormost zone of condensed vitreous framework represents the *anterior border layer*, which is the *anterior hyaloid* of the vitreous body observed clinically. The anterior border layer is composed of filaments (Figs. 7–19 and 7–20) identical to those of the posterior vitreous framework and demarcates the posterior boundary of the posterior aqueous chamber. The layer attaches to the posterior lens capsule in the form of a ring ("Egger's line") ∼ 9 mm in diameter. The strong attachment along this ring is often called the hyaloideocapsular ligament or *ligament of Wieger*. Axially, the vitreous continues more loosely attached to the posterior lens capsule until the vitreous *face* separates from the capsule (thick basement membrane) and turns posteriorly to form the lining border layer of the conical anterior extremity of the canal of Cloquet. A remnant of the embryonic hyaloid system may be seen on the

OCULAR HISTOLOGY

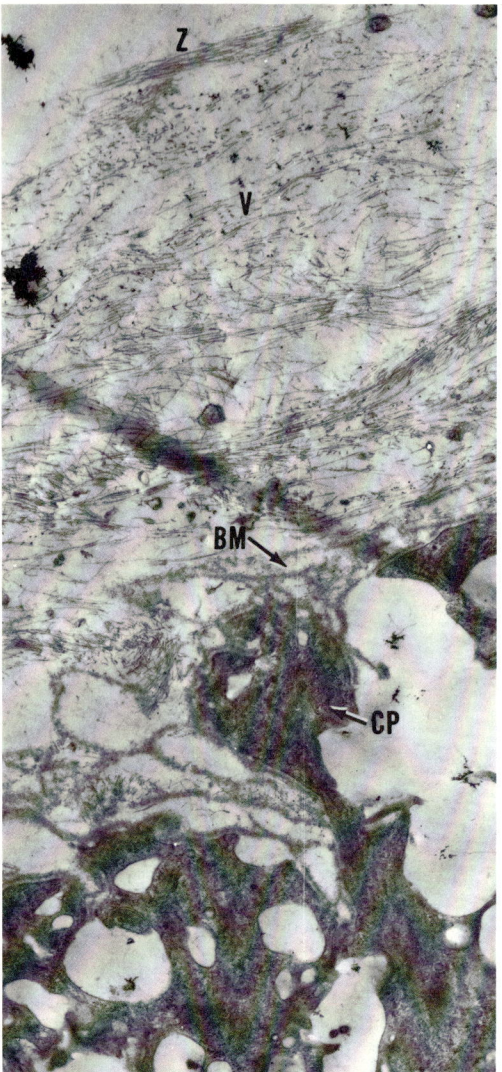

FIG. 7–17. Base of vitreous body in region of ora serrata. *BM*, multi-laminar basement membrane; *CP*, cell processes; *V*, vitreous filaments; *Z*, fragment of zonular fiber. ×8,000. (From Fine, B. S., and Zimmerman, L. E. *Invest Ophthal* 2:105, 1963.)

As with the posterior vitreous, the anterior vitreous face cannot be visualized optically as a distinct entity until it becomes separated from the lens capsule (i.e., detachment of the anterior vitreous). The optically dark zone observed clinically behind the lens (except for the opening to the canal of Cloquet) in a normal eye therefore represents a zone *deep* to the vitreous face, i.e., *within* the anterior vitreous body. This dark zone often is presumably misinterpreted as the *retrolental space of Berger*, a potential space lying between the posterior lens capsule and the vitreous face, which can be demonstrated by the injection of air between the two loosely adherent structures.

The canal of Cloquet can often be recognized in vivo and occasionally is seen to be occupied considerably or partially by a blood-filled vessel from the nerve head representing persistence of hyaloid vasculature. Less often, a small hemorrhage may be observed which may track along the canal so that it is even more clearly demarcated from adjacent, yet clear, vitreous.

ZONES

In the adult eye the vitreous body generally is divided into a condensed interweaving lamellar peripheral zone, the *cortical vitreous*,

FIG. 7–18. Anterior vitreous face (*arrows*) free of anterior pars plana epithelium. Fragments of zonular fibers lie within posterior chamber. H&E, ×80. (AFIP Neg. 62–5292.)

posterior lens capsule just nasal to the visual axis and within the anterior mouth of the canal. This embryonic remnant is called the *Mittendorf dot*.

THE VITREOUS BODY 121

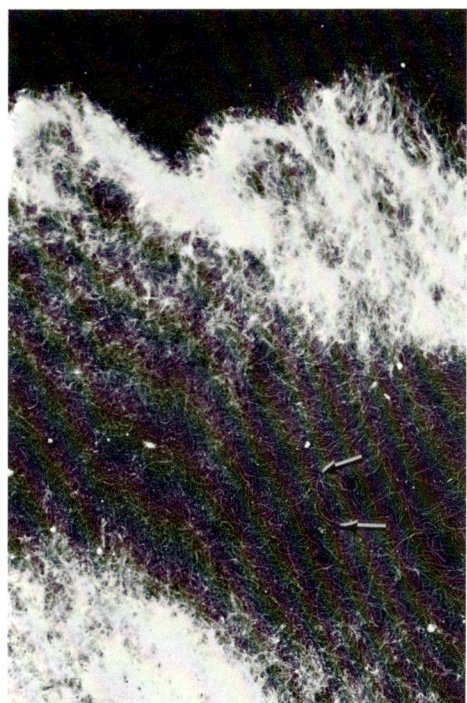

FIG. 7–19. Anterior face of vitreous body. Filament density is too great in some regions of this section to permit their individual resolution. When filaments are in lower concentration (*arrows*) they can be individually observed (see Fig. 7–20). U-shad, ×2,500.

observers, however, attribute their production to cells often lying freely in the posterior vitreous cortex near the retinal surface. These cells, more numerous in bovine eyes, are sometimes termed *hyalocytes*.[8] It is more likely, however, that they are a variety of histiocyte or glial cell that has migrated from surrounding tissues (see retinal vessels and optic nerve head).

MODIFICATIONS

Anteriorly, in the region of the posterior pars plana or vitreous base, the densely aggregated vitreous filaments are aggregated further into "fibers." The fibers pass anteriorly and axially through the vitreous, along its face, and freely into the posterior chamber as the zonular *fibers* of the lens, or *zonule of Zinn* (Figs. 7–21 through 7–24).

The zonular fibers originate as myriad filaments attached to the thin basement membrane covering the markedly infolded basal plasma membranes of the nonpigmented epithelial cells (Fig. 10–54) of the posterior pars

and a looser *central* zone, the central or medullary vitreous. The surfaces are further modified into the dense *anterior border layer* and the somewhat less dense boundary of the canal of Cloquet.

As mentioned previously, the loose arrangement present in fetal life constantly changes with aging, especially in the vitreous periphery, into lamellar aggregations. The loose central vitreous, therefore, represents a more watery "fetal-like arrangement" than the structurally compact cortical zone. The high concentration of acid mucopolysaccharides in the cortical zone can be attributed to their being produced by the ciliary epithelium at the "base" of the vitreous. Some

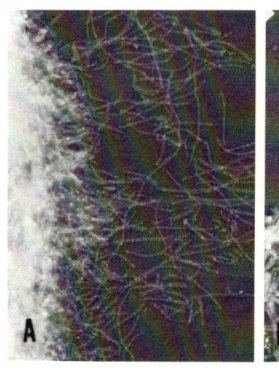 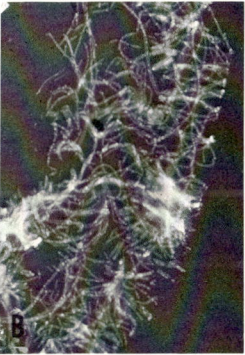

FIG. 7–20. Anterior face of vitreous body. In **A,** filaments are spread out on right; on left, aggregate of filaments is too dense to permit resolution of individual filaments. In **B,** a thin "sheet" of vitreous framework is turned flat to show its filamentous composition. The two cut edges of sheet-like fragment are clearly seen. **A,** U-shad, ×11,000; **B,** U-shad, ×11,500.

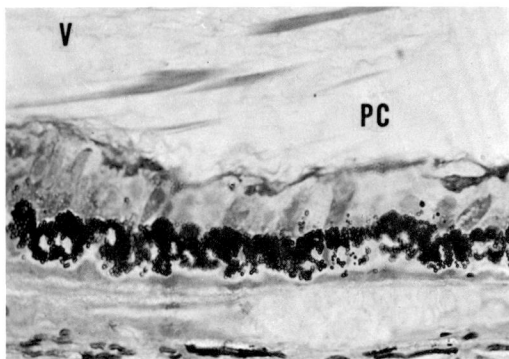

FIG. 7–21. Zonular fibers near their attachments to posterior pars plana epithelium. Zonules lie in posterior chamber (*PC*). Lens is to right and above. *V,* vitreous body. 1.5-μ section, toluidine blue, ×305. (AFIP Neg. 60–2490.)

FIG. 7–22. Shadow-cast section of region over posterior pars plana. A zonular fiber (*Z*) is present within vitreous framework. **Inset** shows cross-section of filamentous zonular fiber (*FZ*) and its close relation to multilaminar basement membrane of nonpigmented ciliary epithelium (*ILM*). (Modified from Fine, B. S., and Tousimis, A. J. *Arch Ophthal (Chicago)* 65:95, 1961.)

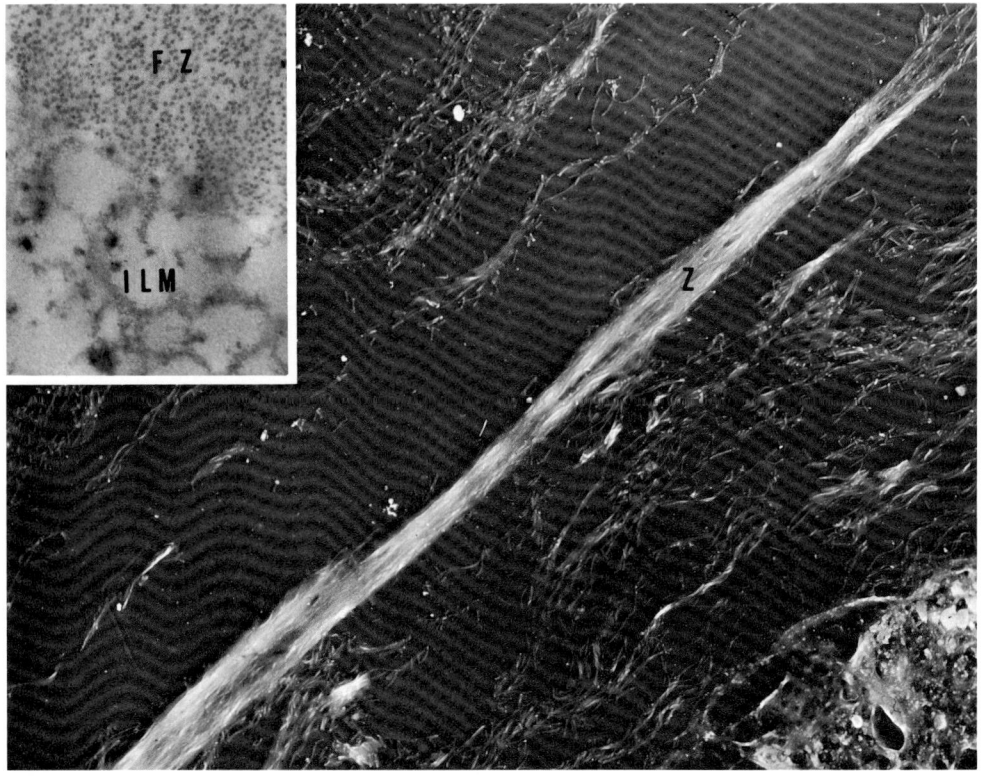

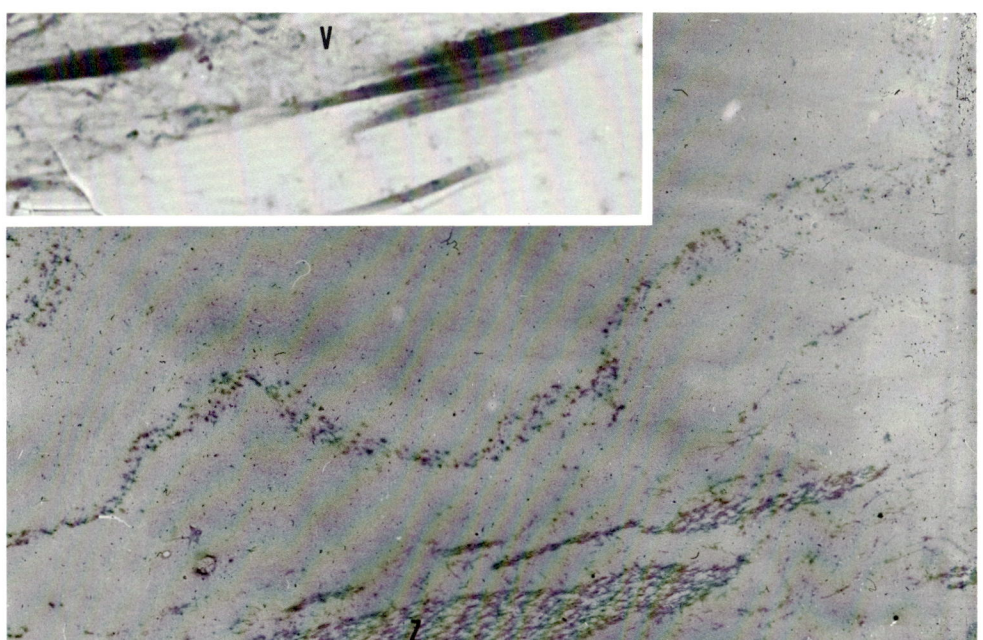

FIG. 7–23. Zonular fibers. Light micrograph (**inset**) shows zonules within vitreous body (V), along anterior face of vitreous body, and within posterior chamber. Electron micrograph, a thin section of a similar region, shows filamentous aggregates of anterior vitreous; similar filamentous aggregates form zonular (Z) fragments. **Main figure,** ×6,900; **inset,** 1.5-μ section, toluidine blue, ×305 (AFIP Neg. 60–4101). (From Fine, B. S., and Tousimis, A. J. Arch Ophthal (Chicago) 65:95, 1961.)

FIG. 7–24. Shadow-cast section showing fragment of an anterior vitreous "sheet" of filamentous composition (V) as well as large zonular fragment (Z). **Inset** a thin-section electron micrograph, clearly shows filamentous composition of a zonular fragment. **Main figure,** ×2,580; **inset,** ×17,000. (Modified from Fine, B. S., and Tousimis, A. J. Arch Ophthal (Chicago) 65:95, 1961.)

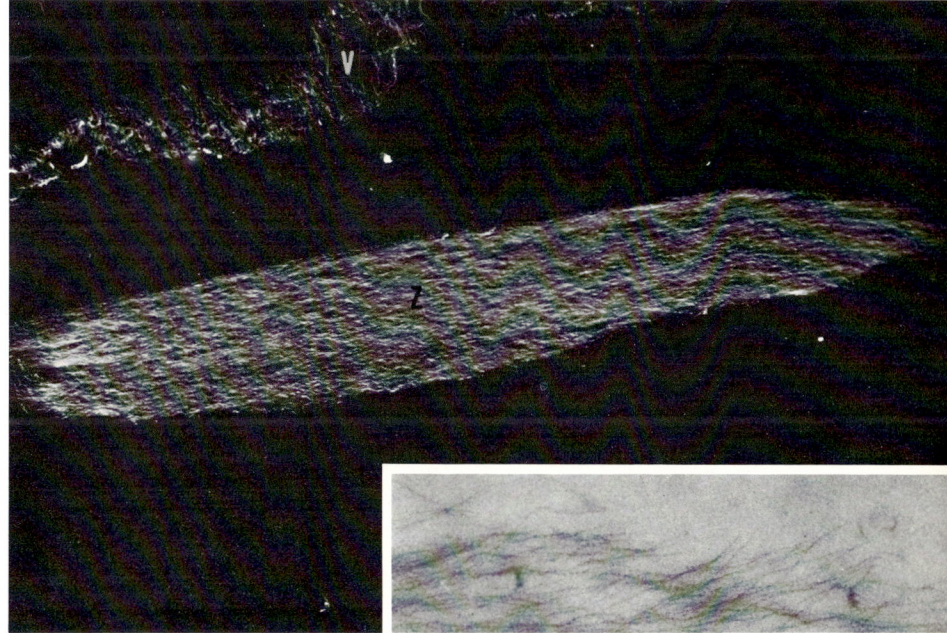

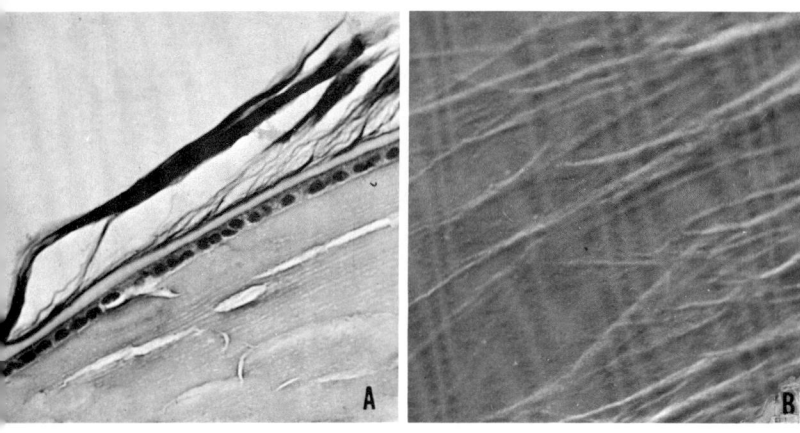

FIG. 7–25. Attachment of orbiculo-anterior zonular fibers to anterior lens capsule seen by light microscopy (**A**) and scanning electron-microscopy (**B**). **A,** bleach-alcian blue, ×350; **B,** ×1,300.

FIG. 7–26. Shadow-cast section from anterior vitreous (V) and adjacent anterior zonule (Z). Both vitreous framework and zonular fibers are composed of morphologically identical filaments. Zonular fibers are too thick at bottom of photograph for their filamentous composition to be visible. This is better appreciated above, where zonule is cut more obliquely as it passes out of plane of section. **Inset** shows some anterior vitreous filaments at higher magnification. **Main figure,** U-shad, ×11,850; **inset,** ×48,000. (From Fine, B. S., and Tousimis, A. J. *Arch Ophthal* (Chicago) 65:95, 1961.)

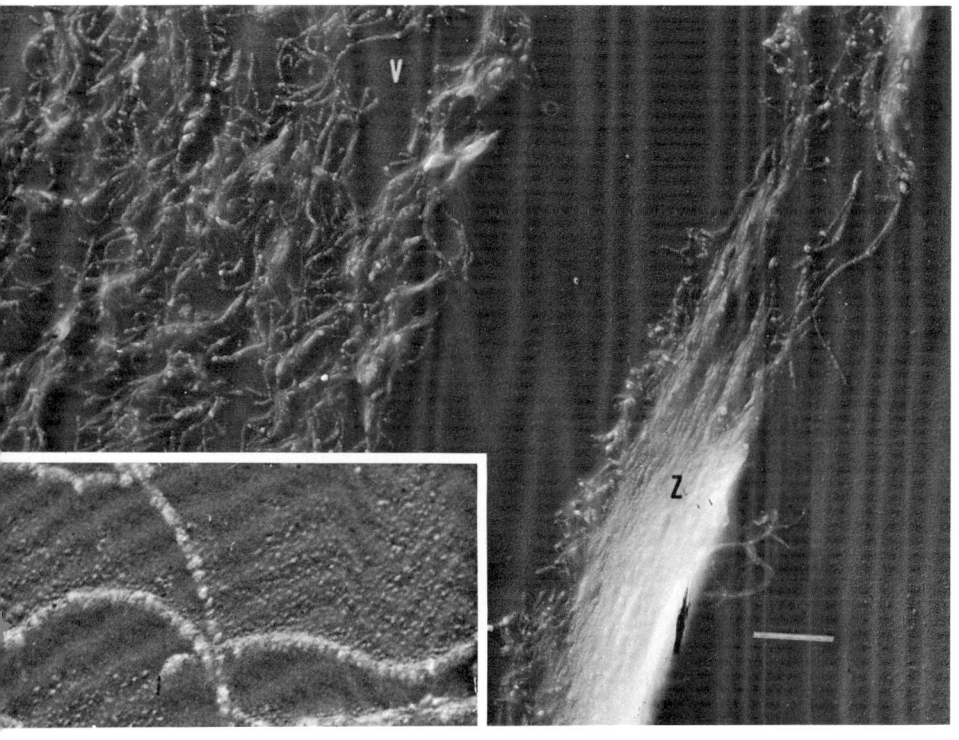

plana or orbiculus ciliaris. Arranged circumferentially, the fibers form essentially two main groups: the *orbiculoanterior* and *orbiculoposterior* zonular fibers. The former attach to the thick basement membrane or anterior lens capsule (Fig. 7–25); the latter attach to the thinner lens basement membrane or posterior lens capsule.

The orbiculoanterior zonular fibers (Fig. 7–26) attach in a ring axial to the lens equator, by spreading out over short regions of the slightly undulating capsular surface (Figs. 8–9 and 8–10). The orbiculoposterior zonular fibers attach in a ring less axial to the equator. The location of the latter ring attachment is identical with the hyaloideocapsular ligament (Figs. 7–27 and 7–28).

A true separate or discrete ligamentous component of the vitreous body in the region of the hyaloideocapsular ligament cannot be identified. The only filamentary aggregates in this zone which are observably different from those of the vitreous anterior border layer and which do not artifactitiously separate from the lens capsule are those of the orbiculoposterior zonular fibers themselves. Since zonular fibers are but specialized portions of the anterior peripheral vitreous framework,[3] the apparent loss of a "ligament" in this region appears explicable on the basis that the hyaloideocapsular ligament is essentially synony-

FIG. 7–27. Region of hyaloideocapsular ligament. Posterior zonular fibers (Z) attach to lens capsule (CAP) together with vitreous body face (V). Melanin granules (arrows) are characteristic of some adult eyes. Note ease of separation of vitreous face from lens capsule. **Inset 1,** higher magnification of lens capsule showing its filamentous arrangement. **Inset 2,** higher magnification of end of zonule (circled area in main figure) showing similarity of zonular and vitreous body filaments. **Main figure,** ×3,800; **inset 1,** ×22,800; **inset 2,** ×22,800. (From Fine, B. S. *Arch Ophthal (Chicago)* 67:689, 1962.)

mous with the attachment ring of the posterior zonular fibers.

Various passageways between anterior and posterior groups of zonular fibers (canal of Hannover) and between the vitreous face and some of the posterior zonular fibers (canal of Petit) have been described on the basis of injection techniques that used such methods as bubbles of air. Such methods, however, because of the surface tensions involved, fail to point up the discrete filamentary zonular bundles which are seen by light and electron microscopy and supported by clinical observations.

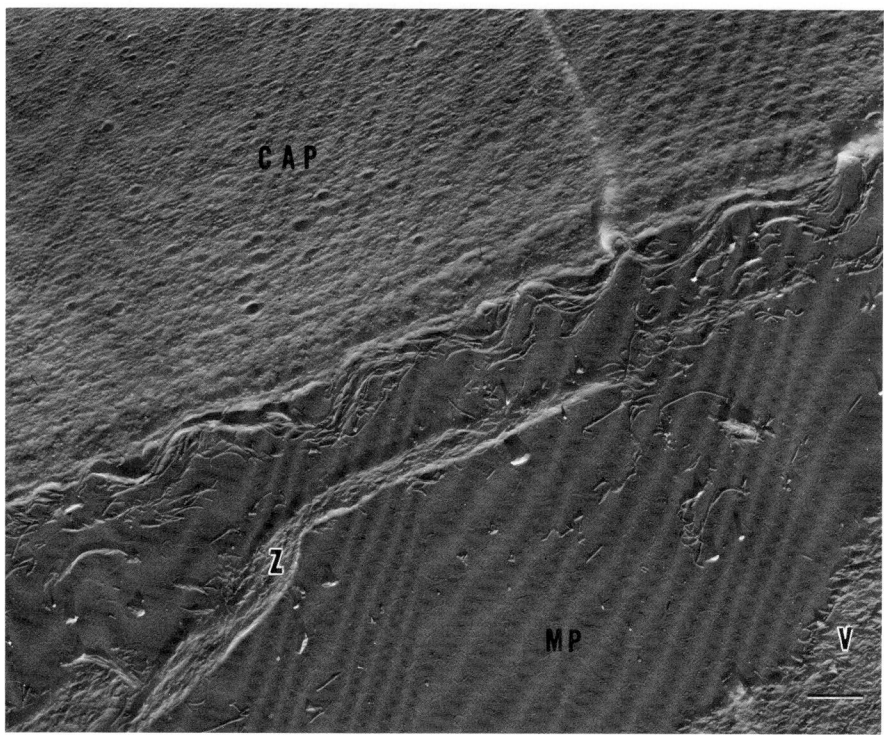

FIG. 7–28. Shadow-cast section of region similar to that in Figure 7–27. Slight undulation of capsular surface (*CAP*) is evident. Zonular filaments and filaments of vitreous body (*V*) appear identical. Mucinous drying patterns are also present (*MP*). U-shad, ×9,200. (From Lerman, S. *Cataracts: Chemistry, Mechanisms, and Therapy.* Springfield, Ill., Thomas, 1964.)

Occasionally the old concept is revived that the zonular fibers are in reality parts of a continuous sheet. No strong histologic evidence exists that these are other than separate groupings of filaments. Clinically, pigment or other deposits are seen *on* but not *between* the zonular fibers.

This specialization of the vitreous framework to become the zonular fibers is in accord with the embryologic observation of a "condensation of tertiary vitreous" to form the zonules. The collagen presumably is secreted by the epithelial cells which impinge upon the lens capsule and spin out the zonular fibers as a spider its web. The fibers become longer as the cells move posteriorly with growth of the globe to come to rest as the epithelial lining of the posterior pars plana. The more anterior, or later, cells apparently lose their ability to synthesize collagen and form only irregular, vestigial, and possibly some *cilioequatorial* zonular fibers, all of which appear to be of limited functional importance. In some pathologic conditions in the adult eye the cells may proliferate and resume their collagen-synthesizing activities; an example is the formation of new vitreous-like material by adenomatous hyperplasias of the nonpigmented ciliary epithelium (see the Ciliary Body, Chap. 10).

REFERENCES

1. Boyer, H. K., Suran, A. A., Hogan, M. J., and McEwen, W. K. Increase of residual protein of bovine vitreous during growth of the eye. *Arch Ophthal (Chicago)* 56:861, 1956.

2. Fine, B. S. Correspondence. *Arch Ophthal (Chicago)* 67:689, 1962.

3. Fine, B. S., and Tousimis, A. J. The structure of the vitreous body and the suspensory ligaments

of the lens. *Arch Ophthal (Chicago)* 65:95, 1961.
4. Fine, B. S., and Zimmerman, L. E. Light and electron microscopic observations on the ciliary epithelium in man and rhesus monkey. *Invest Ophthal* 2:105, 1963.
5. Friedenwald, J. S., and Stiehler, R. D. Structure of the vitreous. *Arch Ophthal (Chicago)* 14:789, 1935.
6. Lerman, S. *Cataracts: Chemistry, Mechanisms, and Therapy.* Springfield, Ill., Thomas, 1964.
7. Meyer, K., and Palmer, J. W. Polysaccharide of vitreous humor. *J Biol Chem* 107:629, 1934.
8. Österlin, S., and Balazs, E. A. Macromolecular composition and fine structure of the vitreous in the owl monkey. *Exp Eye Res* 7:534, 1968.
9. Pirie, A., Schmidt, G., and Waters, J. W. Ox vitreous humour. I. The residual protein. *Brit J Ophthal* 32:321, 1948.
10. Pirie, A., and van Heyningen, R. *Biochemistry of the Eye.* Springfield, Ill., Thomas, 1956, p. 234.
11. Porter, K. R., and Pappas, G. D. Collagen formation by fibroblasts of the chick embryo dermis. *J Biophys Biochem Cytol* 5:153, 1959.
12. Zimmerman, L. E. Demonstration of hyaluronidase-sensitive acid mucopolysaccharides: A preliminary report. *Amer J Ophthal* 44:1, 1957.

chapter 8

The Lens

Embryology
Structure
    Capsule
    Epithelium
    Cortex and nucleus

The lens is an example of a tissue composed entirely of cells completely enveloped by the thickest basement membrane in the body.

EMBRYOLOGY

When the optic vesicle transforms into the optic cup, there is concomitant invagination of the surface ectoderm. The *primordial* lens separates completely from the surface layer to lie suspended within the orifice of the optic cup as a vesicle formed by epithelium, the lens *vesicle* (Fig. 8–1).

The delicate basement membrane associated with the surface ectoderm now encloses the vesicle in which the epithelial cells all project inward.[3,5] This basement membrane, or lens capsule, is so thin that in its early stages it cannot be appreciated by light microscopy. With further development, the posterior epithelial cells elongate to gradually occlude the lumen of the vesicle. These early, elongated lens cells ("fibers") are called the *primary* lens fibers. All subsequent lens cells ("fibers") are derived from the epithelial cell layer at the equator and are termed *secondary* lens cells or fibers (Fig. 8–2).

Presumably it is this narrow band of primary lens fibers that is seen biomicroscopically as the single, narrow central nonrelucent zone of the lens called the *embryonic nucleus* (Fig. 8–3). The early secondary lens fibers or cells presumably form the more relucent, symmetrically paired anterior and posterior portions that "cap" this central zone. These caps, on which the Y-shaped suture is now recognizable, represent the "fetal" nucleus.

In the preequatorial zone the lens epithelium undergoes mitotic division, and at the equator the cells elongate and rotate (apex anterior, base posterior). Further elongation occurs in the postequatorial zone (Figs. 8–2 and 8–4). Successive application of these strap-like cells to the underlying cells results in a layered arrangement, and where the elongated cells meet, a line or lens *suture* is formed (Figs. 8–3 and 8–5).

Because all the lens cells grow equally long, and thus cannot all reach anterior and

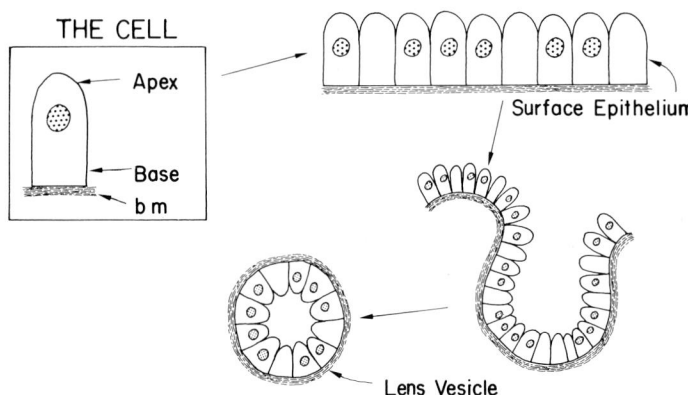

FIG. 8–1. Schematic representation of formation of lens vesicle from surface ectoderm. Polarization of epithelial cells is maintained. *bm,* basement membrane.

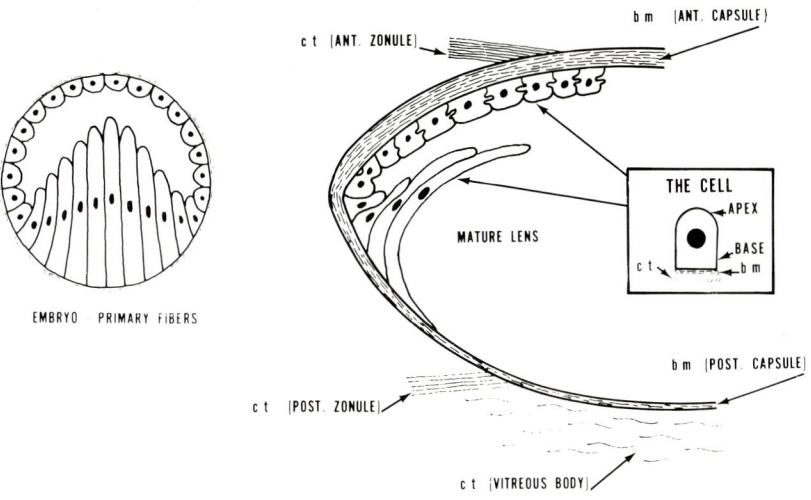

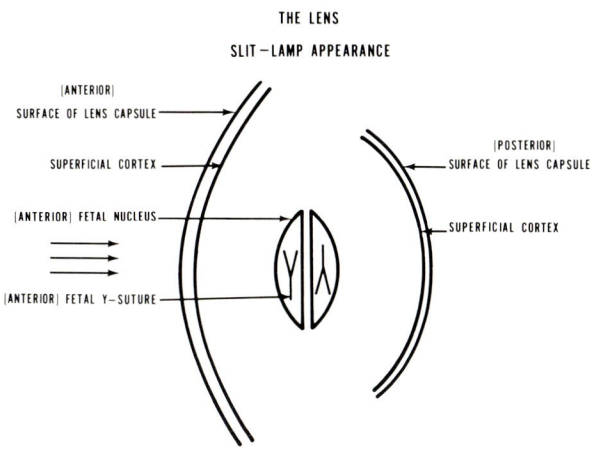

FIG. 8–2. Schematic representation of maturation of primary lens fibers within lens vesicle (**left**) and final arrangement achieved in adult lens (**right**). *bm,* basement membrane; *ct,* connective tissue.

FIG. 8–3. Slit-lamp appearance of major landmarks in lens. The three free arrows indicate direction of light entering eye.

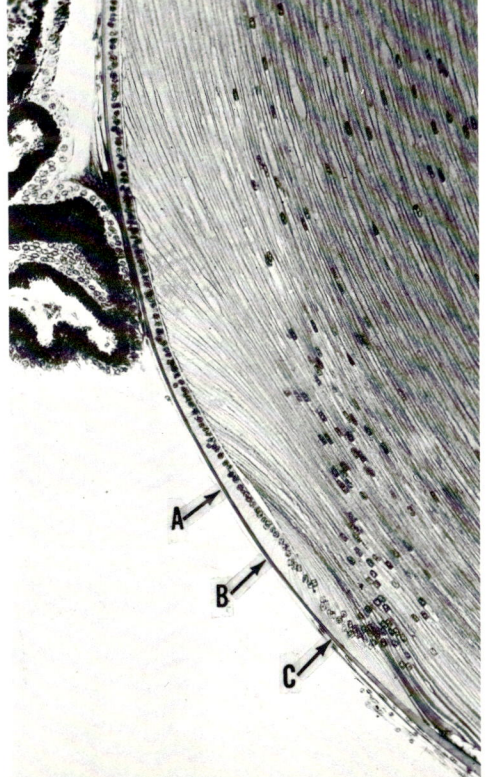

FIG. 8–4. Meridional section from eye of 4-month-old embryo. Preequatorial, equatorial, and postequatorial zones of lens epithelium are indicated by *arrows A, B,* and *C,* respectively. Note persistence of lens cell nuclei deep into cortex and close approximation of ciliary epithelium to lens capsule. Masson, ×115. (AFIP Neg. 61–3077.)

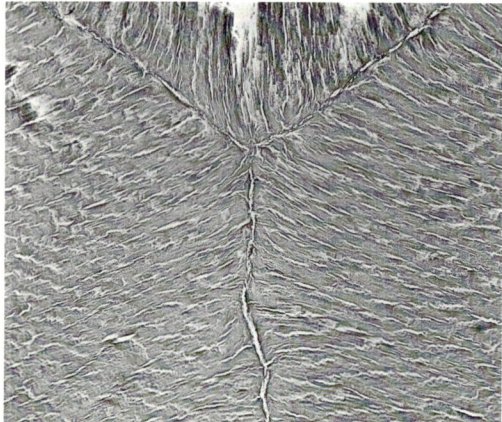

FIG. 8–5. Equatorial section showing formation of Y-suture where lens cells meet. H&E, ×70. (AFIP Neg. 70–1427.)

posterior points, lines of abutment—the anterior and posterior suture lines—are formed. In the human, the lens acquires a biconvex or flattened disc shape so that the suture lines must become quite complex. The result is a Y-shaped suture line which can be observed biomicroscopically as an inverted Y posteriorly and an upright Y anteriorly. With further addition of lens cells, suture lines increase in complexity. All these secondary cells add to the formation of the *lens cortex*.

In spherical lenses (e.g., dogfish, rabbit) the anterior suture is vertical and the posterior is horizontal to accommodate cells of equal length.

The lens epithelium remains over the anterior surface and the equatorial zone. The basement membrane or lens capsule continues to thicken throughout life. Because the thickening occurs mainly where epithelium is present, the posterior lens capsule remains extremely thin (Figs. 8–6 and 8–7).

As the cortical cells are continually layered on from the periphery, the more centrally located begin to lose their recognizable cell organelles; this is most easily noted by observing the loss of a recognizable cell nucleus toward the center of the lens. Forced deeper into the center, the cortical cells are compressed from front to back, becoming increasingly dense and folded to form a central lens mass—the *adult lens nucleus*.

STRUCTURE

After birth the lens appears as a transparent biconvex body enclosed in a capsule, lying directly behind the pupil. The center of its anterior surface is called the anterior pole; the center of its posterior surface, the posterior pole. The radius of curvature of the anterior surface averages 10 mm, but it is

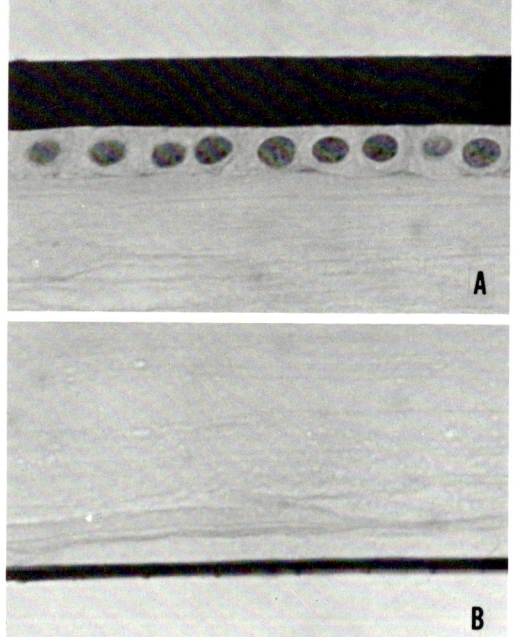

FIG. 8–6. Lens capsule. Thick anterior lens capsule (**A**) is applied to bases of underlying layer of epithelial cells; thinner posterior lens capsule (**B**) is applied to cortical cells only. Both PAS, ×730. (From Scheie, H. G., and Albert, D. M. *Adler's Textbook of Ophthalmology.* Philadelphia, Saunders, 1969.)

THE LENS   133

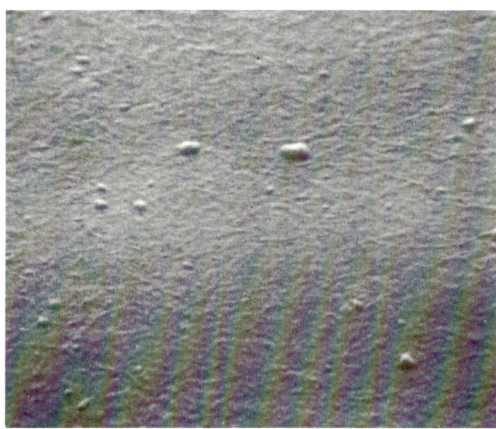

FIG. 8–7. Surface of anterior lens capsule near periphery of pupil as seen by scanning electron microscope. Surface is remarkably smooth in this preparation at this magnification. ×5,500.

FIG. 8–8. Anterior lens capsule. **Main figure.** Electron micrograph of anterior zonule (*Z*), lens capsule (*CAP*), filaments (*F*), and adjacent epithelium (*EP*) in area comparable to that shown in light micrograph (**inset 1**). **Inset 2.** Scanning electron micrograph of surface of anterior lens capsule. Nuclei of underlying epithelial cells protrude slightly because of dehydration. **Inset 3** shows clusters of filaments typical of anterior lens capsule. **Inset 4** shows periodicity that may be observed in anterior lens capsule. **Main figure,** ×5,600; **inset 1,** 1.5-μ section, PD, ×485 (AFIP Neg. 63–6905); **inset 2,** ×250; **inset 3,** ×15,000; **inset 4,** ×40,000.

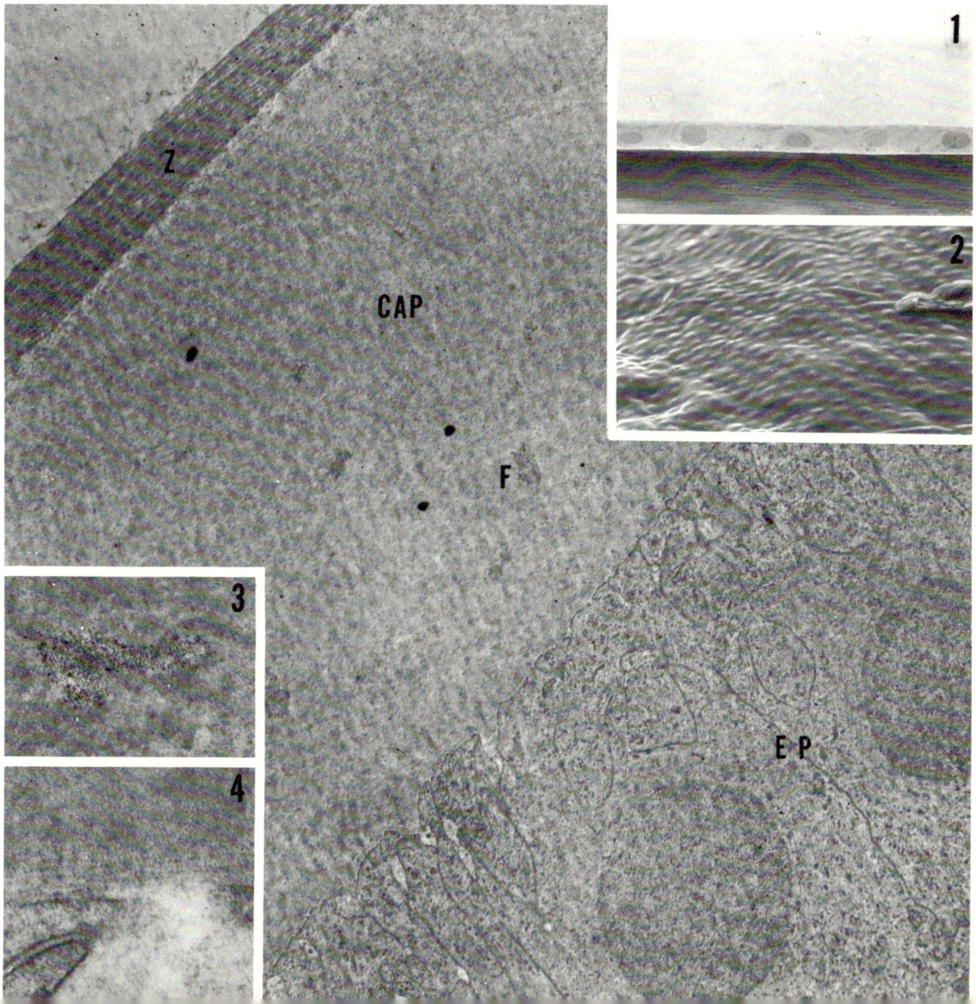

subject to marked changes during accommodation. The radius of curvature of the posterior surface averages 6 mm. The lens equator is not smoothly curved, but shows numerous irregularities resulting from the pull of the zonular fibers.

Capsule

The adult lens is enclosed by the *lens capsule* (*cuticular, glass, or basement membrane*), a transparent, reflective, elastic, indigestible, PAS-positive covering.

Connective tissues are present on the outer, or nonepithelial, side of the thick basement membrane, but are modified in this special case into the anterior and posterior groups of *zonular fibers* and the anterior *face* of the *vitreous body* (Fig. 8–2).

The lens capsule has a distinctly filamentous substructure (Fig. 8–8), with the filaments aligning parallel to the surface. Numerous small patches or clusters of filaments (*inset 3*, Fig. 8–8) lying at varying levels in the anterior capsule differentiate it from the posterior capsule.[4] When appropriately sectioned, this capsular basement membrane occasionally shows a periodicity ($\sim 1{,}000$ Å) (*inset 4*, Fig. 8–8), a characteristic observed in other normal and some pathologic basement membranes.

The outer surface of the lens capsule (to which the connective tissues are attached) is only slightly irregular (Fig. 8–9), with the anterior irregularity slightly coarser than the posterior (see Figs. 8–9 and 7–28).

On biomicroscopic examination, the lens *shagreen* or beaten-metal-like reflex, long believed to represent a reflection from the capsular surface (known in thin optical section as the *first zone of relucency*) is observed in vivo to be coarser anteriorly than posteriorly. Loss of lens shagreen or surface irregularity is a sign that often alerts the cataract surgeon to the possibility of a "weak" capsule and enables him to be more selective in his method of cataract removal. To date (Figs. 8–7 and 8–10), examination of the surface

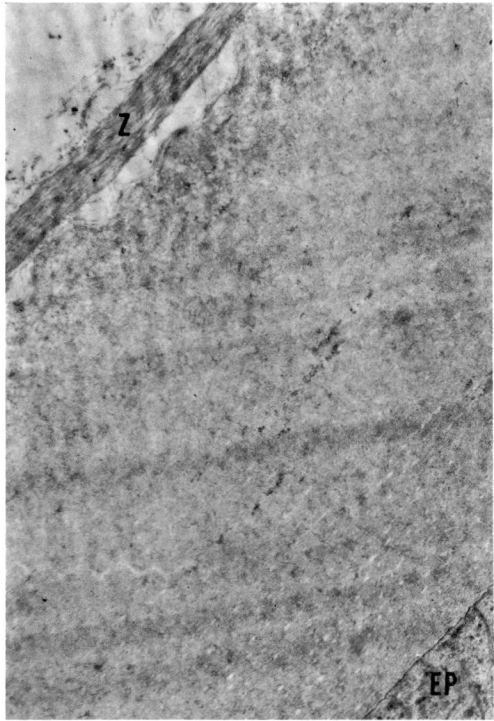

FIG. 8–9. Anterior lens capsule. Note undulation of capsular surface beneath zonule (Z). EP, epithelium. ×5,300. (From Lerman, S. *Cataracts: Chemistry, Mechanisms, and Therapy* Springfield, Ill., Thomas, 1964.)

of the lens capsule by scanning electron microscopy fails to demonstrate a gross surface irregularity. The lens shagreen therefore may represent optical variations *within* the lens capsule.

Although histologic studies show that a thin layer at the surface of the lens capsule can be stained differently from the remainder of the capsule, electron microscopic observations indicate that this pericapsular layer (membrane) differs morphologically only in its comparative looseness from the remainder of the compact lens capsule (Fig. 8–11). That this differential staining probably is due to capsular "looseness" is supported by the observation that when the anterior lens capsule is split or separated mechanically into numerous layers, the loosened structure acquires similarly altered staining properties. On clinical examination a membrane, called pericapsular,

FIG. 8–10. Attachment of anterior zonules to lens capsule as shown by scanning electron microscope. ×680.

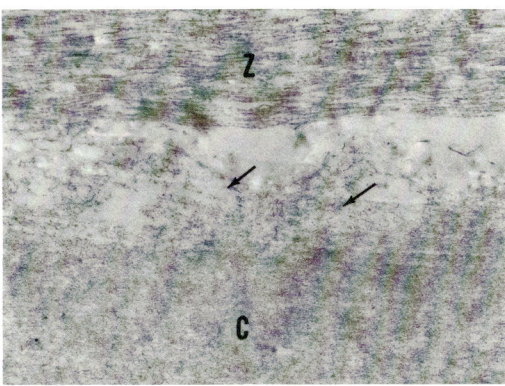

FIG. 8–11. Zone of "looseness" (*arrows*) between zonule (Z) and underlying, more compact anterior lens capsule (C). ×16,000.

can be observed occasionally uniting the ends of zonular fibers that have split off from the pre-equatorial zone of the lens. This membrane is probably composed of the loose surface layer and a variable amount of deeper capsular material.

Apertures in the normal lens capsule have not yet been convincingly demonstrated, but in at least two reports on pathologic tissues[1,2] apparent through-and-through apertures have been observed in the anterior capsule.

Epithelium

Anteriorly, the epithelium is cuboidal, with rather flat cell plasma membranes applied to the equally flat basement membrane. The cells interdigitate laterally in a fashion typical of many epithelia. The intercellular space also is typical except near the basal ends of the cells in the midperiphery, where foci of enlarged intercellular spaces may be observed (Fig. 8–12).

Some of the spaces may be artifacts; some may be physiologic, representing the histologic coun-

FIG. 8–12. Foci of enlarged intercellular spaces (*arrows*) filled with lucent extracellular material at anterior midperiphery. C, lens capsule. ×6,600. (From Lerman, S. *Cataracts: Chemistry, Mechanisms, and Therapy.* Springfield, Ill., Thomas, 1964.)

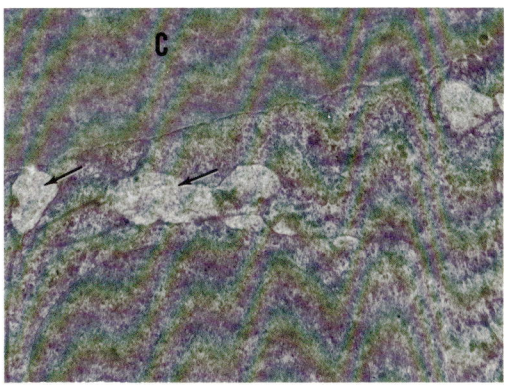

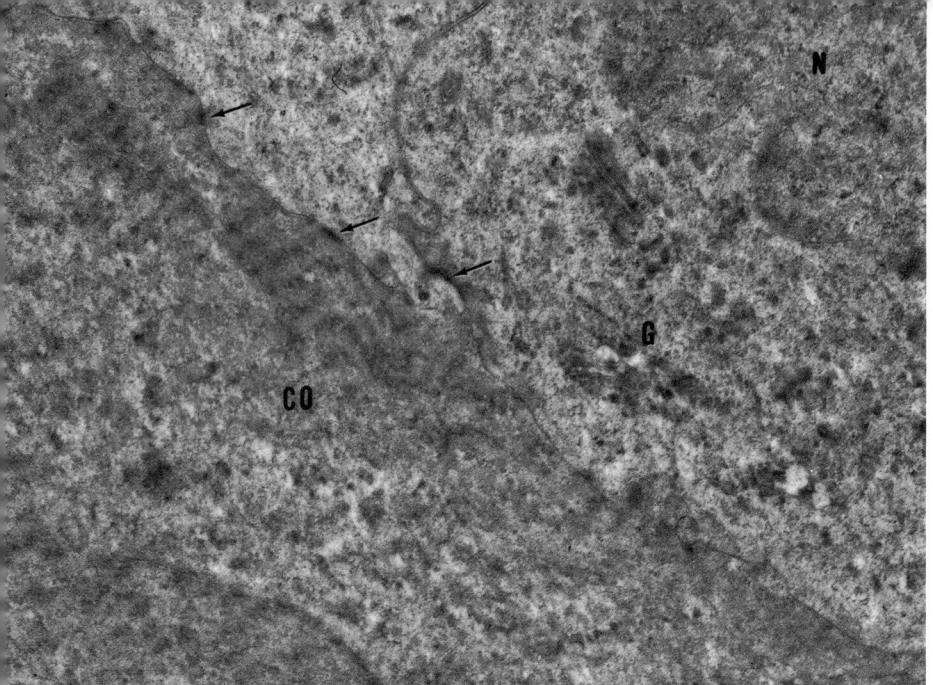

FIG. 8–13. Apex of lens epithelium and adjacent cortical cells (*CO*). *G,* Golgi complex of epithelial cell; *N,* nucleus of epithelial cell. *Arrows* point to adjacent plasma membrane densities. ×16,000. (Modified from Lerman, S. *Cataracts: Chemistry, Mechanisms, and Therapy.* Springfield, Ill., Thomas, 1964.)

FIG. 8–14. Lens cells ("fibers") in cross-section, showing their honeycomb-like arrangement. **Inset** is a representative light micrograph. **Main figure,** ×8,000; **inset,** H&E, ×260. (AFIP Acc. 132413.)

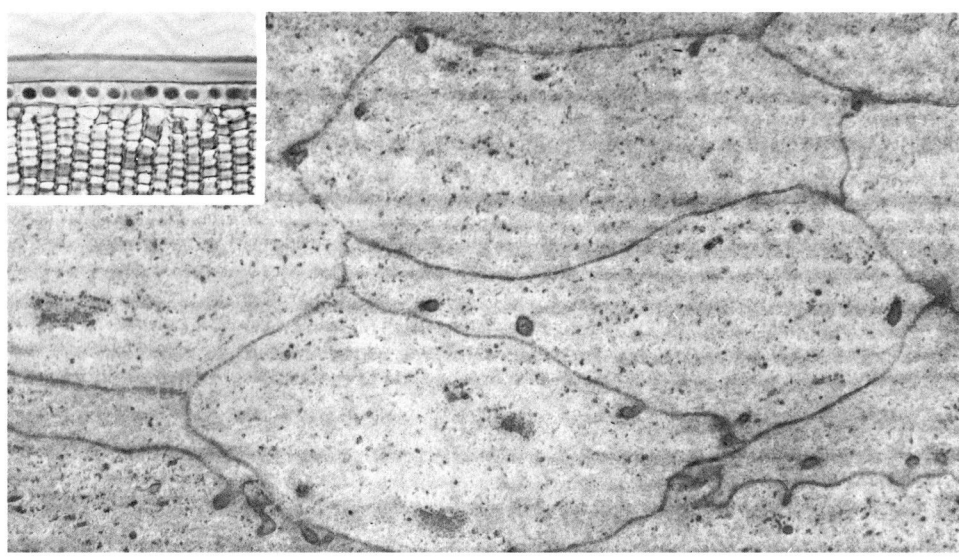

cortical cells so formed are applied to the layers beneath.

Cortex and Nucleus

Because the lens nucleus is formed by increasing density of cortical cells, cortex and nucleus are considered together. Both are made up exclusively of cells derived from the lens epithelium. The common designation of "lens fiber" for the cortical cells is a misnomer for the elongated cells of the lens substance.

Lens cortical cells ("fibers") are elongated strap-like cells which, on cross-section, appear somewhat hexagonal (*inset*, Fig. 8–14). The cortex in this cross-section view resembles the cut surface of a honeycomb. The

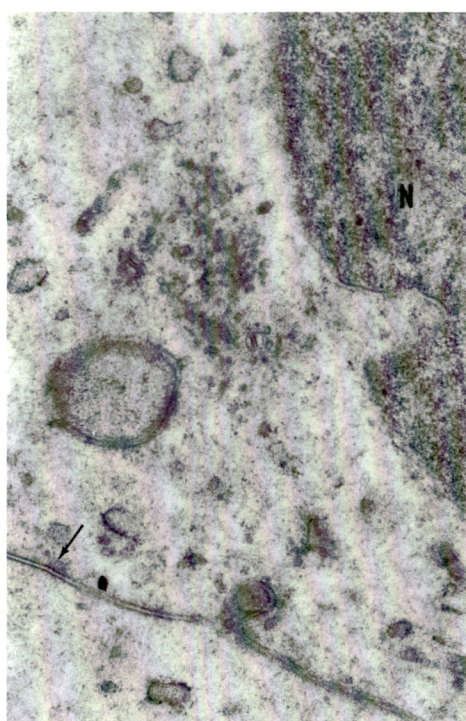

FIG. 8–15. Typical intercellular space (*arrow*) between superficial cortical cells. Cytoplasm contains few organelles. N, nucleus of a cortical cell. ×27,000.

FIG. 8–16. Superficial cortical cell cytoplasm. Note its almost homogeneous appearance and the fragment of a poorly formed Golgi complex. ×27,000.

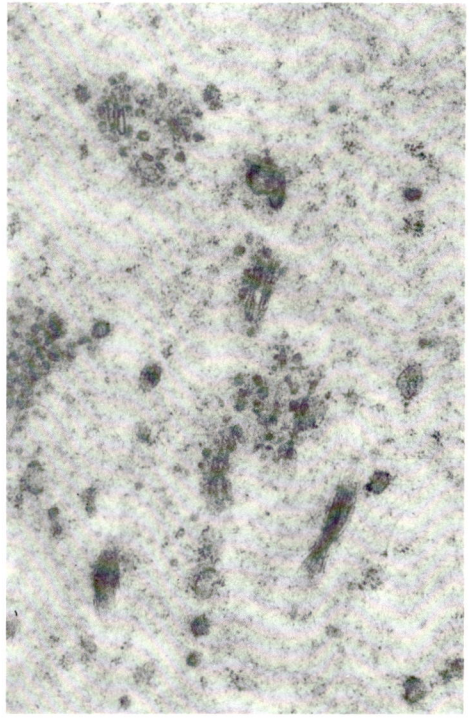

terparts of the globules (i.e., Vogt's "shagreen spheres") that biomicroscopists in the past have noted in the midperipheral zone; and some may represent early pathologic changes.

The organelles within the epithelial cells are not prominent, and the small Golgi complex is found in the apical cytoplasm (toward the lens cortex) (Fig. 8–13.) The apical plasma membranes of the epithelial cells frequently are attached to one another and to the first layer of flattened cortical cells by a series of short desmosome-like densities. (Presumably, removal of lens epithelium is accompanied by a torn superficial layer of cortical cells.)

The cells undergo mitotic division in the preequatorial zone and elongation and rotation in the equatorial zone. The elongated

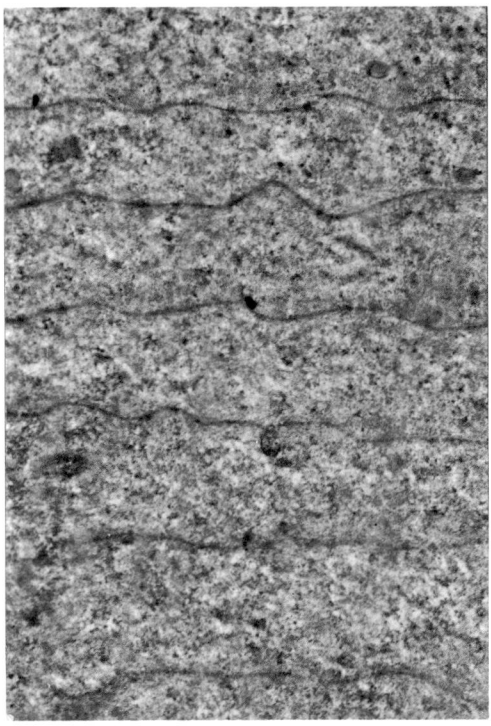

FIG. 8–17. Adult superficial lens cortex. Cells appear homogeneous and possess relatively flat, straight boundaries. Intercellular spaces can be observed. ×7,000.

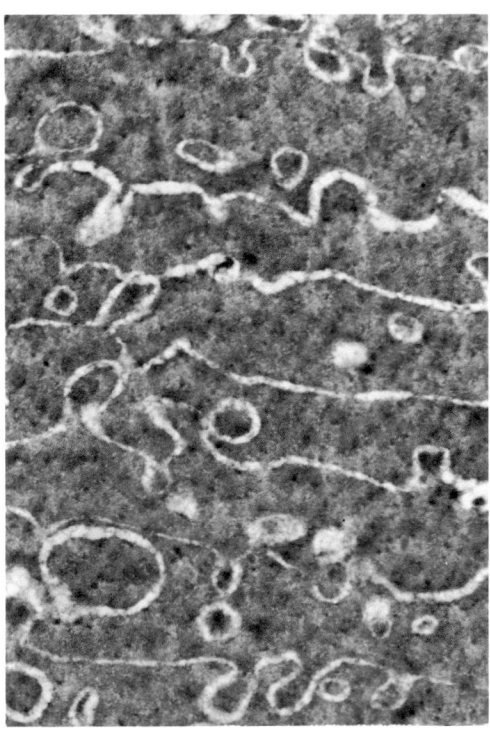

FIG. 8–18. Adult deep lens cortex. Cells are dense and irregularly folded. Cell membranes are indistinct and artifactitious separations are abundant. ×7,000.

superficial cortical cells are relatively flat and are separated by an intercellular space (Fig. 8–15). Cell organelles are almost nonexistent (Figs. 8–14 through 8–16), and the cytoplasm appears more homogeneous than that of the epithelial cells, but of comparable density (Fig. 8–17). Slightly deeper within the cortex (Fig. 8–18) the cell layers may abruptly become dense. These dense cells possess markedly irregular surfaces and the appearance of fusion of adjacent plasma membranes. This abrupt change in cell shape and density is interpreted as due to physiologic compression of the cells with accompanying loss of water and shrinkage.

The abrupt transition of electron microscopic density within the superficial cortex apparently

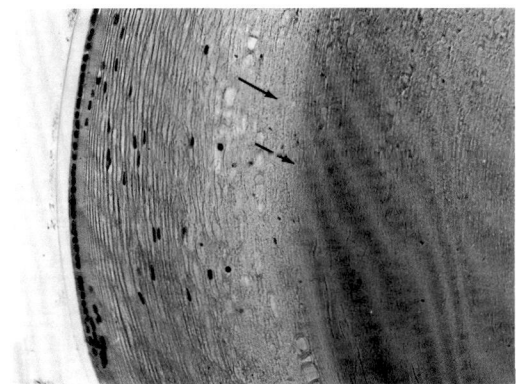

FIG. 8–19. Light micrograph showing normal transition zone (*arrows*) from cortex to nucleus in adult lens. H&E, ×115. (AFIP Neg. 62–2043.)

represents a comparable abrupt change in *optical density* that can be observed in vivo and is known as the *second zone of relucency* (Fig. 8–3). A similar zone is present posteriorly, but is always more closely approximated to that of the posterior lens capsule. The difference in separation between the anterior and posterior zones of relucency is most easily explained by the combination of decreased thickness of capsule and lack of epithelium posteriorly.

In light microscopy, the transition from cortex to nucleus (Fig. 8–19) is characterized by apparent loss of lamellar arrangement and increase in staining density (i.e., eosinophilia). In electron microscopy, as cell density increases, so does the surface irregularity. The plasma membranes of adjacent cells become so tightly adherent to one another that intercellular spaces become impossible to distinguish and the cells may tear through the surface cytoplasm.

These observations are compatible with the increasing density as seen bio- and electron microscopically when deeper layers are examined. In addition, the most dense nuclear cells often show a yellow pigmentation clinically that generally increases with age. When color and density are exaggerated to a degree incompatible with useful vision, it is termed a nuclear *cataract*.

The lens may be compared to a piece of skin that has been turned inside out. The cells which are compressed into the nucleus of the lens are then analogous to those near the surface of the skin epithelium. Comparable densification of the oldest cells in a stratified epithelium also is observed in the superficial cells of corneal epithelium (see Fig. 9–5).

REFERENCES

1. Ashton, N., Shakib, M., Collyer, R., and Blach, R. Electron microscopic study of pseudoexfoliation of the lens capsule. *Invest Ophthal* 4:141, 1965.
2. Bertelsen, T. I., Drablos, P. A., and Flood, P. R. The so-called senile exfoliation (pseudoexfoliation) of the anterior lens capsule: A product of the lens epithelium fibrillopathia epitheliocapsularis. *Acta Ophthal* 42:1096, 1964.
3. Cohen, A. I. Electron microscopic observations on the lens of the neonatal albino mouse. *Amer J Anat* 103:219, 1958.
4. Dark, A. J., Streeten, B. W., and Jones, D. Accumulation of fibrillar protein in the aging human lens capsule (with special reference to the pathogenesis of pseudoexfoliative disease of the lens). *Arch Ophthal (Chicago)* 82:815, 1969.
5. Hunt, H. H. A study of the fine structure of the optic vesicle and lens placode of the chick embryo during induction. *Develop Biol* 3:175, 1961.
6. Lerman, S. *Cataracts: Chemistry, Mechanisms, and Therapy.* Springfield, Ill., Thomas, 1964.
7. Scheie, H. G., and Albert, D. M. *Adler's Textbook of Ophthalmology.* Philadelphia, Saunders, 1969.

chapter 9

The Cornea and Sclera

The Cornea
    Tear film and epithelium
    Bowman's membrane (layer) and Stroma
    Endothelium and Descemet's membrane
    Central compared with peripheral portions
The Sclera
    Emissaria
    Episclera
    Stroma
    Lamina fusca

THE CORNEA

The cornea may be considered a tissue composed almost entirely of extracellular materials and covered on both its anterior and posterior surfaces by a sheet of cells. The two cell layers—a multilayered epithelium anteriorly and a single-layered "endothelium" posteriorly—are apposed to their associated basement membranes and connective tissues in such manner (Fig. 9–1) as to form a three-layered sandwich. The modifications within this simplistic arrangement are described below.

Functionally, the cornea is not only a "window" to the eye but also its *major* refractive element. The corneal diameter in the adult eye averages 12 mm, with the horizontal meridian generally 1 mm larger than the vertical. When viewed from behind (as in an opened eye), the cornea appears more nearly circular. This discrepancy results from the slightly greater overlap of adjoining opaque sclera anteriorly in the vertical than in the horizontal plane.

The cornea is $\sim$ 0.5 mm thick centrally and $\sim$ 1.0 mm thick peripherally.

In addition to the avascularity of the cornea, its transparency appears to be due to the relation of collagen fibrils of small and *uniform* diameter to a *large* bed of acid mucopolysaccharides.

The cornea may be separated into at least six layers (Fig. 9–2): (1) tear film, (2) epithelium (and its basement membrane), (3) Bowman's membrane or layer, (4) stroma, (5) Descemet's membrane, and (6) endothelium.

TEAR FILM AND EPITHELIUM

The tear film covering the corneal surface is made up of three layers: a posterior layer rich in glycoproteins derived from the conjunctival goblet cells, a middle watery layer secreted by the lacrimal tissues, and an anterior oily layer produced by the meibomian glands and the glands of Moll and Zeis in the eyelids.

The nonkeratinized, squamous epithelium of the cornea consists of approximately five layers of cells, which probably are more regularly arranged than any other squamous epithelium in the body. The deepest or basal layer (Fig. 9–3) is composed of columnar cells ($\sim$ 18 μ in height) which characteristically

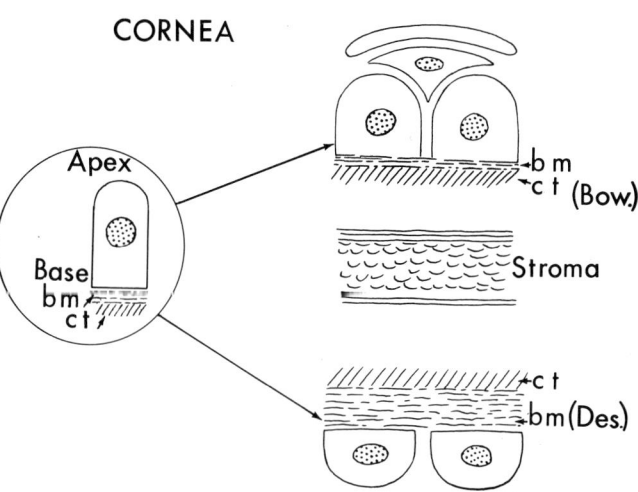

FIG. 9–1. Schematic drawing showing three-layered arrangement of cornea. *bm,* basement membrane; *ct,* connective tissue; *Bow.,* Bowman's membrane (layer); *Des.,* Descemet's membrane.

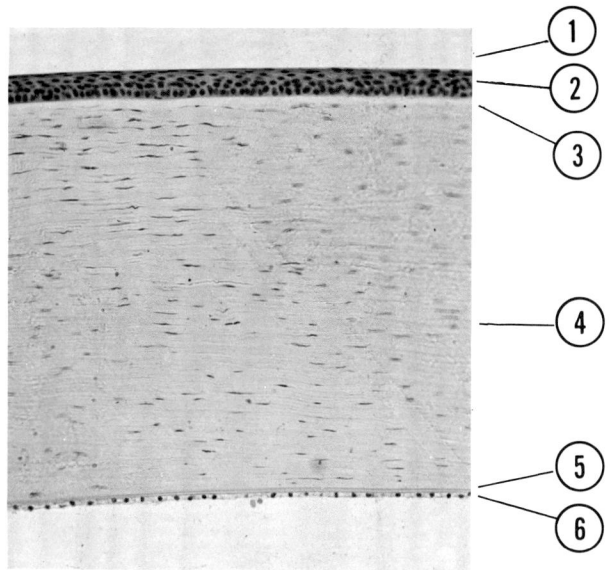

FIG. 9–2. Light micrograph of section of central cornea. Numbers indicate layers: (1) tear film, (2) epithelium, (3) Bowman's membrane, (4) stroma, (5) Descemet's membrane, (6) endothelium. H&E, ×115. (AFIP Neg. 65–238.)

possess a thin, irregular, occasionally diffused (i.e., slightly multilaminar) basement membrane applied to their basal plasma membranes. Focal densities occur regularly along the basal plasma membrane. The densities appear identical to the desmosomes that are attached elsewhere along the cell borders except that here they receive no reciprocity from an adjacent cell to form a complete desmosome; hence this basilar structure is called a *hemidesmosome*.

Although they are structurally unilateral, some evidence exists that the hemidesmosomes may accentuate focally the usual attachment of a cell to its thin basement membrane.[2,12]

The basal cells have a centrally located nucleus as well as the usual complement of mitochondria and Golgi complex. Their cytoplasm contains a high concentration of filaments (characteristic of a number of epithelial cells). As the basal cells undergo mitotic division, they force the daughter cells (which, because of their slightly flattened and curved appearance, are called *wing* cells) outward into the next layer. Persisting in this layer are the myriad desmosomal attachments to adjacent cells (Fig. 9–4) as well as the high concentration of cytoplasmic filaments.*

Beyond this layer the cells become more flattened and more dense (presumably because of increased dehydration) with loss of recognizable cell organelles. The broad, flattened surface cells continue to possess surface irregularities (small folds[1] or villous projections, Figs. 9–5 and 9–6), which are smoothed over by the tear film, especially its mucoprotein and lipoidal components. On occasion, the dried-out remains of the tear film layer can be observed in a micrograph (Fig. 9–5).

The corneal epithelium is continuous with the epithelium of the bulbar conjunctiva at the limbus. Goblet (mucus-secreting) cells appear in the bulbar conjunctival epithelium.

BOWMAN'S MEMBRANE (LAYER) AND STROMA

On the posterior (deep) side of the epithelial basement membrane lies the adjacent connective tissue layer. By light microscopy (inset, Fig. 9–7), this layer $\sim$ 10 to 16 μ thick) is seen as a relatively homogeneous, acellular sheet known as *Bowman's membrane*. In reality, this zone as seen by electron microscopy[7-9,14] is a specialized layer of corneal stroma and not a true membrane. The layer is composed of collagen fibrils (Fig. 9–7) in random distribution, which, except for a slight difference in diameters,[6]† are otherwise

---

* The filamentous aggregates are termed *tonofibrils* which are composed of *tonofilaments*.

† Jakus[6] gives a range of 160 to 240 A diameter (peak $\sim$ 190 A) for fibrils in Bowman's layer and 240 to 280 A diameter in the deeper stroma, with a slight increase in diameter to $\sim$ 340 A near Descemet's membrane.

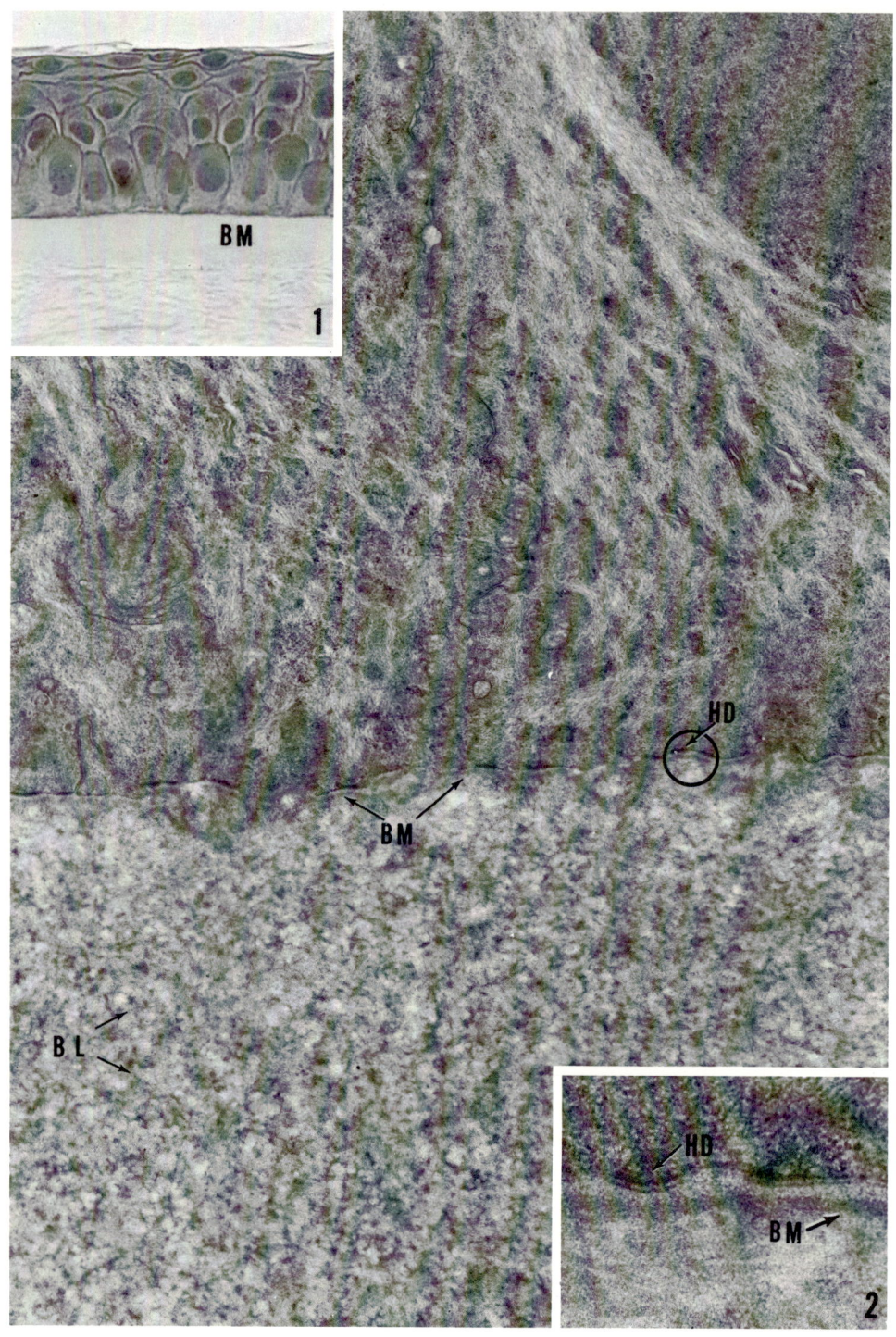

FIG. 9–3. (*Opposite page.*) Basal layer of corneal epithelium. **Inset 1.** Light micrograph showing thin section of corneal epithelium, Bowman's membrane (*BM*), and adjacent corneal stroma. **Main figure.** Electron micrograph showing portion of basal epithelial cells, their hemidesmosomes (*HD*), their basement membrane (*BM*), and a portion of Bowman's layer (*BL*). **Inset 2.** Enlargement of basal cell hemidesmosome (circled in main figure) and adjacent basement membrane (*BM*). **Main figure,** ×16,000; **inset 1,** 1.5-μ section, PD, ×485 (AFIP Neg. 63–6904); **inset 2,** ×48,000. (Modified from McTigue, J. W. *Trans Amer Ophthal Soc* 65:591, 1967.)

FIG. 9–4. Wing cell layer of corneal epithelium. Adjacent cells are attached by myriad desmosomes (*arrows*), shown enlarged in **inset.** *FIL,* intracellular filaments. **Main figure,** ×16,000; **inset,** ×80,000.

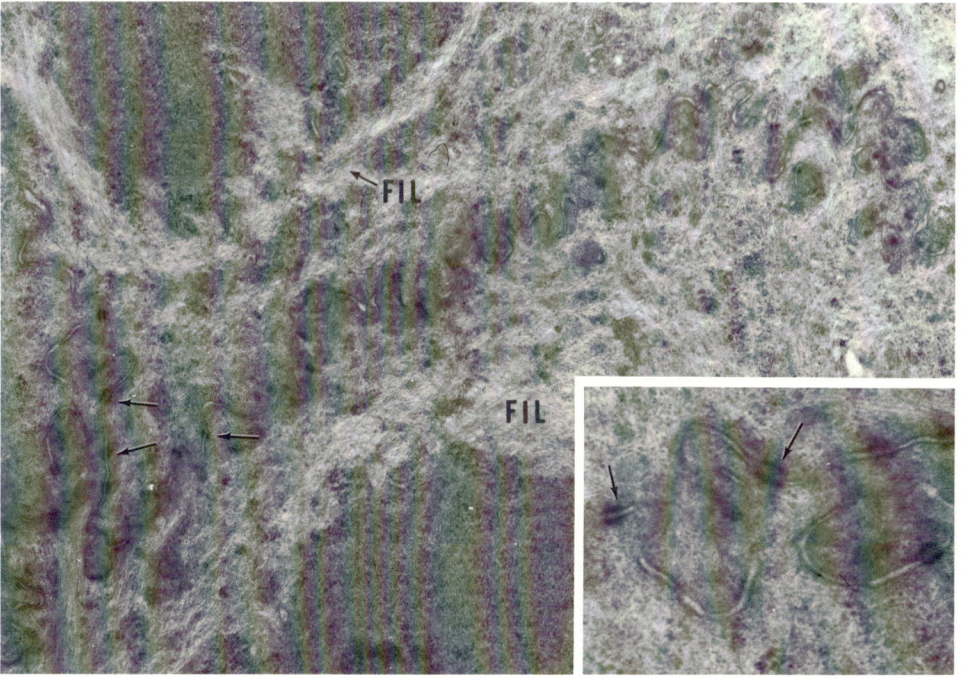

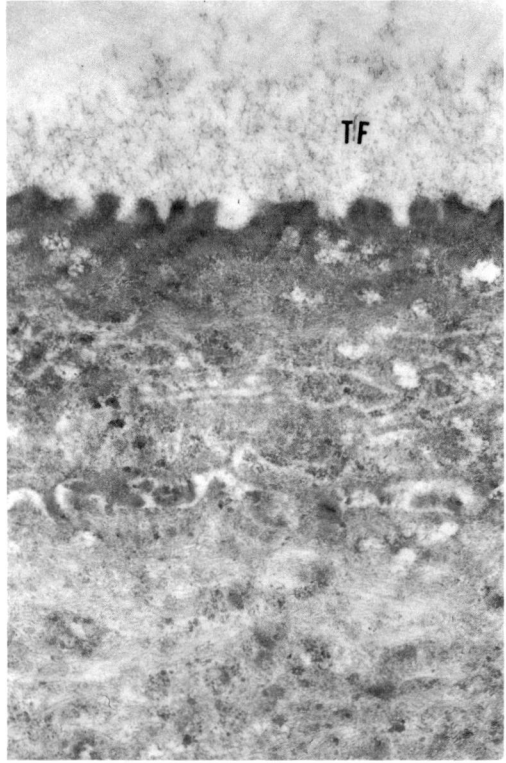

FIG. 9–5. Surface cells of corneal epithelium. Anteriormost cell possesses many surface projections or folds. *TF,* dried remains of tear film. ×14,500. (From McTigue, J. W. *Trans Amer Ophthal Soc* 65:591, 1967.)

similar to those of the deeper stroma. Where it joins underlying *lamellar* stroma, the "membrane" alters via a narrow transition zone in which the collagen fibrils aggregate into the first recognizable, obliquely arranged, collagenous lamellas of the superficial corneal stroma.

To avoid confusion in terminology and to specify the method used for a particular description, the term Bowman's *membrane* refers to this structure's appearance by light microscopy, whereas Bowman's *layer* refers to its appearance by electron microscopy. The transition into obliquely arranged stromal lamellas prevents the layer from being easily separated from the remainder of the stroma.

The membrane or layer contains apertures (*inset,* Fig. 9–8), pores or canals to allow the corneal nerves to pass through to reach the epithelial cells. Groupings of these nonmyelinated neurites often are observed between adjacent basal epithelial cells (Fig. 9–8), but their canals through Bowman's layer and the appearance of their endings within the epithelium have not as yet been seen by electron microscopy.

Because the layer is normally *acellular,* one of the earliest pathologic alterations that may be noted is the appearance of one or more cell nuclei. Some cell migration between superficial stroma and epithelium may be aided by the presence of the normal apertures or pores which allow for passage of the corneal nerves.

FIG. 9–6. Corneal surface (similar to Fig. 9–5) as viewed by scanning electron microscope. Tear film (above) obscures surface microvilli or plicae that are more easily observed below. *Arrow* indicates free edge of a flattened surface epithelial cell that has begun to peel away. ×10,000.

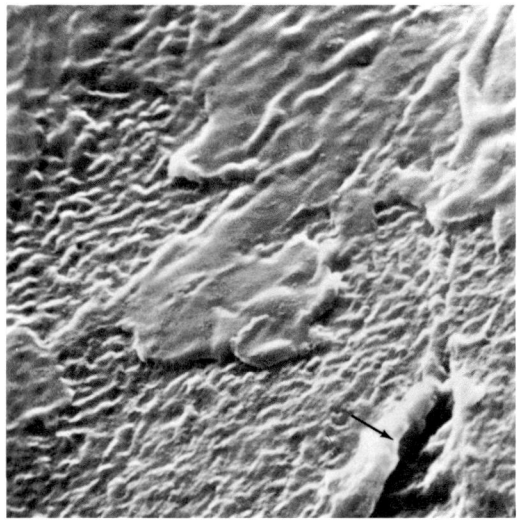

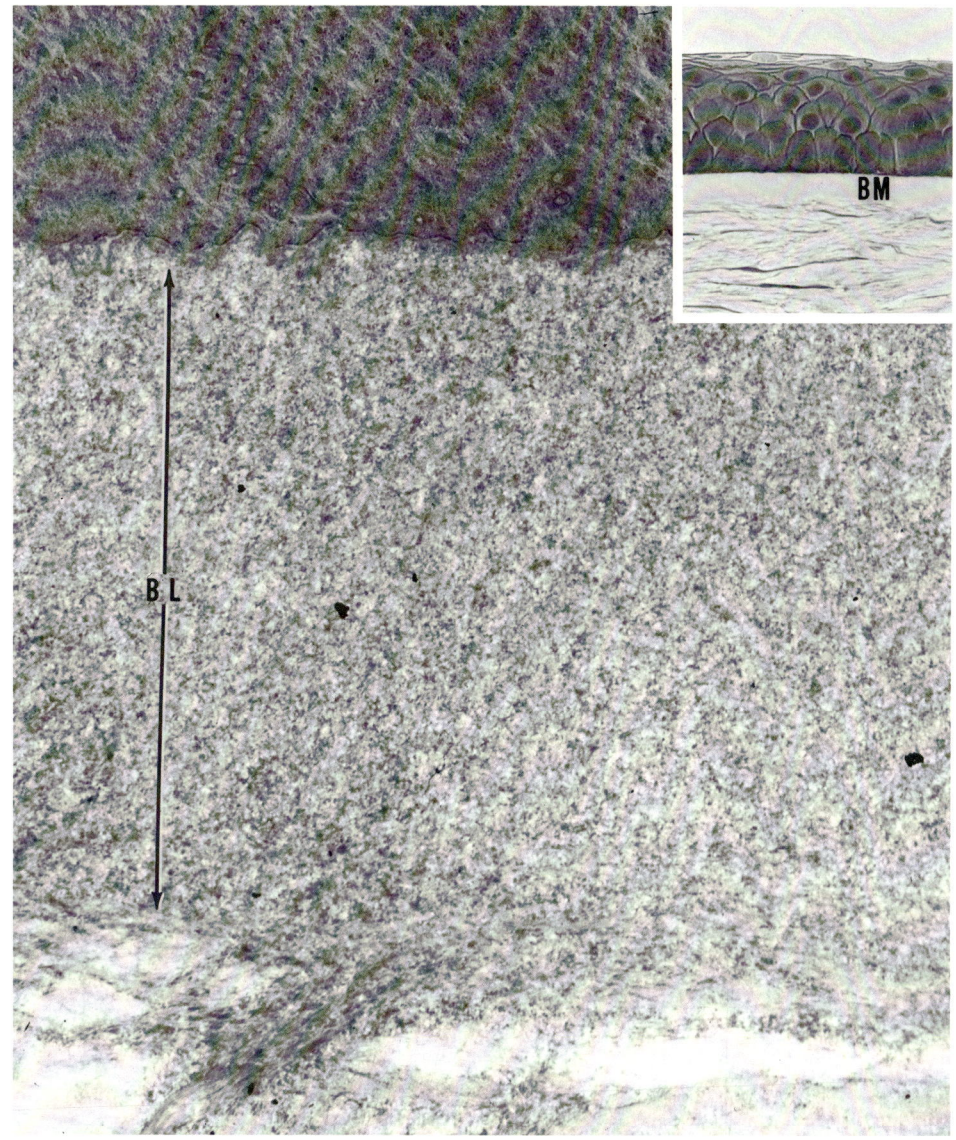

The adjacent corneal stroma (approximately the anterior one-third) is arranged much like a flattened honeycomb into obliquely oriented bundles of collagen fibrils (Fig. 9–9) known as lamellas. Lying between the lamellas are the keratocytes (Figs. 9–9 and 9–10), believed to represent fibrocytes that can become fibroblastic after appropriate stimulation (e.g., trauma).[18]

The keratocytes generally are recognized in H&E-stained sections by their very basophilic nuclei. Their cytoplasmic extensions are

FIG. 9–7. Bowman's membrane or layer. By light microscopy (**inset**) structure appears as a membrane (*BM*). By electron microscopy it is a thick acellular layer of collagen fibrils in random distribution (Bowman's layer, *BL*). **Main figure,** ×8,400; **inset,** 1.5-μ section, PD, ×300.

seen by electron microscopy to extend for enormous distances.

Artifactitious "clefting" of the stromal lamellas usually occurs from separation in the plane of

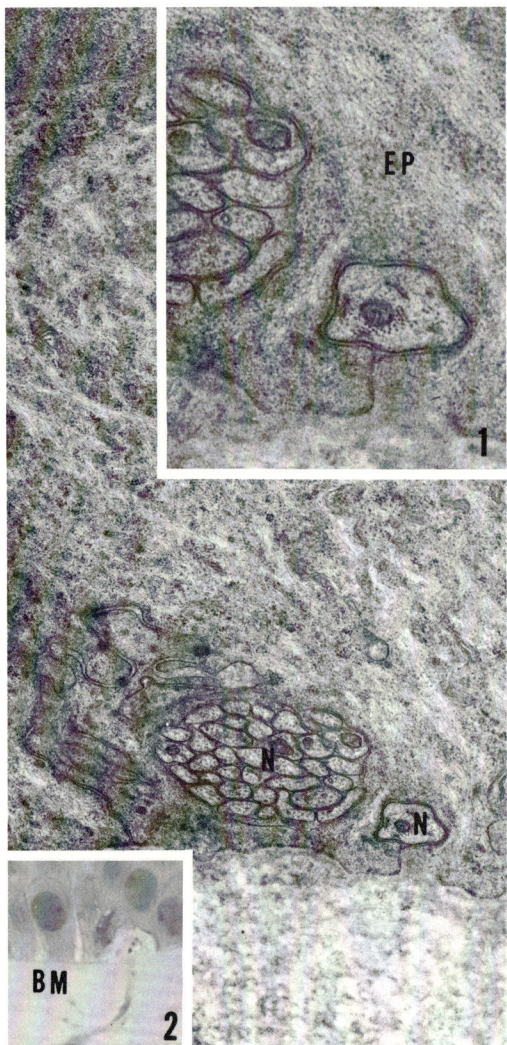

FIG. 9–8. Main figure. Electron micrograph showing groups of neurites (N) present between adjacent epithelial cells. **Inset 1** shows that outer neurites in each group are enveloped only by adjacent epithelial cells (EP). **Inset 2** is a light micrograph showing passage of presumed nerve fiber bundle through Bowman's membrane (BM). **Main figure,** ×16,000; **inset 1,** ×32,000; **inset 2,** 1.5-μ section, PD, ×575 (AFIP Neg. 66–6701).

the keratocytes. The lamellas themselves do not readily come apart. The concept of a flattened honeycomb-like arrangement (Fig. 9–11) in this region is further supported by the observation that artifactitious clefting does not extend for any great distance in any single plane. A corneal surgeon attempting a lamellar dissection or *keratectomy* at these superficial levels also appreciates this arrangement.

A single stromal lamella is made up of collagen fibrils of uniform diameter (∼ 300 Å) in highly ordered array (Fig. 9–12). The various acid mucopolysaccharides, which occupy the interfibrillar spaces, are best visualized by applying to a conventional tissue section such light microscopic histochemical staining procedures as the colloidal iron (AMP) or alcian blue techniques. In electron micrographs the delicate web-like material that frequently may be observed spanning the interfibrillar spaces is interpreted as the drying patterns of these same acid mucopolysaccharides.

Patches of basement-membrane-like material are frequently observed in direct contact with the plasmalemma of keratocytes, especially in the peripheral cornea (Fig. 9–13). The material is occasionally observed to be filamentous. Close association of similar material with collagen fibrils produces a periodic structure which may also be observed in the uveal meshwork (see Chap. 11).

Deeper within the stroma (at the approximate junction of the anterior one-third and posterior two-thirds) a transitional zone occurs. The lamellas become arranged more parallel to the corneal surface (Fig. 9–14), and therefore the long cytoplasmic extensions of the keratocytes can be traced with greater ease. The parallelism continues posteriorly to Descemet's membrane.

It has been generally and widely believed that the corneal layer equivalent to skin was limited to the epithelium. It appears, from the combined evidence of phylogeny,[17] embryology (Fig. 9–15), histology, and biomicroscopy, that all the layers of the skin are represented in the anterior cornea.

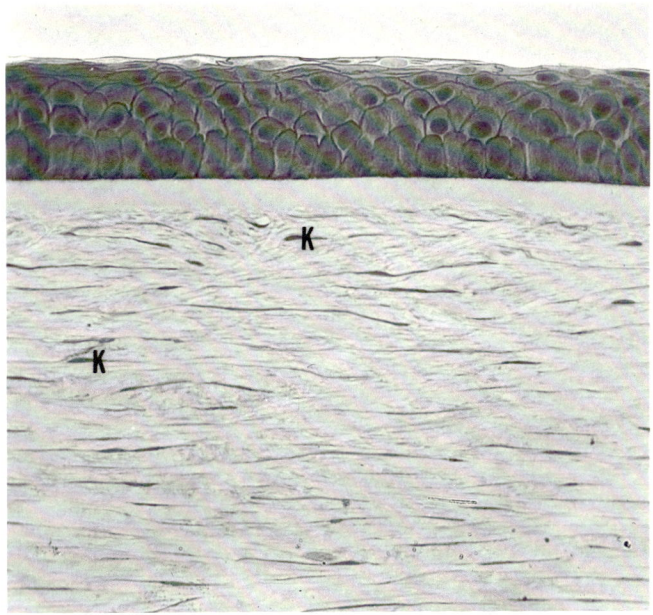

FIG. 9–9. Anterior corneal stroma composed of obliquely arranged collagenous lamellas. *K*, keratocytes. 1.5-μ section, PD, ×970. (AFIP Neg. 67–5144.)

FIG. 9–10. Keratocytes lying between corneal stromal lamellas. Presence of two nuclei in **main figure** indicates either presence of two adjacent cells or sectioning of a bilobed or grossly folded nucleus. **Inset,** a higher magnification of area circled in main figure, shows typical intercellular space which indicates that the two nuclei may belong to two separate but adjacent cells or a single cell folded. **Main figure,** ×16,000; **inset,** ×25,000. (Modified from McTigue, J. W. *Trans Amer Ophthal Soc* 65:591, 1967.)

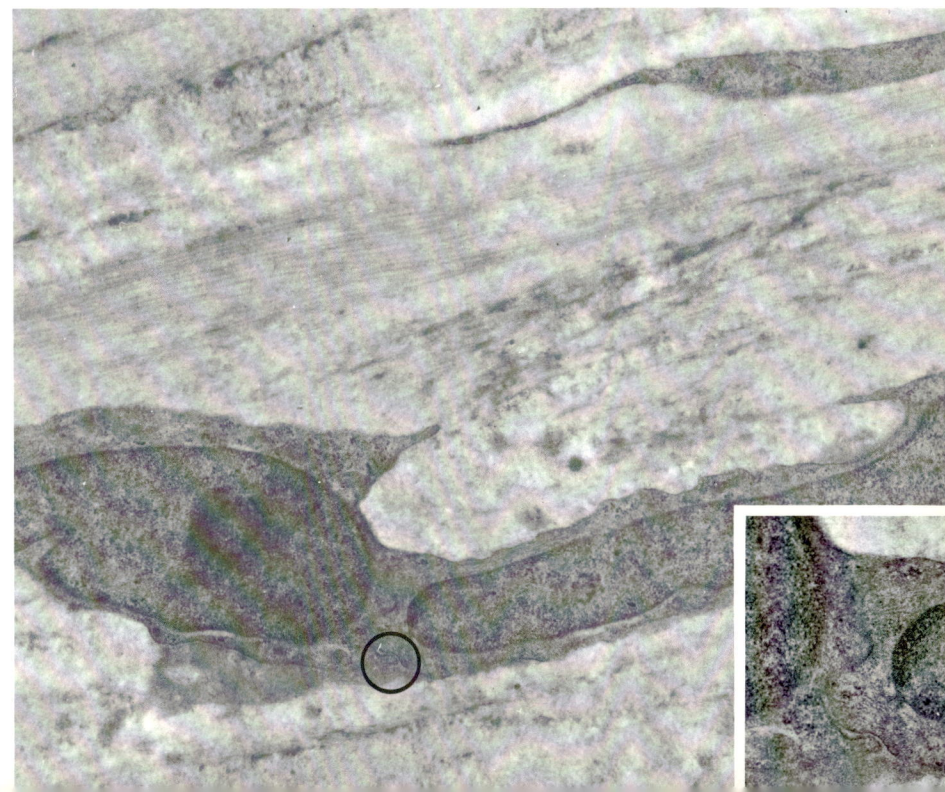

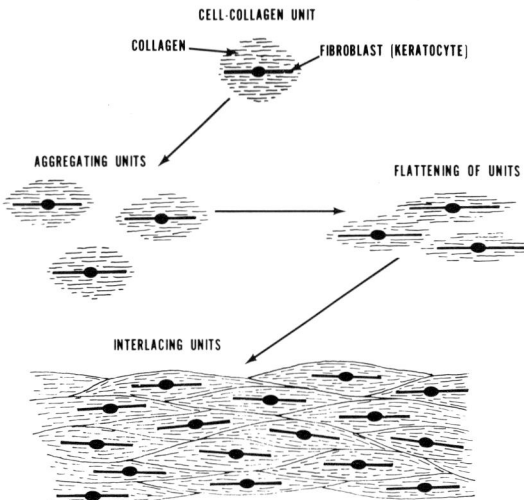

FIG. 9–11. Schematic representation of morphogenesis of corneal stroma. Stroma is formed by collagen-producing keratocytes which surround themselves with collagen. This produces a flattened honey-comb-like arrangement which is more prominent in anterior than in posterior corneal stroma.

The epithelium, therefore, represents the epidermis, and the anterior third of corneal stroma, the dermis. The analogy can be carried even further, with Bowman's layer being represented by the loosely arranged collagen of the papillary layer of dermis, and the obliquely arranged lamellas of anterior corneal stroma by the densely arranged bundles of collagen that compose the "reticular layer" of the dermis. The deeper layers of corneal stroma would, therefore, represent the subcutaneous tissue. Vasculature, present in the skin, is, of course, normally absent in the cornea. The histologic difference between the arrangement of the anterior one-third and the posterior two-thirds of the corneal stroma (Fig. 9–16) is also well reflected by the normal biomicroscopic differences in relucency of these two zones (Fig. 9–17). These differences in relucency are exaggerated in early stromal edema. The *increase* in relucency anteriorly with swelling was apparently confusing because of experiments which indicated that swelling occurred more easily and to a greater degree in the posterior stroma.[13] The anatomic arrangements of lamellar obliquity anteriorly and lamellar parallelism posteriorly appear to explain both the experimental and clinical observations.

ENDOTHELIUM AND DESCEMET'S MEMBRANE

Lining the posterior surface of the cornea (Fig. 9–18) and forming the anterior boundary of the anterior aqueous chamber is a single layer ($\sim$ 5 to 6 μ thick) of flattened hexagonally arranged cells generally known as the corneal endothelium.

The "endothelium" is not a true endothelium (i.e., lining a vascular or lymphatic channel), but a *mesothelium* (i.e., resembling those cells that line such body cavities as peritoneum and pleural or pericardial clefts or spaces). Furthermore, cells called *mesothelia* frequently are considered to possess the property of readily undergoing alteration into fibroblasts, a property that is not widely attributed to true endothelia. If such is the case, then the concept becomes readily acceptable that this layer of cells produces connective tissue membranes, either normal or pathologic. On this basis, the peculiar changes recently observed in certain disease processes (Fuchs' endothelial dystrophy[10,14] or congenital stationary hereditary dystrophy[11]) as well as the mysterious formation of *some* retrocorneal fibrous membranes are more easily explained.

Like some epithelia, the endothelial cells are attached by terminal bars to one another circumferentially in a plane near the cell apex (Fig. 9–18). The adjacent cells, separated

FIG. 9–12. (*Opposite page.*) Corneal stromal lamellas, showing arrangement of collagen fibrils. **Inset,** at lower magnification, shows attenuated cytoplasm of a keratocyte (*single free arrow*) lying between lamellas, a keratocyte cell body (*CE*) with adjacent homogeneous substance (*HS*), and banded appearance of some collagenous aggregates (*free double arrows*). **Main figure,** ×18,000; **inset,** ×11,000.

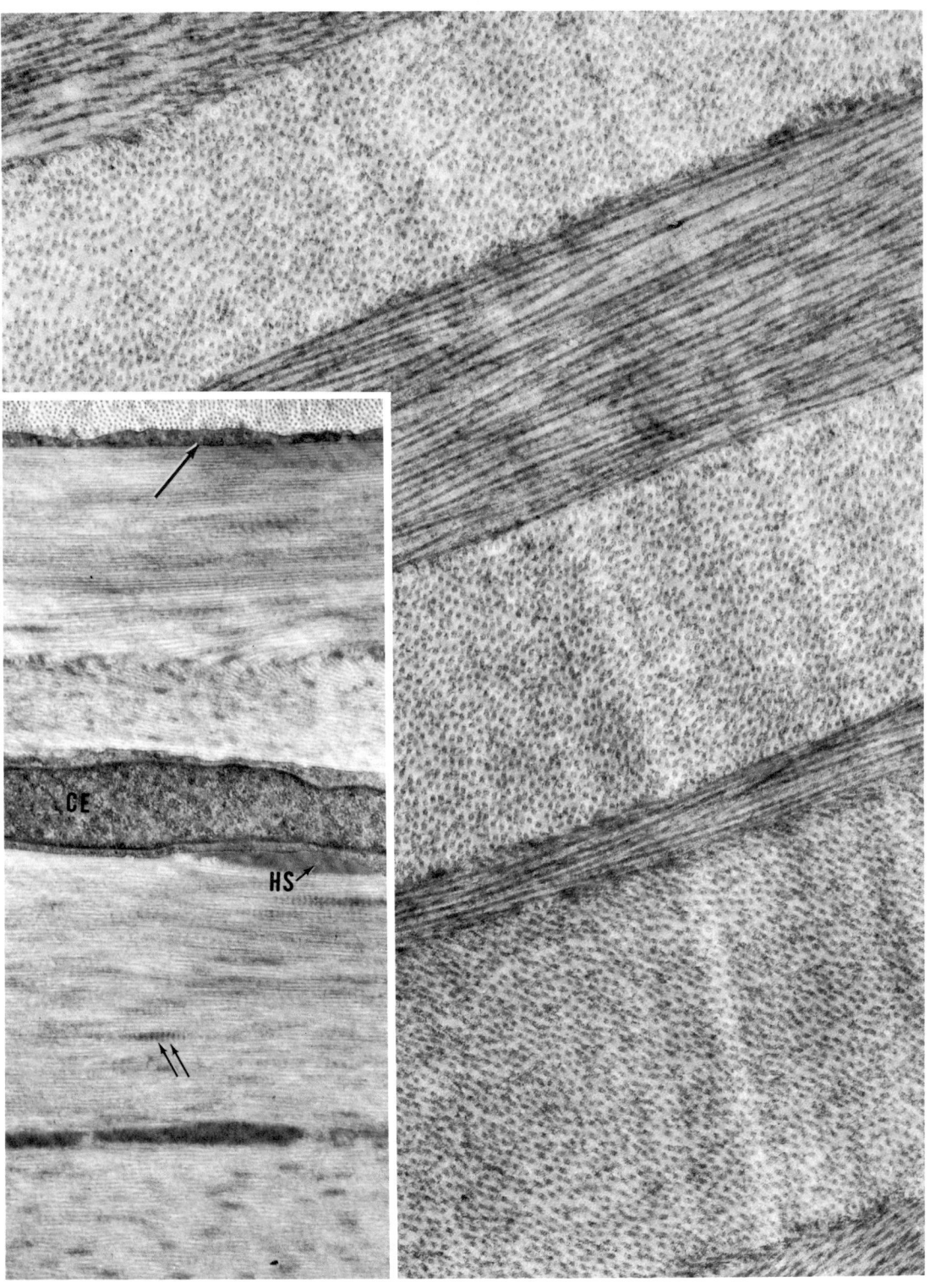

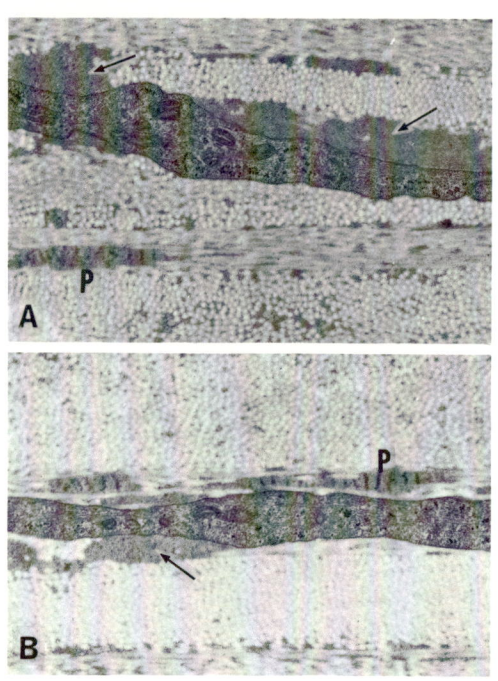

FIG. 9–13. A. Extracellular masses of homogeneous basement-membrane-like material closely applied to keratocyte (*arrows*). Occasionally homogeneous extracellular material appears filamentous. Patches of similar material involving several collagen fibrils produce a pattern of periodicity (*P*). **B.** Another keratocyte with adjacent patches of basement membrane-like material (*arrow*) and periodic (*P*) arrangements. Peripheral cornea. **A,** ×13,500; **B,** ×16,000.

FIG. 9–14. Light micrograph of posterior cornea showing more parallel arrangement of stromal lamellas. 1.5-μ section, PD, ×300. (AFIP Neg. 68–5145.)

FIG. 9–15. Fetal cornea at 12 weeks' gestation. Stroma is clearly separable into anterior layer (one-third) and posterior layer (two-thirds). 1.5-μ section, toluidine blue, ×220. (AFIP Neg. 71–876.)

MATURE CORNEA

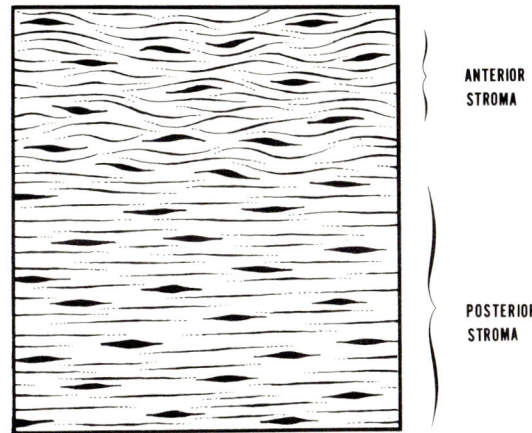

FIG. 9–16. Arrangement of anterior and posterior corneal stromal lamellas.

FIG. 9–17. Slit-lamp differences in relucency of anterior and posterior corneal stroma.

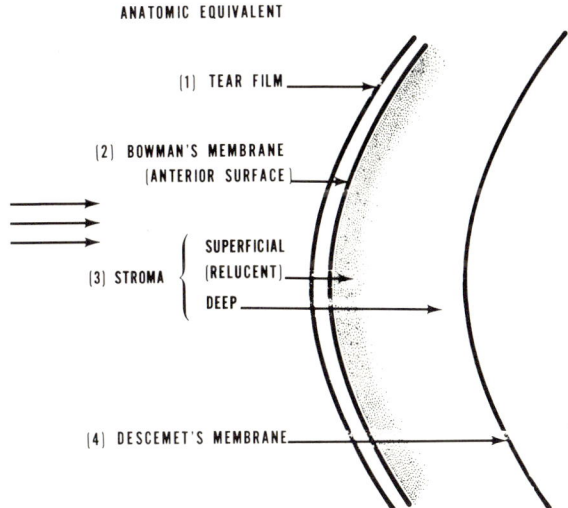

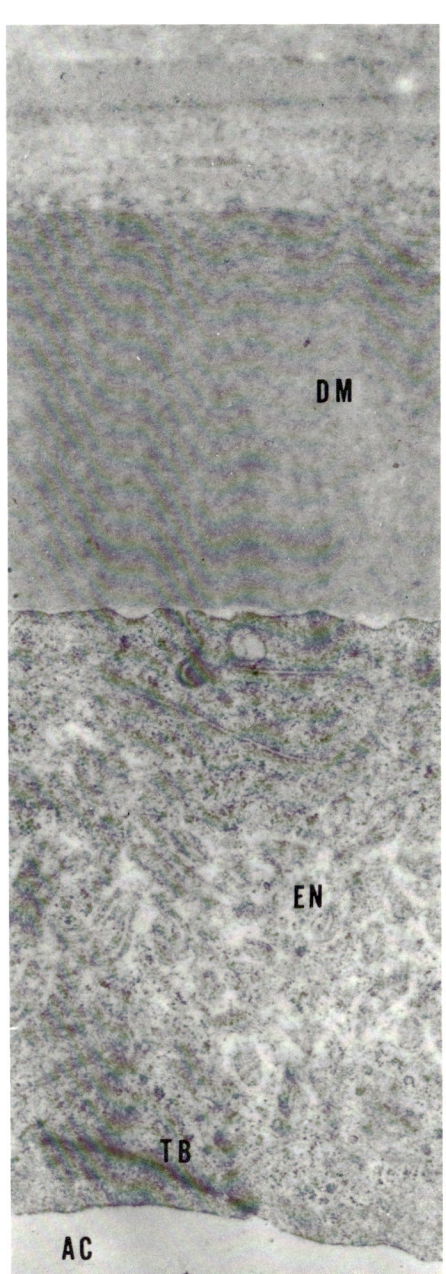

FIG. 9–18. Corneal endothelium (*EN*) applied to Descemet's membrane (*DM*) and lining anterior chamber (*AC*). Endothelial cells are attached to each other by terminal bars (*TB*). ×24,000.

FIG. 9–19. Mitochondria of corneal endothelium with characteristically dilated, lucent intercristal spaces. ×16,000.

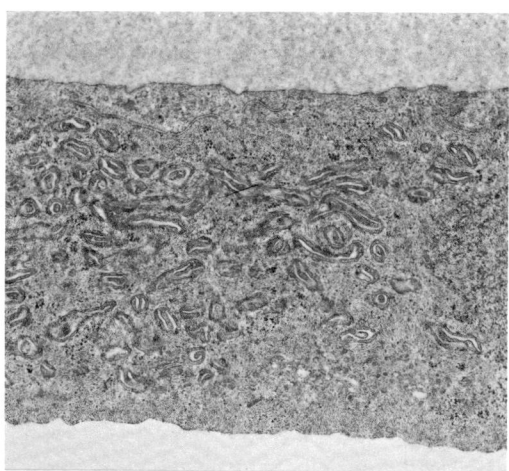

by intercellular spaces, are well interdigitated. No desmosomes are present. The basal plasma membranes are quite flat, are closely applied to a thick basement membrane (Descemet's membrane), and are lacking in hemidesmosomes.

The only peculiarity of the cellular organelles is that of the mitochondria, which appear to possess a matrix of increased intracristal density accompanied by slight dilation of their intercristal spaces (Fig. 9–19) and so generally differ from mitochondria in other ocular tissue (except keratocytes) in their response to fixation.

Descemet's membrane, which closely resembles the lens capsule and the lamina vitrea of Bruch's membrane, has long been known as a typical ocular cuticular, glass, or PAS-positive membrane. Produced by the endothelium, the membrane is thin in infancy, increases in thickness to ~ 5 μ in childhood, and then to ~ 8 to 10 μ in adulthood (Fig. 9–20).

Where Descemet's membrane joins the flat collagenous stromal lamellas, the deepest lamellas are slightly infiltrated by patches of basement-membrane-like material.

The anterior third or so of this thick basement membrane (the "oldest" portion and therefore the earliest to be laid down during embryonic or fetal life) possesses a peculiar structural arrangement.[5] On sectioning, it presents an appearance of periodicity that measures approximately 1,000 A. Similar periodicity can be observed in other basement membranes under certain conditions and planes of sectioning (see Chap. 11).

The innermost (most posterior, "youngest," or latest) portion of Descemet's membrane appears quite homogeneous (except for a fine filamentous appearance that occurs under suitable conditions of fixation, sectioning, and resolution). Separation of this refractile "elastic," indigestible membrane into two or more layers not only is possible but also can be observed on occasion by light microscopic examination of the frayed ends of a break in it.

That Descemet's membrane is a secretory product of the endothelium is supported by the obser-

FIG. 9–20. (*Opposite page.*) Descemet's membrane. **Inset 1.** Light micrograph of posterior corneal stroma, Descemet's membrane, and corneal endothelium. **Main figure.** Electron micrograph of region comparable to that circled in inset 1. Stromal lamellas are flat and arranged relatively parallel to one another. Attenuated keratocyte processes (*KP*) extend for long distances. Characteristic periodicity of anterior portion of Descemet's membrane can be seen at *free arrows*. **Inset 2** shows slight blending (at *arrow*) of this thick basement membrane with adjacent collagenous fibrils. **Main figure,** ×18,000; **inset 1,** 1.5-μ section, PD, ×300; **inset 2,** ×53,700. (Modified from McTigue, J. W. Trans Amer Ophthal Soc 65:591, 1967.)

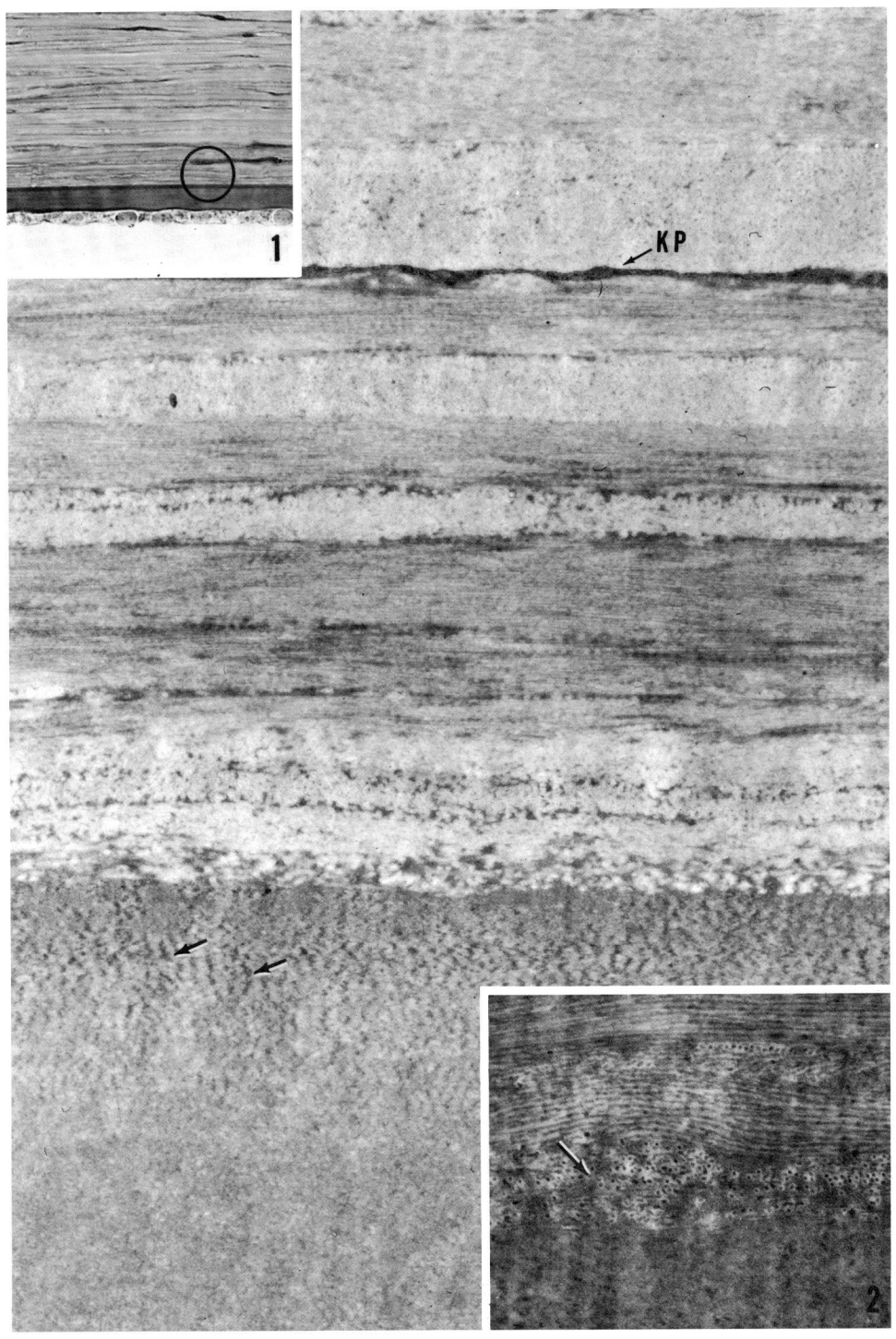

vations that a small, clean traumatic break in the membrane is bridged posteriorly by a new cuticular layer and the old cut ends are left curled slightly outward and separated. The location of the bridge together with lack of interposition of another cell type (e.g., keratocyte) between it and the broken ends of the original membrane clearly support the concept of synthesis and deposition by the overlying endothelium.

CENTRAL COMPARED WITH PERIPHERAL PORTIONS

The preceding description has been limited for the most part to a typical example of *central* cornea examined from front to back.

The epithelial basement membrane increases in thickness from the center to the periphery of the cornea.[15] This can be appreciated by PAS reaction in appropriately cut tissue sections for light microscopy (Fig. 9–21) as well as by electron microscopic examination.

At the periphery (Fig. 9–22), i.e., just central to the termination of Bowman's layer, the basal plasma membranes of the basal epithelial cells become elongated into villi which continue to possess their complement of hemidesmosomes (Fig. 9–23). The villi project into a markedly thickened multilaminar basement membrane in which peculiarly banded fibrils are present.[3,15,16] Such interweaving of the long cytoplasmic processes bearing hemidesmosomes with a thickened, interwoven basement membrane suggests the anatomic basis for the strong adherence of corneal epithelium in this region.

Posteriorly, proceeding from the center of the cornea to its periphery, there is again a change in the relation of cell to basement membrane (Fig. 9–24). As the corneal periphery is approached (Fig. 9–25), basal villi of the endothelial cells are seen projecting into the adjacent basement membrane (Descemet's membrane). These villi lack any form of "attachment" structure. The endothelial cells become slightly attenuated and the basement membrane is thicker focally. At the

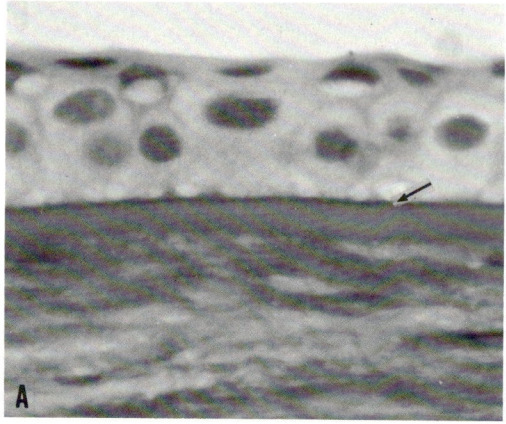

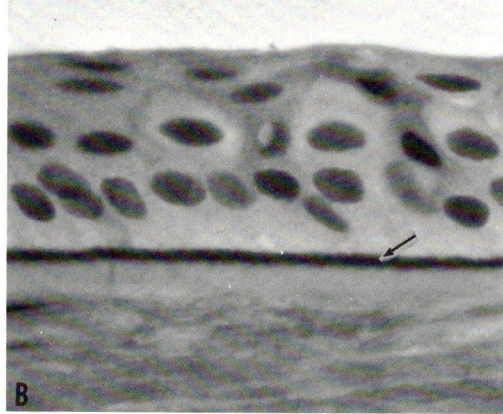

FIG. 9–21. Variation in basement membrane (*arrows*) thickness of central cornea (**A**) compared with peripheral cornea (**B**) from same section. PAS, ×900. (AFIP Negs. 65–12184 and 65–13102. Modified from McTigue, J. W. *Trans Amer Ophthal Soc* 65:591, 1967.)

periphery, the foci of basement membrane thickening are so large (*inset*, Fig. 9–25) that they are easily appreciated by light microscopy as the *warts of Hassall-Henle*. The cell bodies containing their nuclei are located along the slopes or in the valleys between these excrescences, which are covered by an extremely attenuated endothelial cytoplasm (Figs. 9–26 and 9–27). Myriad endothelial villi project into the fissures of these base-

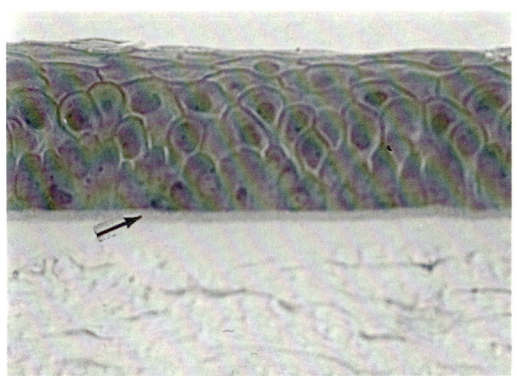

FIG. 9–22. Light micrograph showing thick epithelial basement membrane (*arrow*) present at corneal periphery. 1.5-μ section, PD, ×395. (AFIP Neg. 66–3320. From McTigue, J. W., and Fine, B. S. In Uyeda, R. (ed.). *Proceedings, Sixth International Congress for Electron Microscopy, Kyoto, Japan, 1966*, vol. 2. Tokyo, Maruzen, 1966.)

FIG. 9–23. Electron micrograph showing long villi of basal epithelial cells projecting into thick, expanded (multilaminar) basement membrane at corneal periphery. *BL,* collagen fibrils of Bowman's layer. **Inset** shows peculiar banded fibrils scattered within this basement membrane. **Main figure,** ×16,000; **inset,** ×32,000.

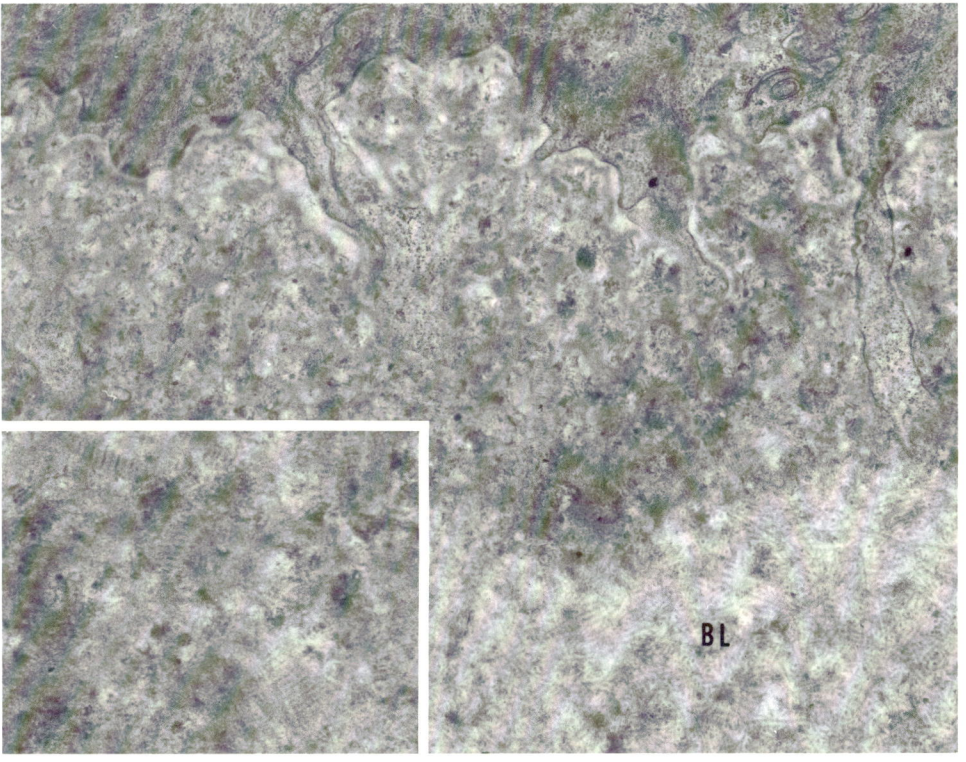

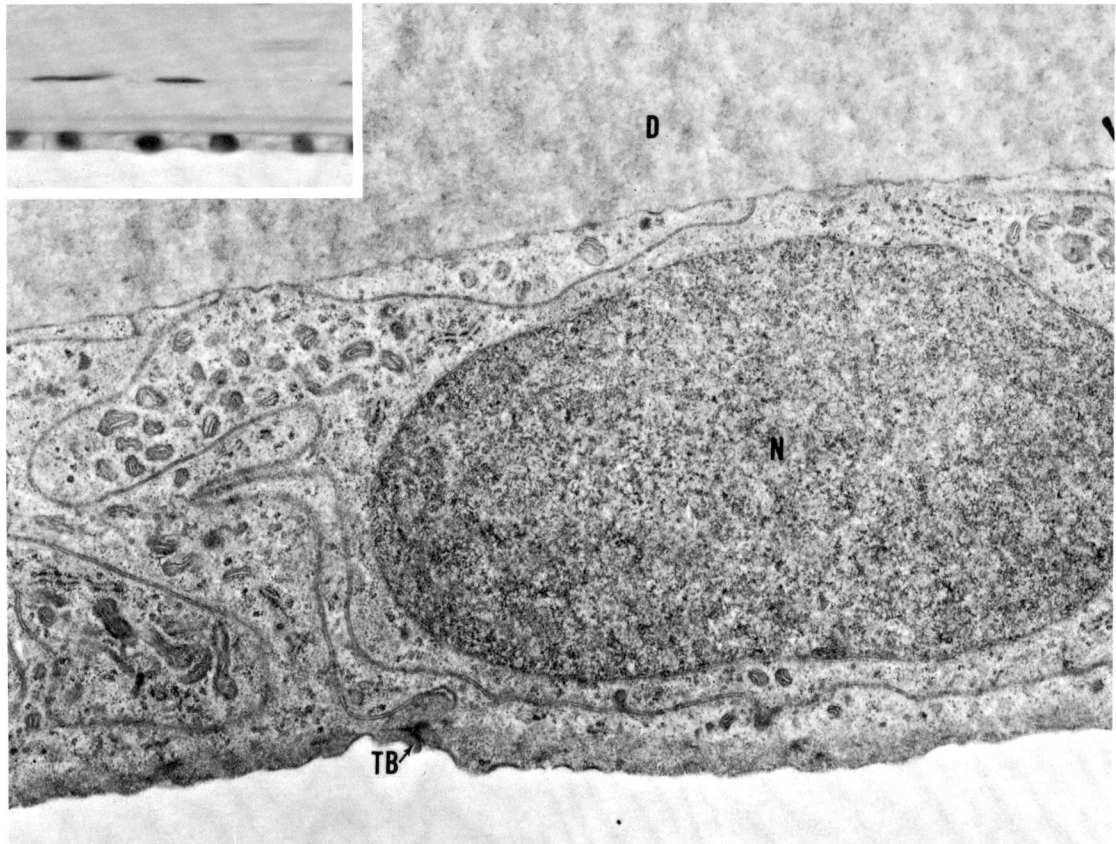

FIG. 9–24. Typical relation of corneal endothelium to Descemet's membrane in central cornea. *D,* Descemet's membrane. *N,* nucleus of endothelial cell; *TB,* terminal bar. **Main figure,** ×16,000; **inset,** H&E, ×575 (AFIP Neg. 65–237). (Modified from McTigue, J. W. *Trans Amer Ophthal Soc* 65:591, 1967.)

ment membrane accumulations (Fig. 9–26).[4,8,10,14] In some instances the fissure appears to be occupied by dense globules which may be interpreted as degenerated fragments of pinched off cytoplasmic villi.

Exaggeration of the "physiologic" or aging process at the periphery of Descemet's membrane into central cornea gives rise to the entity known clinically as *cornea guttata*. This condition can be identified clinically in its earliest stages by defects in the endothelial pattern noted on biomicroscopic examination (specular reflection). The defects result from displacement of the endothelium posteriorly from the plane of reflection by the enlarging foci of wart-like material.

There is evidence that these pathologic warts or excrescences differ in morphologic composition from those normally present in the corneal periphery. The pathologic warts contain more prominent patches of basement membrane material of 1,000-A periodicity, a material reminiscent of the "oldest" portions of Descemet's membrane, and suggest a reversion of the cells to some of their "fetal-like" activities. The combination of cornea guttata and epithelial changes is known clinically as *Fuchs' combined dystrophy*.

The extreme periphery of Descemet's membrane forms a ring or line, if observed in three dimensions (in vivo), known as *Schwalbe's ring* or *line*. Not infrequently this line is thickened by basement membrane, collagenous connective tissue, or a mixture of the two. Since endothelium (i.e., mesothelium) here has the potential to elaborate these materials, other sources need not be strongly considered (see The Drainage Angle, Chap. 11).

A similar arrangement of mesothelium ("endothelium") basement membrane and col-

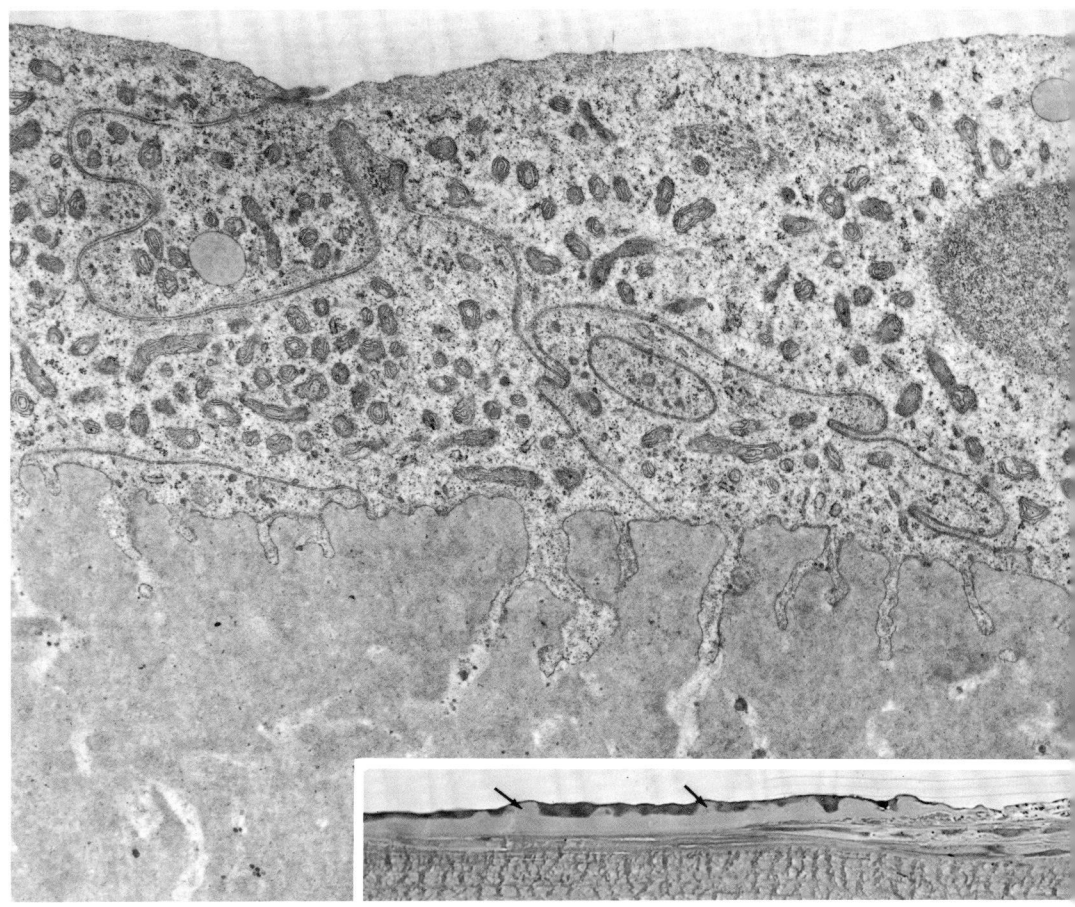

lagenous connective tissue is carried over into the adjacent trabecular meshwork (Fig. 9–28).

THE SCLERA

The white, collagenous sclera forms approximately five-sixths of the outer coat of the eye, with a radius of curvature of ~12 mm. As previously mentioned, the anteriormost one-fifth of this ocular layer is highly modified into the transparent cornea. Uniting the sclera to the cornea is a transition zone, the limbus, described in Chapter 11. Like the cornea, the sclera is composed mostly of extracellular materials.

EMISSARIA

Vessels and nerves pass through canals or *emissaria* in varying degrees of obliquity en

FIG. 9–25. Relations of corneal endothelium to Descemet's membrane in peripheral cornea. Light micrograph (**inset**) shows Descemet's membrane with a number of excrescences, i.e., Hassall-Henle warts (*arrows*). Electron micrograph shows fissuring of Descemet's membrane. Many processes of endothelial cells lie within fissures. Debris of cytoplasmic processes remain within distal portions of fissures. **Main figure,** ×10,500; **inset** 1.5-μ section, PD, ×220 (AFIP Neg. 66–1291).

route to and from the interior of the eye. In addition to vessels and nerves, the emissaria contain uveal tissue which may be heavily pigmented in deeply pigmented individuals.

It is possible to have uveal nevi located partly or wholly within scleral canals. The emissaria

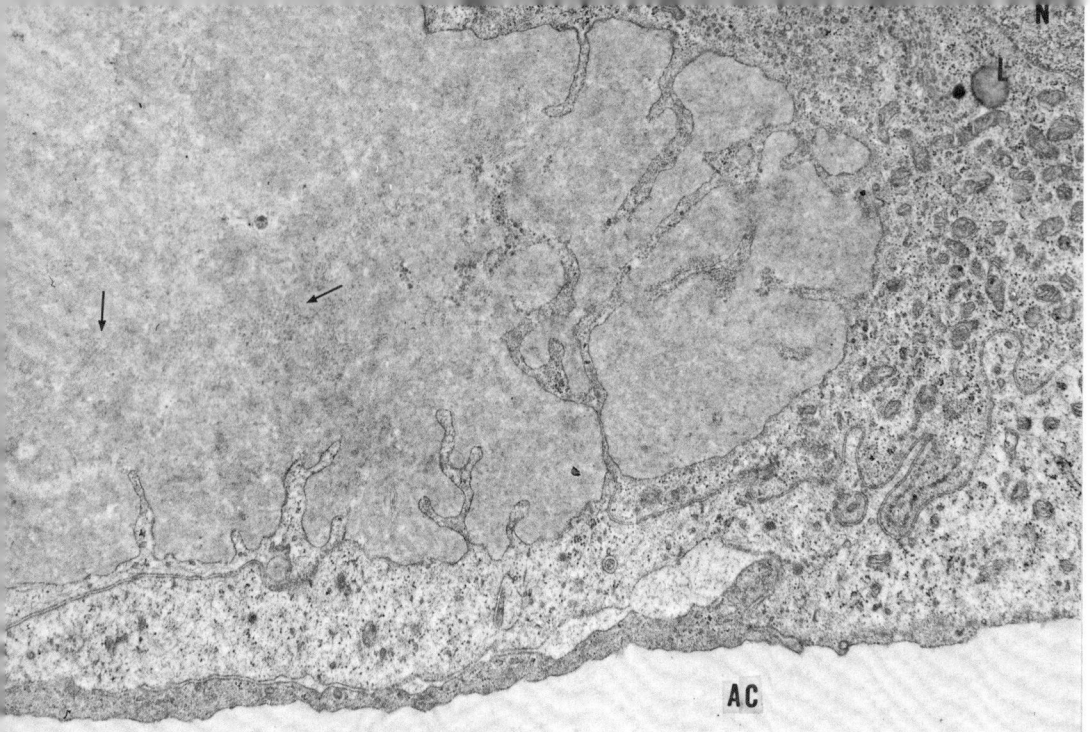

FIG. 9–26. Portion of Hassall-Henle wart at extreme corneal periphery. Endothelial cells are attenuated over wart. Many endothelial cytoplasmic processes lie within more recently elaborated basement membrane excrescences. A suggestion of periodicity within basement membrane is present (*arrows*). A lipofuscin granule (*L*) is present within endothelial cell. *AC,* anterior chamber; *N,* nucleus of endothelial cell. ×11,000.

FIG. 9–27. Hassall-Henle warts as seen by scanning electron microscope. Schwalbe's line (*S*) is seen above at extreme corneal periphery. *Arrow* indicates isolated red blood cell lying on corneal endothelium. ×310.

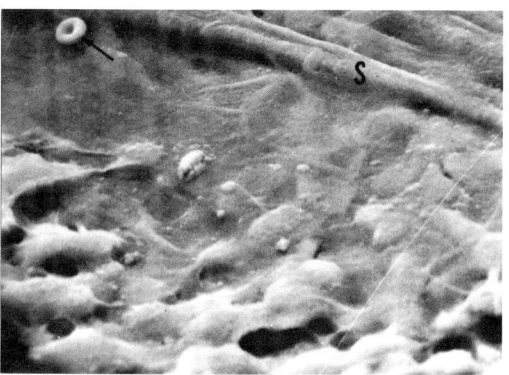

may provide an avenue by which neoplasms such as malignant melanomas of the uvea, and inflammations such as sympathetic uveitis, may emerge from the globe.

Melanocyte-containing uveal tissue may extend out of the emissaria onto the scleral surface or episclera and in the living eye be clearly viewed anteriorly through the transparent conjunctiva. Episcleral uveal tissue, called *pigment spots* of the sclera,[19] may be present 3 to 4 mm from the limbus. The spots are most commonly seen in eyes with darkly pigmented irises, usually in the superior episclera. Inferior, temporal, and nasal episcleral spots occur in decreasing frequency.

Largest among the apertures for the passage of nerves, arteries, and veins in the sclera is the one for the optic nerve. It lies just nasal to the posterior pole (see Fig. 5–5, on Color Plate I) and is surrounded by a cluster of smaller apertures for passage of the adjacent short posterior ciliary arteries to the choroid. In the horizontal meridian are two additional

approximately 14 in number, at least 2 of which accompany each artery. Pigment spots are generally associated with perforating anterior ciliary vessels.

Branches or loops of the long ciliary nerves (nerve loop of Axenfeld, Fig. 9–29) often

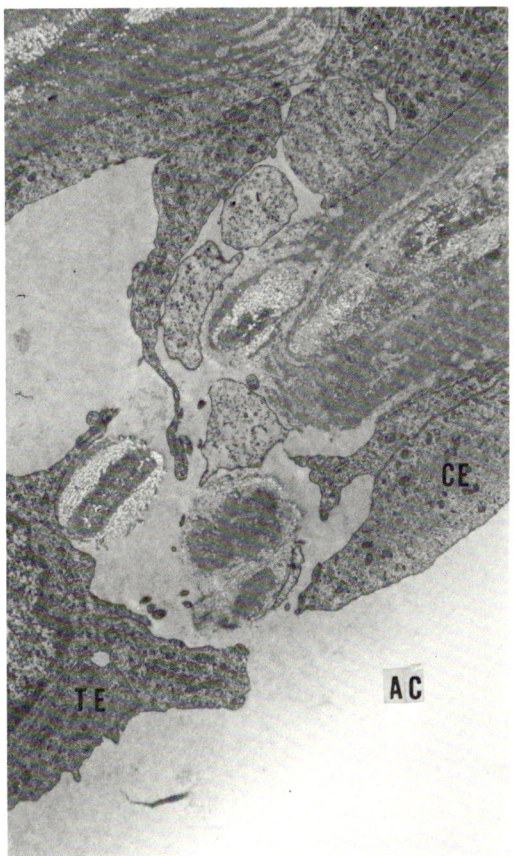

FIG. 9–28. Region of Schwalbe's line or ring where corneal endothelium (*CE*) alters to become endothelium of trabecular meshwork (*TE*). *AC*, anterior chamber. ×5,400.

oblique apertures for the passage of the two long posterior ciliary arteries and accompanying nerves. The corresponding veins (vortex veins) that drain the posterior uveal tract traverse the sclera in the four posterior quadrants.

Anteriorly, just behind the limbus but still well within the sclera, the anterior ciliary arteries perforate to reach the ciliary muscle. Approximately 7 anterior ciliary arteries are derived from the blood supply to the 4 rectus muscles; the lateral rectus has the single vessel. The associated anterior ciliary veins are

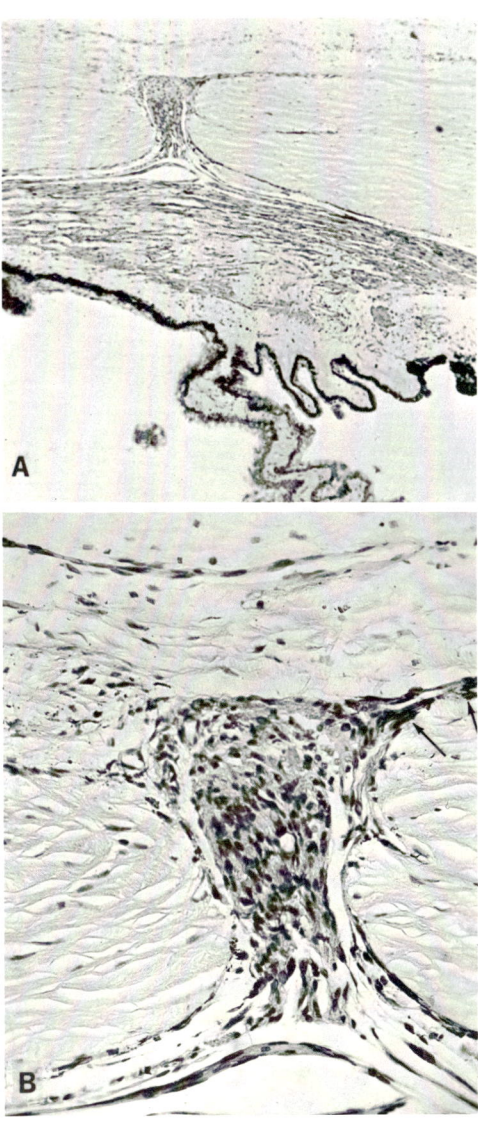

FIG. 9–29. Nerve loop of Axenfeld. Higher magnification (**B**) shows uveal tissue (*arrows*) in episclera. **A,** H&E, ×47; **B,** H&E, ×125.

penetrate or even perforate the sclera just behind the limbus to form a small, often pigmented episcleral nodule.

These pigment spots and intrascleral nerve loops may be confused with nevi, cysts, or extensions of adjacent or underlying malignant melanomas. On clinical examination, the conjunctiva is freely movable over episcleral pigment spots, and the intrascleral nerve loops remain painful to touch even after the instillation of a topical anesthetic.

The collector channels from the canal of Schlemm perforate the sclera via a tortuous path through the region of the limbus. As mentioned in Chapter 11, some of these unite with anterior ciliary veins within the substance of the sclera (intrascleral plexus), whereas others reach the surface of the limbus (episclera) to appear clinically as aqueous veins.

EPISCLERA

The term episclera refers to the few loose, vascularized surface layers of scleral collagen beneath Tenon's capsule. Its vascularity differentiates it from the avascular Tenon's capsule and its looser texture from scleral stroma.

STROMA

The stroma (Figs. 9–30 and 9–31) consists of obliquely arranged, interlacing bundles of collagen fibrils with a paucity of cells. By electron microscopy the fibrils possess the 640 Å axial periodicity (Fig. 9–32) similar to corneal collagen but differ in their *nonuniformity* of fibril diameter (Fig. 9–33).[8] A combination of oblique arrangement, variability in collagen fibril diameter,[20] and relative deficiency in water-binding substances probably helps account for scleral opacification.

It long has been known that white, opaque sclera, if dehydrated in vitro, becomes significantly translucent but not transparent.

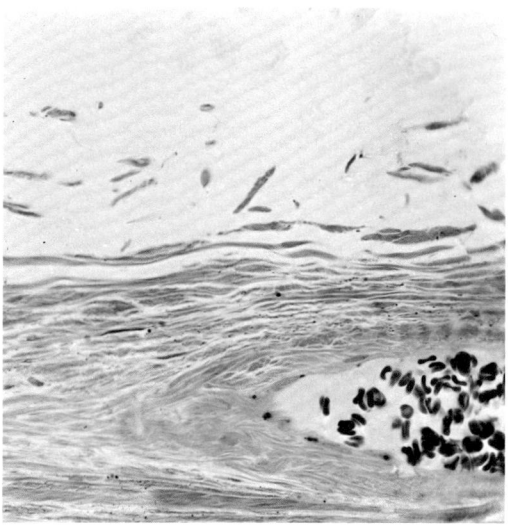

FIG. 9–30. Outer layers of scleral lamellas. A large blood vessel is passing through sclera. 1.5-μ section, PD, ×305. (AFIP Neg. 63–6902.)

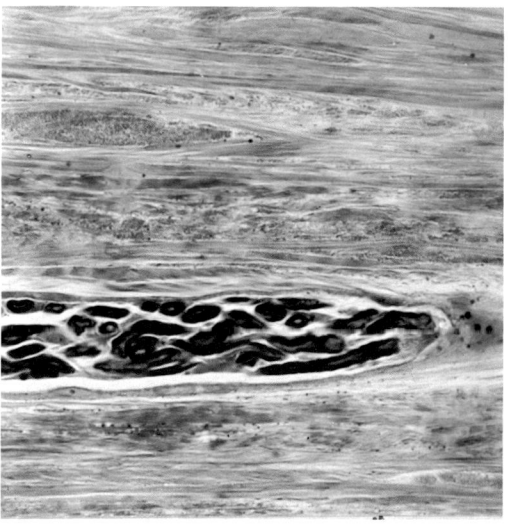

FIG. 9–31. Dense, interlacing collagenous lamellas of sclera. A large intrascleral nerve is present. 1.5-μ section, PD, ×485. (AFIP Neg. 63–6900.)

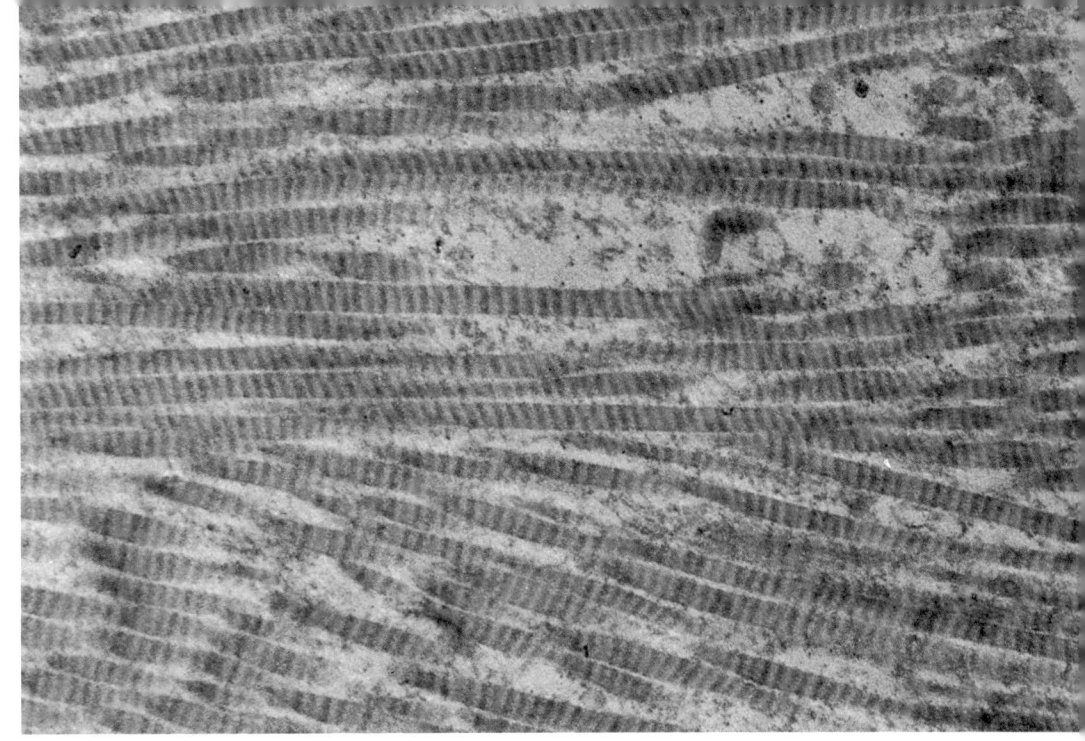

FIG. 9–32. Scleral collagen fibrils (longitudinal). ×27,200.

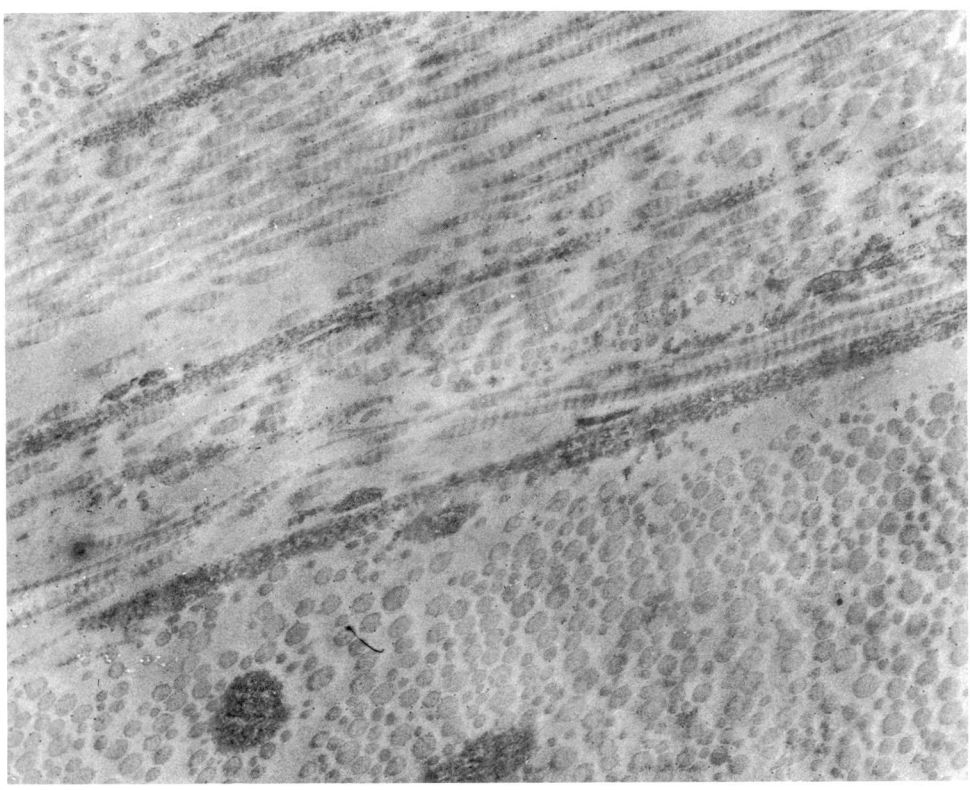

FIG. 9–33. Scleral collagen fibrils (oblique and cross-section). ×24,000.

FIG. 9–34. Doubling of thickness of sclera at attachment of a rectus muscle. H&E, ×35. (AFIP Neg. 70–1422.)

(~ 1.0 mm), thinner at the equator (~ 0.4 to 0.5 mm), and thinnest (~ 0.3 mm) just beneath the attachments for the rectus muscles (Fig. 9–34). With addition of the rectus muscle tendons, the thickness doubles to ~ 0.6 mm, and this thickness continues to the limbus, where the sclera is again thinned, and thereby weakened, by the two circumferential furrows (outer and inner scleral sulci).

The two "thin" regions of the sclera are of clinical importance. The surgeon engaged in muscle surgery must be constantly aware of the thinness of the sclera behind the muscle attachment to avoid penetrating the globe. Because of the two circumferential furrows present in the limbic area, that part of the globe is vulnerable to rupture in a contusion injury.

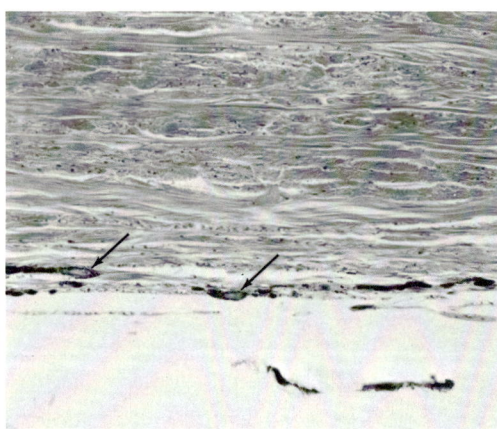

FIG. 9–35. Lamina fusca of sclera, formed by the few pigmented cells (*arrows*) that remain adherent to sclera after separation of choroid. 1.5-μ section, PD, ×305. (AFIP Neg. 63–6901.)

FIG. 9–36. Lamina fusca of sclera (*arrows*). Lamina is difficult to appreciate in sections where choroid has not separated from sclera. *SC*, interlacing bundles of scleral collagen. H&E, ×265. (AFIP Neg. 68–9953.)

A special arrangement is present at the posterior margin of the inner scleral sulcus (scleral roll); this is described in detail in Chapter 11.

The sclera is thickest at the posterior pole

LAMINA FUSCA

If all the loose inner layers of an opened eye are stripped away from the sclera, the inner scleral surface remains pigmented. In sections (Fig. 9–35), these few pigmented inner layers (*lamina fusca*) appear less conspicuous. They contain a number of scattered uveal melanocytes which appear less flattened than those of the adjacent choroidal layers (Fig. 9–36). This appearance, in sections, may give the false impression that this surface in the gross specimen should be less pigmented than it actually shows.

REFERENCES

1. Blümke, S., and Morgenroth, K., Jr. The stereo ultrastructure of the external and internal surface of the cornea. *J Ultrastruct Res* 18:502, 1967.
2. Blümke, S., Niedorf, H. R., Rode, J., and Kudszus, G. Feinstrukturelle Veränderungen des Corneaepithels in der Gewebekultur. III. Die Desmosomen. *Z Zellforsch* 84:189, 1968.
3. Bruns, R. R. A symmetrical extracellular fibril. *J Cell Biol* 42:418, 1969.
4. Feeney, M. L., and Garron, L. K. Descemet's Membrane in the Human Peripheral Cornea: A Study by Light and Electron Microscopy. In Smelser, G. K. (ed.). *The Structure of the Eye* New York, Academic Press, 1961, p. 367.
5. Jakus, M. A. Studies on the cornea: II. The fine structure of Descemet's membrane. *J Biophys Biochem Cytol* 2. (Suppl.) 243, 1956.
6. Jakus, M. A. The Fine Structure of the Human Cornea. In Smelser, G. K. (ed.). *The Structure of the Eye*. New York, Academic Press, 1961, p. 343.
7. Jakus, M. A. Further observations on the fine structure of the cornea. *Invest Ophthal* 1:202, 1962.
8. Jakus, M. A. *Ocular Fine Structure: Selected Electron Micrographs*. Boston, Little, Brown, 1964.
9. Kayes, J., and Holmberg, A. The fine structure of Bowman's layer and the basement membrane of the corneal epithelium. *Amer J Ophthal* 50:1013, 1960.
10. Kayes, J., and Holmberg, A. The fine structure of the cornea in Fuchs' endothelial dystrophy. *Invest Ophthal* 3:47, 1964.
11. Kenyon, K. R., and Maumenee, A. E. The histological and ultrastructural pathology of congenital hereditary corneal dystrophy: A case report. *Invest Ophthal* 7:475, 1968.
12. Khodadoust, A. A., Silverstein, A. M., Kenyon, K. R., and Dowling, J. E. Adhesion of regenerating corneal epithelium: The role of the basement membrane. *Amer J Ophthal* 65:339, 1968.
13. Kikkawa, Y., and Hirayama, K. Uneven swelling of the corneal stroma. *Invest Ophthal* 9:735, 1970.
14. McTigue, J. W. The human cornea: A light and electron microscopic study of the normal cornea and its alterations in various dystrophies. *Trans Amer Ophthal Soc* 65:591, 1967.
15. McTigue, J. W., and Fine, B. S. The Basement Membrane of the Corneal Epithelium. In Uyeda, R. (ed.). *Proceedings, Sixth International Congress for Electron Microscopy, Kyoto, Japan, 1966*, vol. 2, p. 775. Tokyo, Maruzen, 1966.
16. Palade, G. E., and Farquhar, M. G. A special fibril of the dermis. *J Cell Biol* 27:215, 1965.
17. Walls, G. L. The vertebrate eye and its adaptive radiation. *Cranbrook Institute Science Bulletin* No. 19. 1942.
18. Weimer, V. The transformation of corneal stromal cells to fibroblasts in corneal wound healing. *Amer J Ophthal* 44:173, 1957.
19. Yanoff, M. Pigment spots of the sclera. *Arch Ophthal (Chicago)* 81:151, 1969.
20. Zinn, K. M. Changes in corneal ultrastructure resulting from early lens removal in the developing chick embryo. *Invest Ophthal* 9:165, 1970.

chapter 10

The Uveal Tract

The iris
  Uveal (mesodermal) portion
    Anterior border layer
    The stroma
  Retinal (neuroepithelial) portion
    Sphincter muscle
    Clump cells
    Dilator muscle
    Pigment epithelium
The choroid
  Suprachoroidal space and lamina fusca
  Choroidal stroma
    Pigmented and nonpigmented cells
    Vessels
    Bruch's membrane
The ciliary body
  Zones
  Components
    Retinal (neuroepithelial) portion
      Nonpigmented epithelium and internal basement membrane
      Pigment epithelium and external basement membrane
    Uveal (mesodermal) portion

The uveal tract is the pigmented vascular layer between the corneosclera and the neuroepithelial layers. It is conveniently divisible into three parts: the iris stroma anteriorly, the choroid posteriorly, and an intermediate part, the ciliary body.

The uveal tract is a tissue composed of a relatively even mixture of cells and extracellular materials.

Although the two-layered epithelium lining the inner surface of the ciliary body and posterior surface of the iris belongs embryologically to the neuroepithelia of the optic cup, the close relation of the epithelia with the uveal tract in the iris and ciliary body makes it convenient to describe them as associated structures.

THE IRIS

The iris is a circular diaphragm separating the aqueous compartment into anterior and posterior chambers. Its central aperture, the pupil, is slightly nasal to center and controls the amount of light entering the eye. The iris diaphragm is thickest not far from the pupillary zone in the region of the *collarette* and thinnest at its periphery, the iris *root* (Fig. 10–1).

The thin periphery of the iris is subject to traumatic tearing or separation, *iridodialysis*, upon injury. Vascularity of the iris root may cause hemorrhage into either the anterior or posterior chamber following ocular trauma. On the other hand, the pupillary margin rests on the lens anterior to the frontal plane of the iris root. This contact by the iris pigment epithelium may produce pigment cell adhesions to the lens capsule, as in inflammations or contusions. Absence of the supporting lens, as in subluxations, dislocations, or surgical removal due to cataract, allows the iris diaphragm to fall back to the plane of the iris root. The anterior chamber deepens and the iris looks flat and acquires a tremulousness (*iridodonesis*) with movement of the eye.

The color of the iris is determined for the most part by its content of pigmented cells in the stroma. Many irises appear blue at birth because the uveal tract is not maximally pigmented then. By 3 to 6 months of age, however, many blue irises have changed color because of increased stromal pigmentation. If the collagenous stroma of the iris lacks pigmented cells, but its double layer of pigment epithelium is normally pigmented, the iris appears blue. In the albino, however, in whom the pigment is deficient not only in the stroma but also in the pigment epithelium, the red glow of the illuminated fundus shows through, giving the iris a pink color.

Stromal pigmentation (i.e., green or brown irises) obscures the finer morphologic detail of the iris that may be observed with the slit lamp.

UVEAL (MESODERMAL) PORTION

The iris stroma consists of two parts: the *anterior border layer* and the *stroma* proper (Fig. 10–2), which is vascularized.[12] Two cell types, *pigmented cells* and *nonpigmented cells*,[16] are found in the iris stroma (Figs. 10–3 through 10–5). In the adult uvea, pigmentation appears to be almost all or none in any single cell (see Figs. 10–4 and 10–31); in the pigmented cells every cell seems to have a maximum number of pigment granules.

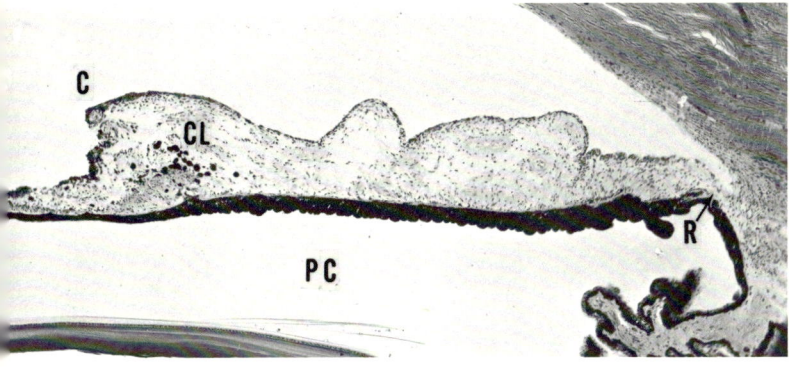

FIG. 10–1. Meridional section of one leaf of iris diaphragm. *C*, edge of collarette; *CL*, clump cells anterior to sphincter muscle; *P*, pupillary border; *PC*, posterior chamber; *R*, iris root. H&E, ×35. (AFIP Neg. 70-1424.)

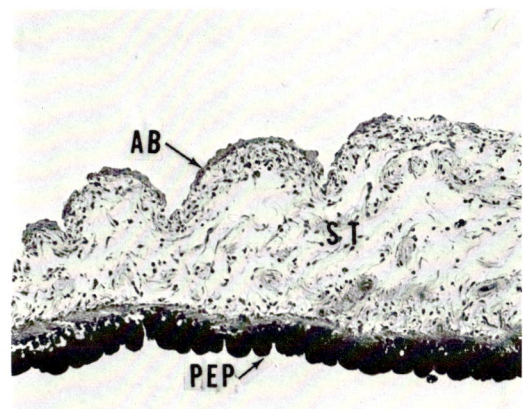

FIG. 10–2. Iris stroma and associated pigment epithelium (*PEP*). *AB*, anterior border layer; *ST*, stroma. 1.5-μ section, PD, ×60. (AFIP Neg. 63-6907.)

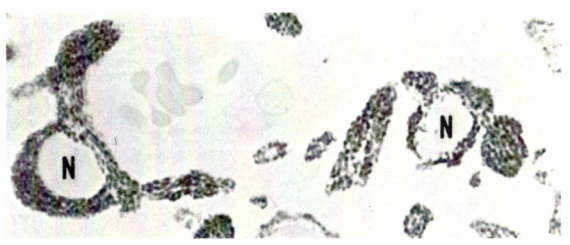

FIG. 10–3. Light micrograph of pigmented iris stromal cells. Large, circular nonpigmented portions of cells (*N*) represent nonstained nuclei. PD, ×1,000. (AFIP Neg. 68-9945.)

The blue iris (Fig. 10–5) possesses amelanotic cells having a few small dense granules in their cytoplasm. The granules are not melanin and are typically found in blue irises, although their actual color in vivo is not yet defined. The cells containing the granules may represent potential melanocytes.

Anterior Border Layer

The anterior border layer (Fig. 10–6) consists of a dense packing of pigmented or nonpigmented cells (Fig. 10–7) similar in appearance to the cells present throughout the remainder of the stroma. Cell density in this surface layer varies from iris to iris. Within any particular iris there may be considerable variation from location to location. Absence of cells appears as *crypts* or "apertures" in the border layer (Figs. 10–5 and 10–8). The crypts are usually more prominent in the thick pupillary zone and toward the iris root.

The iris surface, therefore, varies widely, from the smooth, heavily pigmented surface of a dark-brown iris to the irregular, crypt-marked surface of green or blue irises. In the latter irises, groupings of surface cells can be observed which bridge some of the spaces in the anterior border layer. These bridge-like tissue clusters are called *iris trabeculae*.

No continuous cell layer that can be defined as an endothelium lines the surface of the iris. The anterior border layer of the iris is a loose connective tissue whose surface has become visible due in part to the very specialized cleft, the anterior aqueous chamber, that has been formed embryonically *entirely* within the predominantly mesodermal, i.e. middle ocular, layer (Fig. 10–9).

Peripherally, the anterior border layer ends abruptly at the iris root, except where spoke-

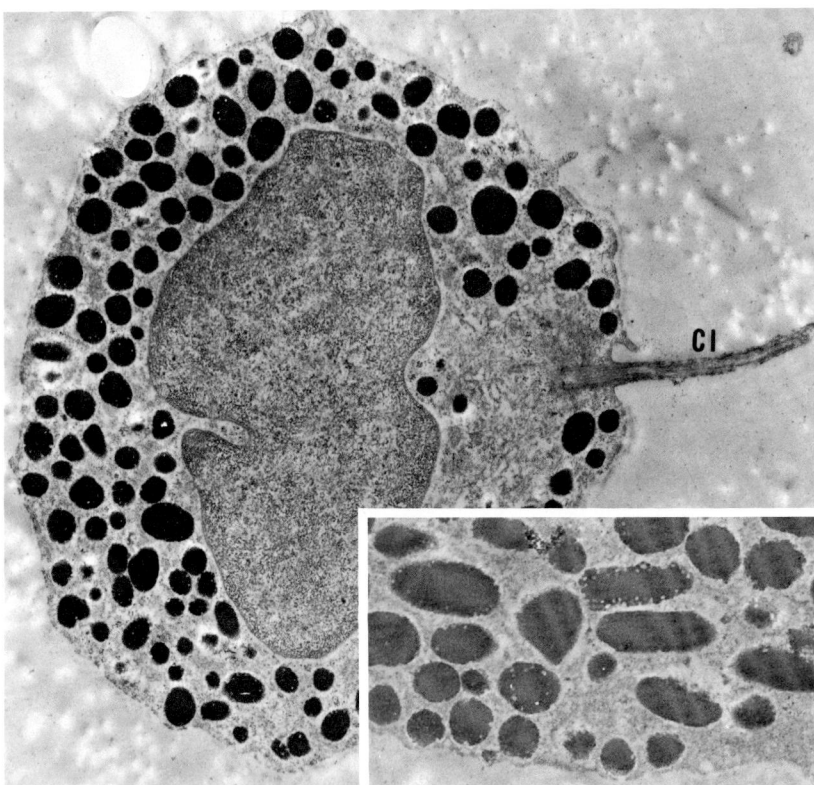

FIG. 10–4. Electron micrograph of typical pigmented iris stromal cell. A cilium (CI) is seen projecting from cell. **Inset** shows characteristic stromal melanin granules, sectioned in various planes at higher magnification. **Main figure,** ×8,500; **inset,** ×23,400.

like portions continue toward Schwalbe's line as *iris processes* (Fig. 11–32).

These processes are most noticeable when pigmented. The novice at gonioscopy sometimes confuses them with *peripheral anterior synechias*.

The Stroma

The loose stroma consists of pigmented and nonpigmented cells and extracellular materials, mainly bundles of collagen fibrils (*inset 2*, Fig. 10–10) in their matrix of hyaluronidase-sensitive mucopolysaccharides. The fibrils are generally arranged in cylindrical groupings, columns, or bundles around a number of nerves or a blood vessel (Fig. 10–10).[17] The collagenous *columns* have a characteristic arrangement, tending to radiate sinuously from the back of the iris toward the front and from the iris periphery toward the pupil (Fig. 10–11). When iris blood vessels surrounded by the columns have been observed in histologic sections, they have been inaccurately referred to as "thick-walled" blood vessels (*inset 1*, Fig. 10–10, and Figs. 10–12 and 10–13).

The collagenous columns enclosing the cells, blood vessels, and nerves may be seen biomicroscopically in blue irises as white stromal columns (Fig. 10–11). Occasionally a vessel deviates from its axial course and may be observed protruding from the white column. The nerves, lacking color, are not recognizable unless perhaps they are represented in vivo by the translucent central line that can be observed in many of the columns at higher magnification (i.e., 40×). Because the space between adjacent *columns* is filled with transparent acid mucopolysaccharides, the entire thickness of the blue iris stroma as well as the

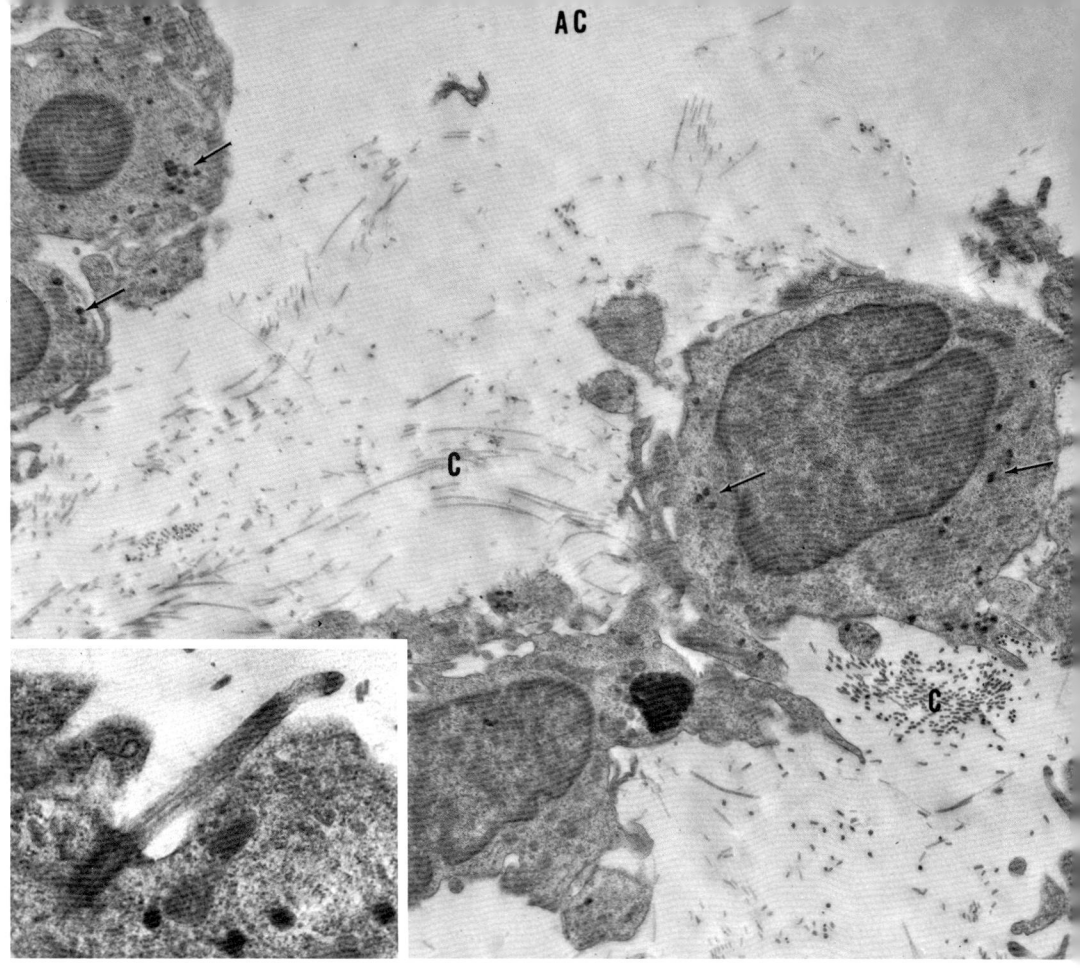

FIG. 10–5. Anterior border layer in blue iris. Note large space between two of surface cells. Dense granules are present (arrows) within nonpigmented stromal cells. Collagen fibrils (C) occupy much of intercellular space. AC, anterior chamber. Inset shows cilium projecting from nonpigmented stromal cell at higher magnification. Main figure, ×7,000; inset, ×15,000.

FIG. 10–6. Anterior border layer of iris. This layer may be lightly pigmented (main figure) or heavily pigmented (inset). Main figure, 1.5-μ section, PD, ×145 (AFIP Neg. 63–6906); inset, 1.5-μ section, PD, ×180 (AFIP Neg. 65–6680).

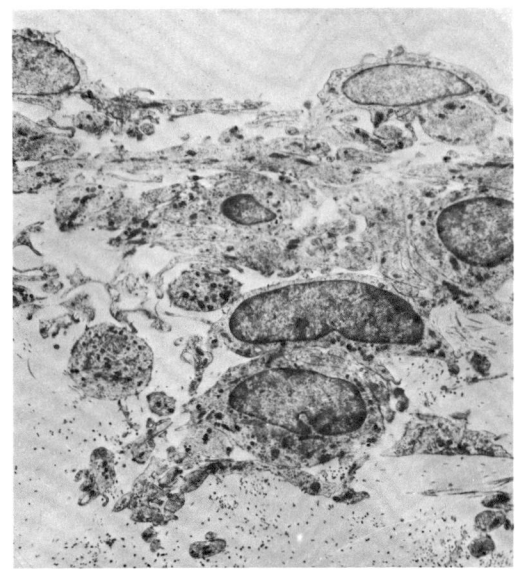

FIG. 10–7. Low power electron micrograph of nonpigmented anterior border layer of iris composed of a dense packing of stromal cells. ×1,500.

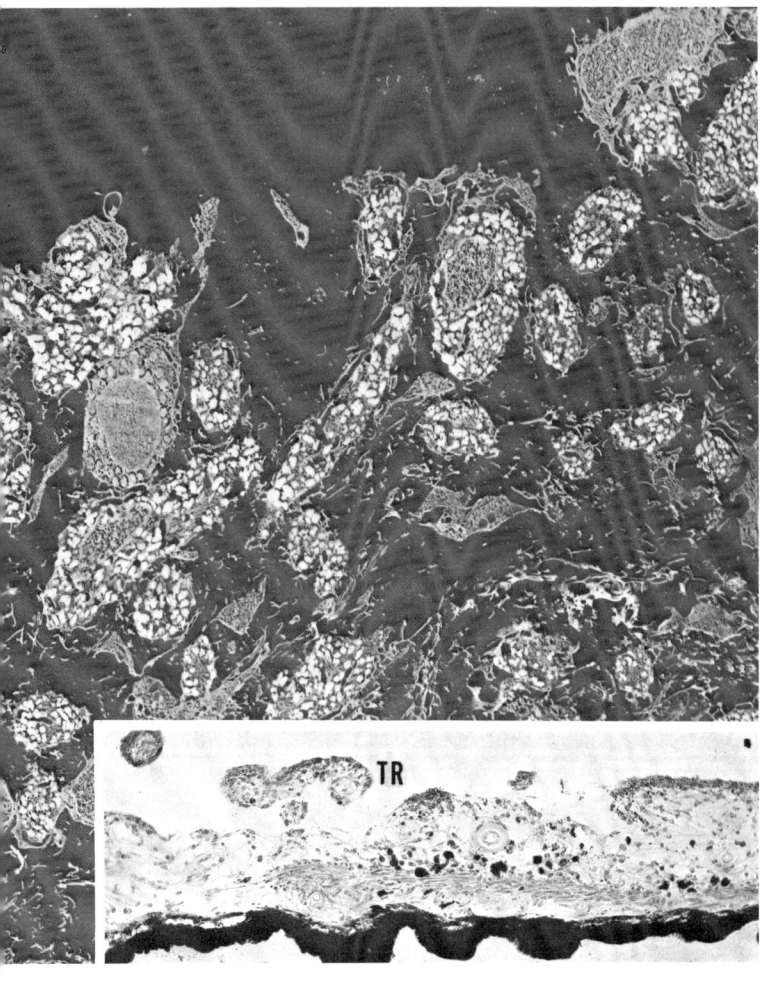

FIG. 10–8. Shadow-cast tissue section of anterior border layer of iris, showing many surface apertures. Collagen fibrils are distributed in extracellular spaces. **Inset** shows many larger crypts within anterior border of pupillary zone as well as sections of surface trabeculae (*TR*). **Main figure,** U-shad, ×3,000; **inset,** H&E, ×85. (AFIP Neg. 70–1418). (Modified from Tousimis, A. J., and Fine, B. S. *Arch Ophthal* (Chicago) 62:974, 1959.)

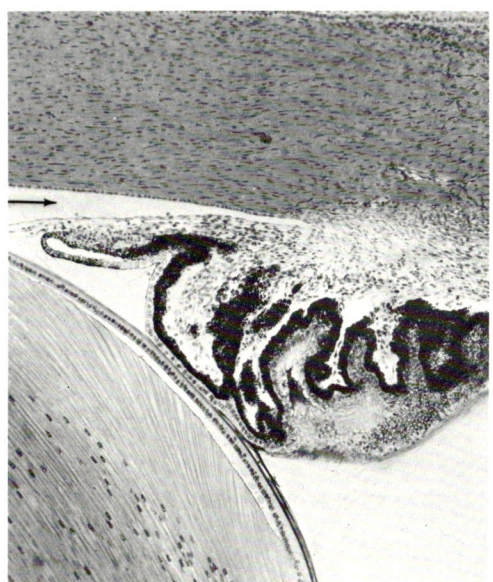

FIG. 10–9. A cleft, the anterior chamber (*arrow*), formed within mesodermal tissue. Masson, ×80. (AFIP Neg. 61–3078.)

pigmentation of the pigment epithelium can be examined directly. Defects in the pigment epithelial layer of light-colored irises can also be detected by transpupillary retroillumination.

These biomicroscopic observations may be clinically useful, as when determining if a retroiridal mass is simply pressing forward as a cyst may do, is focally reducing the stromal thickness as a more solid tumor may do, or is altering the stromal composition as an infiltrating tumor may do.

Except for the great differences in cell and pigment density and surface trabeculae of the anterior border layer, the structure of the iris stroma appears identical in irises of all colors. Iris variation, therefore, is a property of stromal pigmentation and anterior border layer arrangement.

That the subsurface stromal arrangement of a blue iris is identical to that of a brown iris can be appreciated in vivo in a brown-eyed patient with *early* stromal atrophy. The white columnar arrangement of the collagen bundles can then be seen to be similar.

Many stromal vessels are large (of the diameter of the choriocapillaris), endothelium-lined structures with the morphology typical of a nonfenestrated capillary. A basement membrane is present which blends with the surrounding matrix, except where it joins the basement membrane of an associated *pericyte* (Fig. 10–12). The endothelium possesses apical junctional complexes (terminal bars; see Chap. 6) as well as the pinocytotic vesicles, micro and macro (morphologic as well as functional), similar to those present in other nonfenestrated capillaries. To date a *minor* arterial circle analogous to the major arterial circle within the ciliary body has not been observed.

RETINAL (NEUROEPITHELIAL) PORTION

Sphincter Muscle

Lying in the pupillary zone of the iris stroma is a ring of typical smooth muscle almost 1 mm wide, the *iris sphincter* (Fig. 10–1 and 11B). The smooth muscle cells are concentric with the pupil (Fig. 10–14), so that when the muscle contracts it causes the pupil to constrict. The muscle plate is separated from the pigment epithelium of the iris posteriorly by a layer of connective tissue which becomes increasingly dense with age.

It is probably this dense plate of retrosphincter connective tissue, occupying the area posterior to the muscle, that contributes most toward the ease with which the "periphery of the sphincter muscle" in the blue iris is observed biomicroscopically (see Fig. 11B).

The muscle cells satisfy all the histologic criteria for smooth muscle (see discussion of ciliary muscle, later in this chapter). In the iris sphincter, however, they are unusual, for they are considered to be derived *wholly* from neuroectodermal cells that have migrated into the stroma from the neuroepithelium during embryonic life. Adding support to this belief is the presence within these smooth muscle cells of melanin granules that are characteristic in size and shape of neuroepithelial melanin.[16] A series of pigmented projections from the plane of the dilator muscle remains in the adult (Fig. 10–15). These projections or "spurs" are known as Fuchs' or Michel's spurs in the vicinity of the sphincter muscle and as Grunert's spurs at the periphery of the iris (Fig. 10–16).

This muscle of neuroectodermal origin is innervated[11] by parasympathetic nerve fibers that originate in the Edinger-Westphal nucleus at the anterior aspect of the third nerve nucleus. The fibers are carried into the orbit by the branch of the third nerve that goes to the inferior oblique muscle. They branch off like small twigs to arrive at the ciliary ganglion, where a synapse occurs. Fibers from the ciliary ganglion enter the eye by way of the short ciliary nerves. There is some evidence that the sphincter muscle is supplied with inhibitory fibers by way of sympathetic nerves.

Clump Cells

Heavily pigmented, rounded cells are often present just anterior to the sphincter muscle (Fig. 10–17).[12,16] Known as the *clump cells of Koganei*, they were once believed to have migrated into the stroma as did the precursor cells of the sphincter muscle, but without having undergone further modification. More recent observations,[20] however, supported by morphologic, embryologic, and experimental evidence, indicate that most of these cells are in reality macrophages, having engulfed melanin granules of both neuroepithelial and stromal varieties. A second type of cell grouping is less frequently present in these regions of stroma (Fig. 10–18). This latter type consists of a small grouping of cells completely surrounded by a basement membrane. The melanin granules are uniformly distributed within the cell cytoplasm and are all of the neuroepithelial type in size and shape. The macrophage-like cell has been designated the type 1 clump cell and the migrated neuroepithelial cells the type 2 clump cell.

Type 1 clump cells (Fig. 10–17) may be best observed in vivo in blue irises near the pupillary margin as tiny, spherical deep-brown dots. They are best seen under 40× magnification and can be easily distinguished from typical stromal melanocytes by their deeper brown color (identical with that of the cells of the pupillary pigment seam) and by their spherical shape (unlike the

FIG. 10–10. (See page 175.)

COLOR PLATE IV. A. Radial, frequently sinuous, collagenous columns are evident in the blue iris. An enclosed blood vessel can often be observed biomicroscopically. **B.** Patches of pigmented anterior border layer focally screen the underlying stromal columns. The zone of the sphincter muscle can be observed in the thinner pupillary zone beyond the collarette. **C.** One sector of a blue iris is heavily pigmented effectively concealing the underlying stromal columns. **D.** The brown iris is the result of diffuse and dense pigmentation. The peripheral concentric rings are contraction furrows. **E.** The topography of the iris may be irregular because of trabeculae formed within the anterior border layer. **F.** Pigmentation of the anterior border layer clearly outlines the ring of the collarette. **G.** Broad trabeculae of the anterior border layer may course widely over the iris surface. Variation in topography combined with variation in pigmentation produces the assorted clinical appearances of the iris.

Color Plate IV

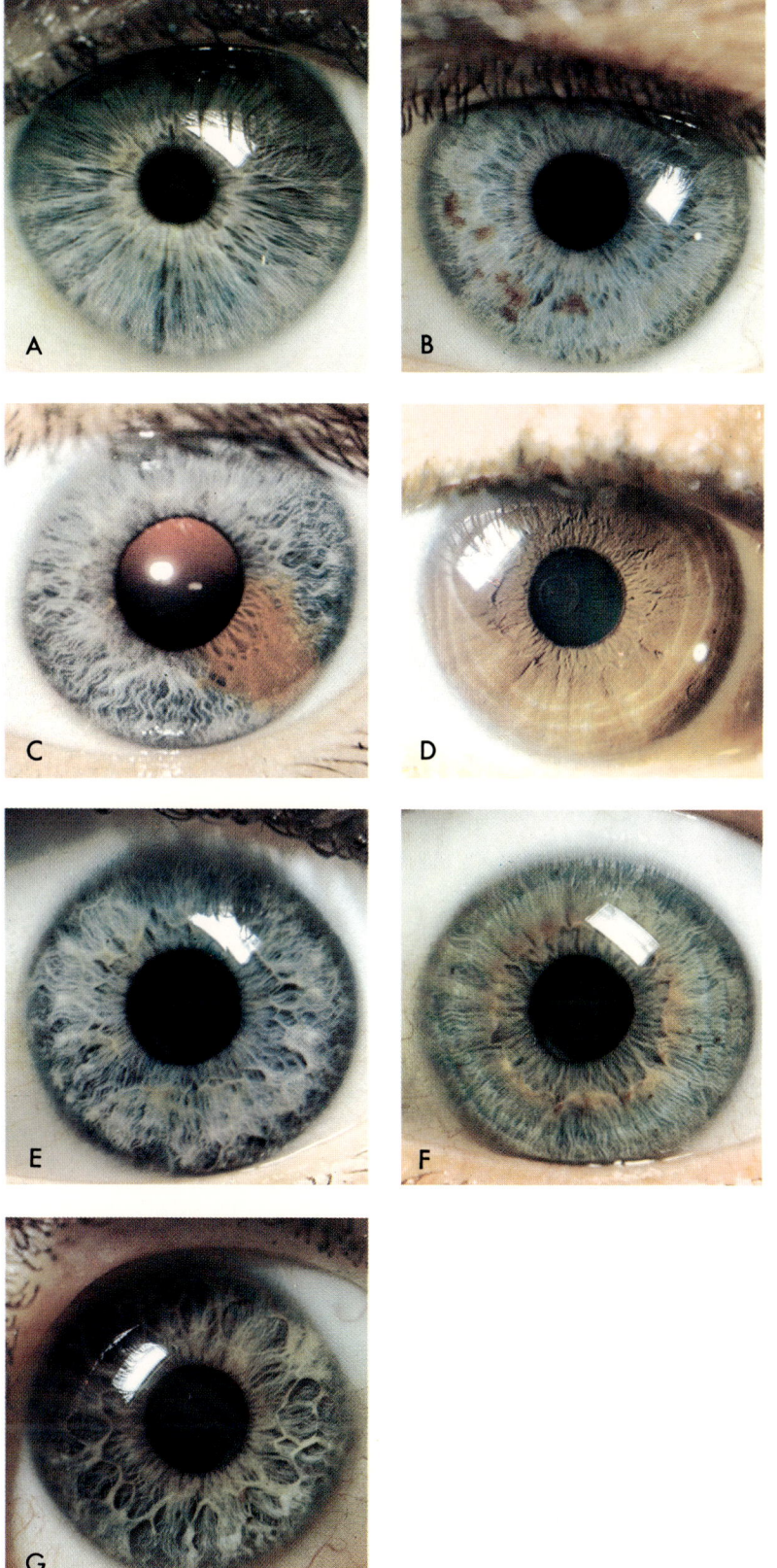

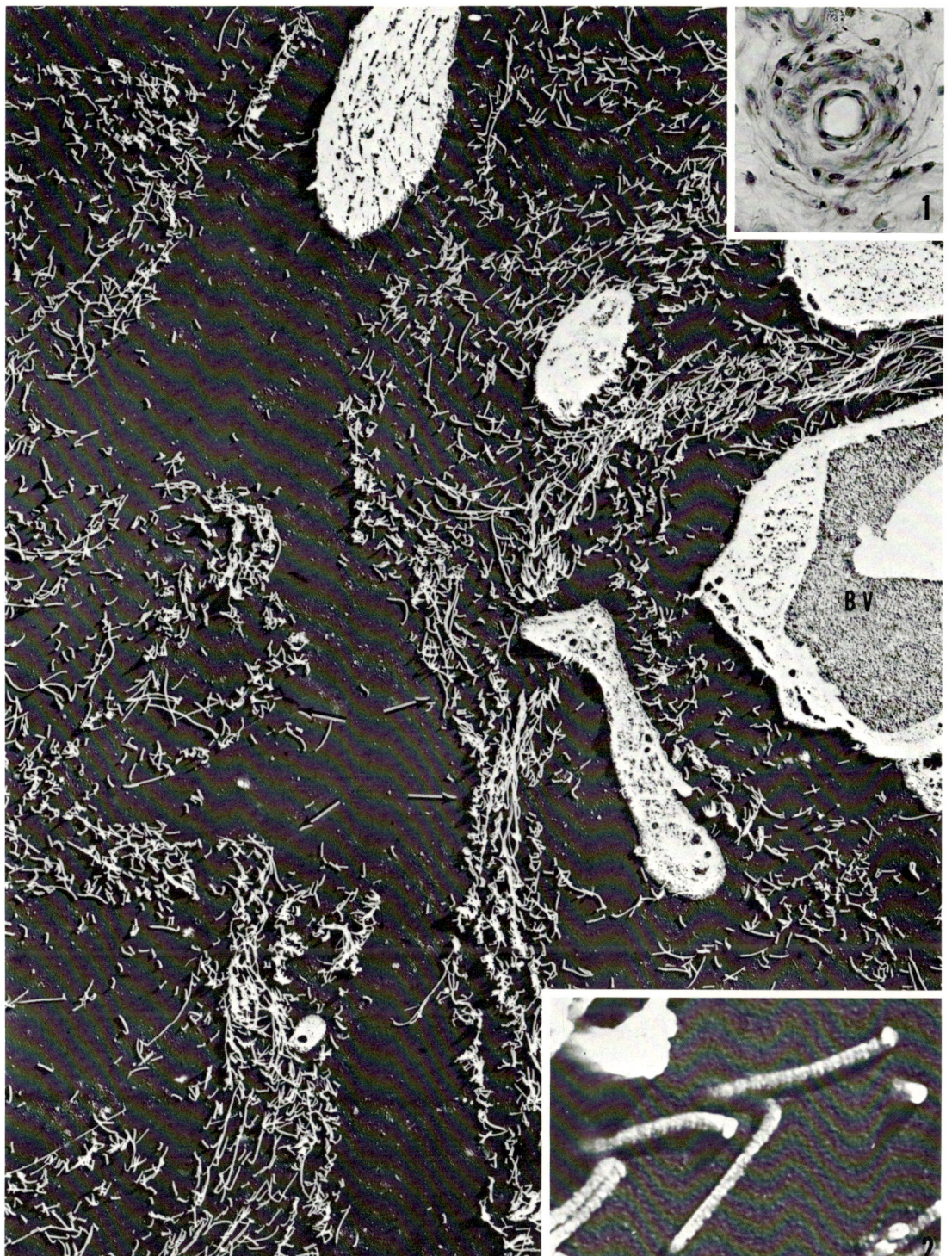

FIG. 10–10. Shadow-cast tissue section of iris stroma. Collagen fibrils, acid mucopolysaccharides, and cells form columns or aggregates around blood vessels (*BV*) or nerves. Edges of two adjacent columns are marked by *arrows*. Space between is occupied by mucinous drying patterns. **Inset 1.** Light micrograph of "thick-walled" blood vessel. **Inset 2** shows typical banded collagen fibrils of iris stroma. **Main figure,** U-shad, ×5,250; **inset 1,** H&E, ×220 (AFIP Neg. 70–1420); **inset 2,** ×28,000.

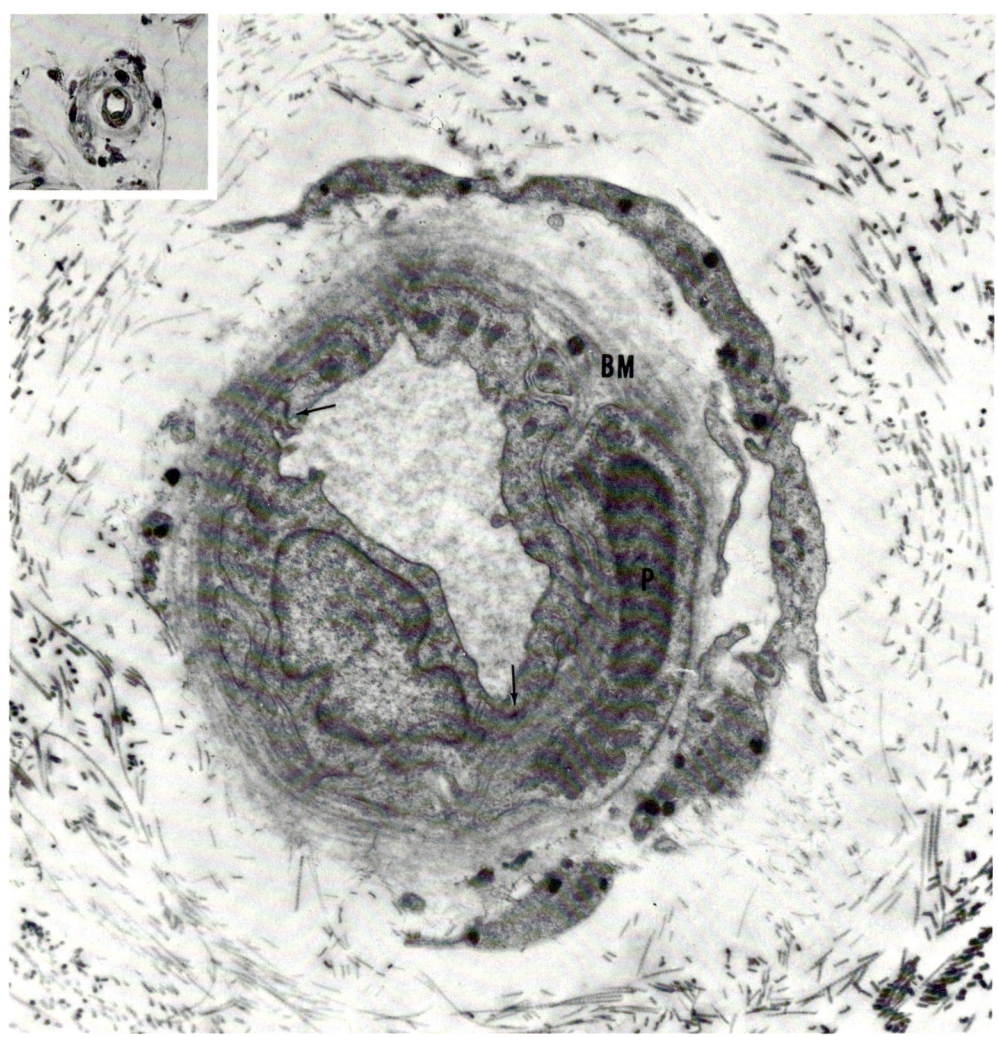

FIG. 10–12. Iris blood vessel. Electron micrograph shows small stromal vessel. Apical junctional complexes attach adjacent endothelial cells (*arrows*). A pericyte (*P*) is present. *BM*, basement membranes. Arrangement of cells and collagen makes up "thick walls" of blood vessel seen in light micrograph (**inset**). **Main figure,** ×10,000; **inset,** H&E, ×220.

linear appearance of the stromal melanocytes). They are more common in older individuals and most noticeable when changes are also observable in the nearby pigment seam.

Dilator Muscle

The posterior boundary of the iris stroma, peripheral to the sphincter muscle, is demarcated by another sheet of smooth muscle, the *dilator muscle*. The fibers of the dilator muscle are derived from, and remain in continuity with, the cuboidal pigmented cell bodies which make up the anterior layer of iris pigment epithelium (Fig. 10–19). The dilator is therefore a *partial* specialization of cytoplasmic processes of the anterior layer of iris pigment epithelium into smooth muscle (neuroectodermal). The dilator processes are arranged in an overlapping manner somewhat like shingles on a roof. When the muscle ele-

FIG. 10–13. Shadow-cast tissue section of large iris stromal vessel occupied by preserved plasma proteins in which red blood cells (*RBC*) are suspended. Collagen and stromal cells surround vessel. U-shad, ×5,000. (From Tousimis, A. J., and Fine, B. S. *Arch Opthal* (Chicago) 62:974, 1959.)

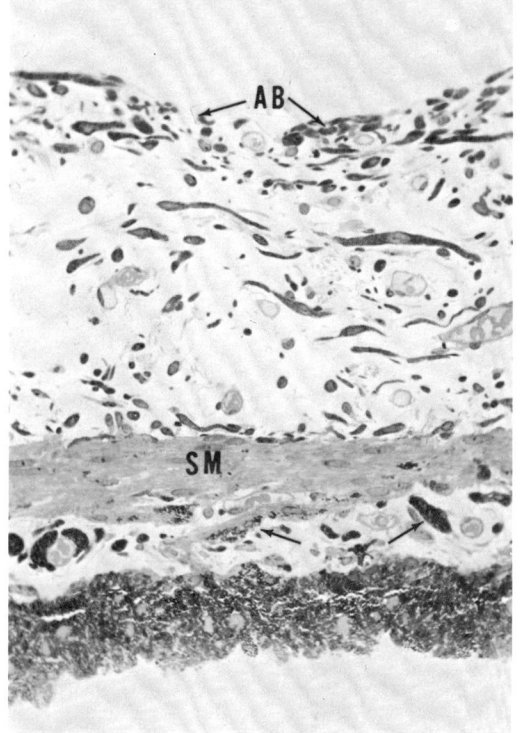

FIG. 10–14. Light micrograph of young rhesus monkey iris cut along long axis of sphincter muscle (*SM*). Minimal connective tissue is present here between muscle and pigment epithelium. Pigmented septa (*arrows*) passing toward pigment epithelium are called Fuchs' spurs. *AB*, anterior border layer. 1.5-μ section, toluidine blue, ×305 (AFIP Neg. 59–2588). (From Tousimis, A. J., and Fine, B. S. *Amer J Ophthal* 48:397, 1959.)

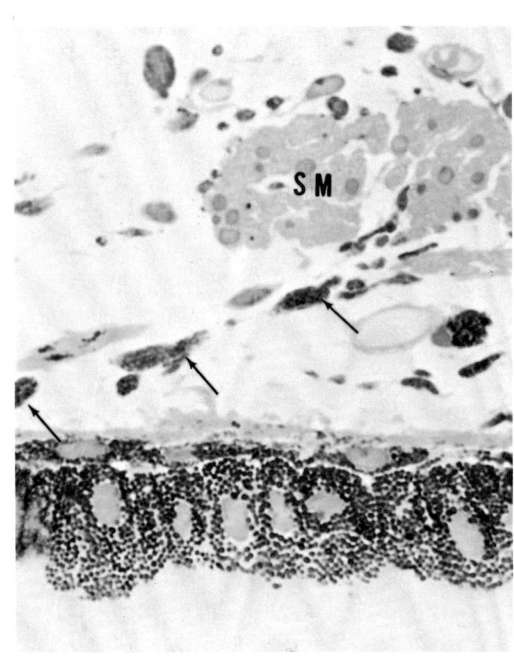

FIG. 10–15. Meridional section of young rhesus monkey iris showing peripheral edge of sphincter muscle (*SM*). A line of pigmented cells (*arrows,* Michel's spur) passes from this edge to pigment epithelium. 1.5-μ section, toluidine blue, ×820. (AFIP Neg. 59–2552.)

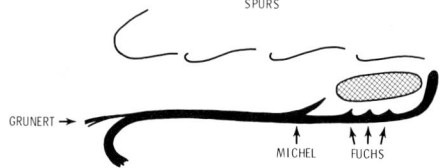

FIG. 10–16. Schematic drawing showing location of often pigmented "spurs" that project from region of dilator muscle of iris. (From Wobmann and Fine.[20])

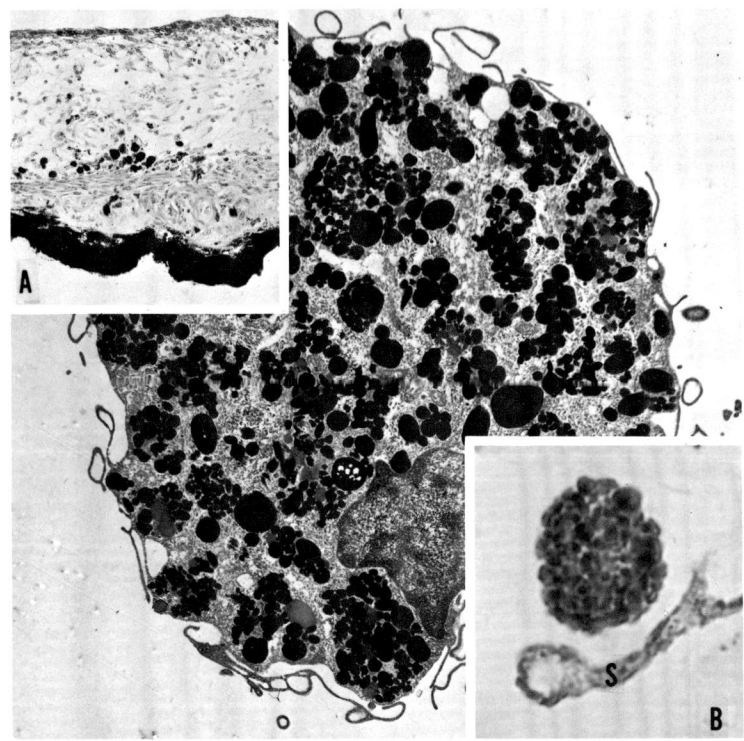

FIG. 10–17. Type 1 iris clump cells. **Inset A.** Light micrograph of clump cells in region of sphincter muscle. **Main figure.** Electron micrograph of typical type 1 clump cell with its delicate villi and clusters of pigment granules which vary in size and shape. **Inset B.** Light micrograph showing mulberry-like appearance of cell. A nearby branching stromal cell (*S*) is easily differentiated by its content of smaller granules of uniform size. **Main figure,** ×4,200; **inset A,** H&E, ×85 (AFIP Neg. 60–1421); **inset B,** 1.5-μ section, PD, ×1,040 (AFIP Neg. 70–8848). (From Wobmann and Fine.[20])

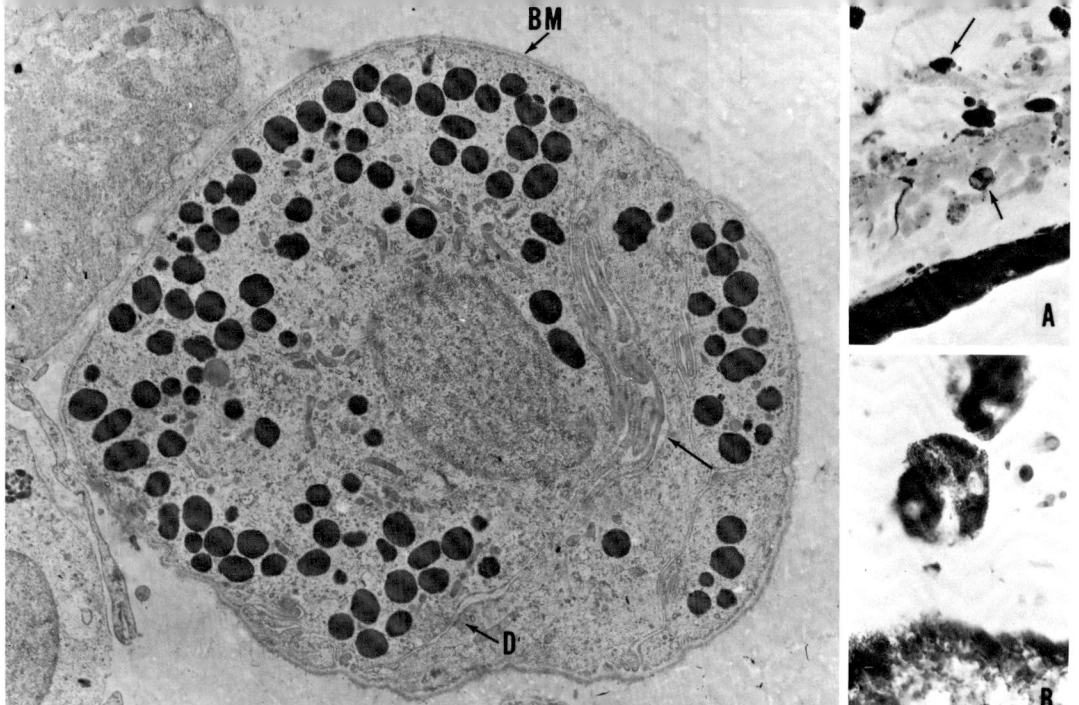

FIG. 10–18. Type 2 iris clump cells. Less frequently observed in region of sphincter muscle, this type is recognized in light microscopy (**insets A and B**) by its multinucleated appearance and by uniformity of distribution of its melanin granules. By electron microscopy (**main figure**), type 2 clump cell is seen to consist of group of cells surrounded by continuous basement membrane (*BM*). *D,* desmosome. *Free arrow* points to interdigitating apical villi. Melanin granules are all of neuroepithelial type. In **inset A,** upper arrow indicates a type 1 clump cell anterior to sphincter muscle; lower arrow, a type 2 clump cell. **Inset B** shows multinucleate type 2 clump cell near iris root. **Main figure,** ×5,600; **inset A,** ×145 (AFIP Neg. 70–8836); **inset, B,** ×305 (AFIP Neg. 70–8850). (From Wobmann and Fine.[20])

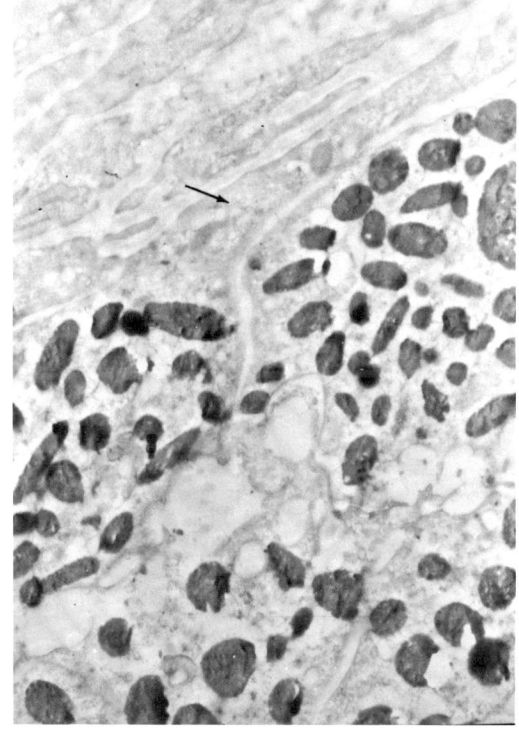

FIG. 10–19. Continuity of anterior pigment epithelium with dilator muscle of iris (*arrow*). ×12,300. (From Tousimis, A. J., and Fine, B. S. *Amer J Ophthal* 48:397, 1959.)

ments contract, their radial direction causes pupillary dilation. Melanin granules typical of neuroepithelium may often be seen within the smooth muscle portions of the anterior pigment epithelial cells.

As in other typical smooth muscle cells, the dilator cells have an investing membrane, cytoplasmic filaments, densities, and surface-connected vesicles.

Pigment Epithelium

The double layer of pigment epithelium that covers the posterior iris surface results from apposition of the two epithelial layers forming the anterior extremity of the embryonic cup (see Figs. 6–2 and 6–3, on Color Plate II).

The anterior border layer is separated from the pupil by a ridge of more heavily pigmented cells, which is the visible portion of iris pigment epithelium, the *pigment seam* (Fig. 10–20) or *ruff*.[2] Under higher magnification this ridge can be observed as an apparent chain of little beads. With wide excursions of the pupil the arrangement can be observed to indeed resemble a ruffle which folds up like an accordion on pupillary constriction, opens to reveal more clearly the ruffle-like appearance on semidilation, and finally stretches to form an almost smooth ridge lining the pupillary margin on wide dilation. Why it has this appearance can be better appreciated on gross examination of the pigment epithelium lining the posterior surface of the iris.

In a meridional section of a dilated iris the posterior epithelial cells* appear to be arranged in small clusters (Fig. 10–21). Each cluster represents a cross-sectioned ridge lying between two *circumferential furrows* on the posterior iris surface (Fig. 10–22). The wedge-shaped configuration of the cells in each ridge often appears as an incomplete triangle because the cells are not only arranged radially across each ridge but are also staggered along its length (Figs. 10–23 and 10–24).

The circumferential ridges do not form complete circles around the pupil but taper repeatedly to blend into adjacent furrows (Figs. 10–23 and 10–25). This arrangement allows the posterior iris surface to be properly covered during contraction and dilation of the iris. The circumferential furrows are least prominent in the region behind the pupillary zone of the sphincter muscle and are most prominent at the iris root (Fig. 10–22).

The posterior layer of pigment epithelium is continuous with the *nonpigmented* epithelium of the ciliary body and ultimately with the neural retina.

The prominence of the circumferential furrows at the iris root may be related in some way to the high incidence of pigment cell proliferation in this region to form folds and processes (Fig. 10–21B) leading to the formation of pigment cysts or pigmented "balls,"[3,22] which may be observed clinically in the anterior chamber. On rare occasions such pigment epithelial cysts or clusters may even be confused with a malignant melanoma of the uveal tract.[22]

FIG. 10–20. Termination of iris pigment epithelium in pupil as pigment seam (*arrow*). H&E, ×50. (AFIP Neg. 62-5287).

* For convenience, in descriptions of the iris, the terms "anterior" and "posterior" replace the terms "inner" and "outer" that are used for most other ocular tissues.

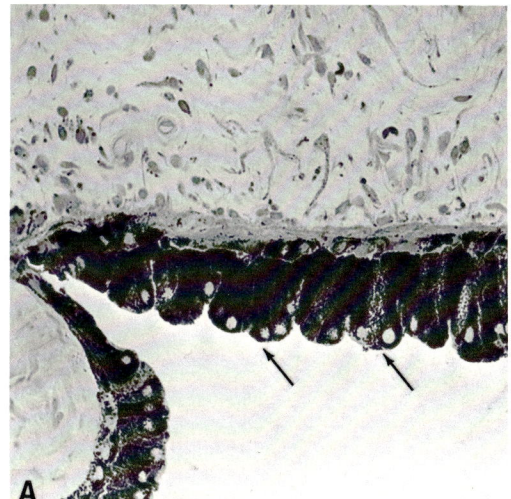
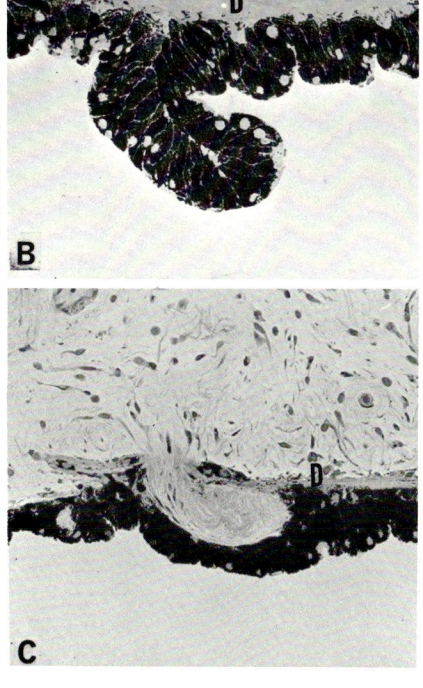
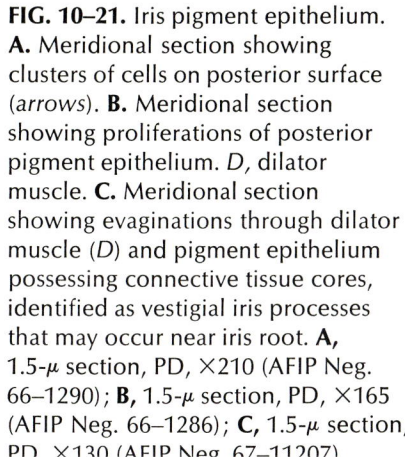

FIG. 10–21. Iris pigment epithelium. **A.** Meridional section showing clusters of cells on posterior surface (*arrows*). **B.** Meridional section showing proliferations of posterior pigment epithelium. *D*, dilator muscle. **C.** Meridional section showing evaginations through dilator muscle (*D*) and pigment epithelium possessing connective tissue cores, identified as vestigial iris processes that may occur near iris root. **A,** 1.5-μ section, PD, ×210 (AFIP Neg. 66–1290); **B,** 1.5-μ section, PD, ×165 (AFIP Neg. 66–1286); **C,** 1.5-μ section, PD, ×130 (AFIP Neg. 67–11207).

In the region of the sphincter muscle many short *radial* or *longitudinal* furrows replace the circumferential furrows (Fig. 10–22). They can be observed in vivo where they terminate anteriorly, arranging the pigment epithelial cells of the pupillary border into the ruffle-like arrangement seen clinically.

This blind ending of the optic vesicle (pupillary pigment seam) projects slightly over the adjacent anterior border layer of the iris stroma, and the slight overhang is sometimes known clinically as the *physiologic ectropion* of the pigment epithelium. This appearance is in contrast to *pathologic ectropion*, often termed *ectropion uveae*, in which the pupillary border of the iris is sufficiently everted that much more pigment epithelium from the posterior surface becomes exposed to view. In histologic sections the pupillary margin of the sphincter muscle generally is also everted in pathologic ectropion.

Peripheral to the sphincter muscle, the delicate pupillary radial furrows expand into the *structural furrows*, which are less well demarcated and rapidly diverge. They traverse the remainder of the iris to its root and become continuous with the valleys between the ciliary crests (see Fig. 10–22).

The large melanin granules of the pigment epithelium (spherical, ~ 0.8 μ in diameter; ovoid, ~ 0.5 by 1.3 μ) permit adequate resolution by oil-immersion light microscopy (see Chap. 2) (Fig. 10–26). They contrast with the smaller (~ 0.2 by 0.6 μ) uniformly ovoid granules of the pigmented cells of the iris

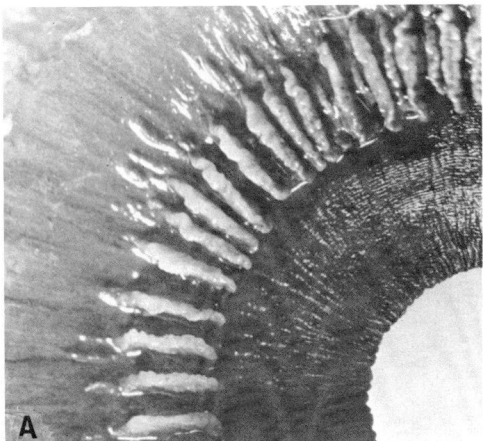

FIG. 10–23. Posterior iris ridges and furrows as seen by scanning electron microscopy at higher magnification. Note tapering and blending of adjacent ridges as well as staggered arrangement of cells along each ridge. ×500.

FIG. 10–22. Posterior iris surface. **A.** Gross photograph of posterior surface and adjacent ciliary crests. Coarse longitudinal or radial furrows are clearly seen; finer circumferential furrows are seen with difficulty. **B.** Scanning electron micrograph of posterior surface (infant) showing gross longitudinal furrows at pupillary margin (below, in photograph) and along posterior iris surface. Circumferential ridges and furrows are also clearly seen. Anterior edges of infantile ciliary crests are seen above in the photograph. **B,** ×50.

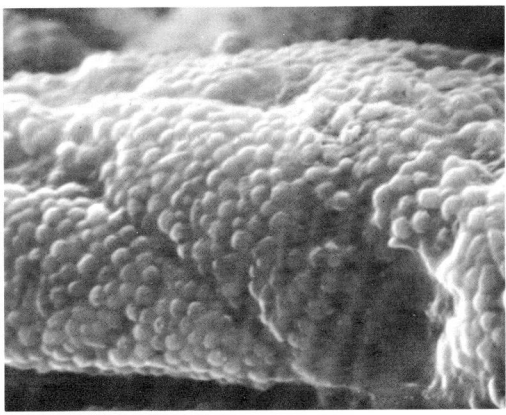

FIG. 10–24. Iris epithelial ridge as seen by scanning electron microscopy at still higher magnification. Note staggered arrangement of cells (outlined by deep grooves). Myriad small bumps are produced by melanin granules. Adult, ×2,800.

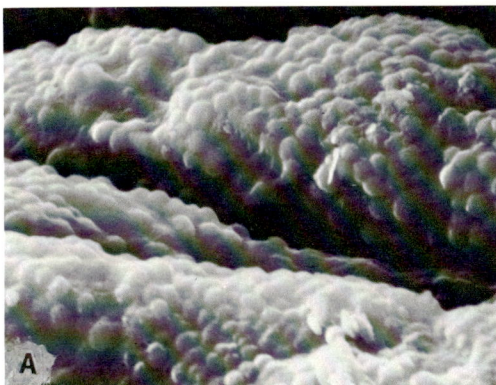

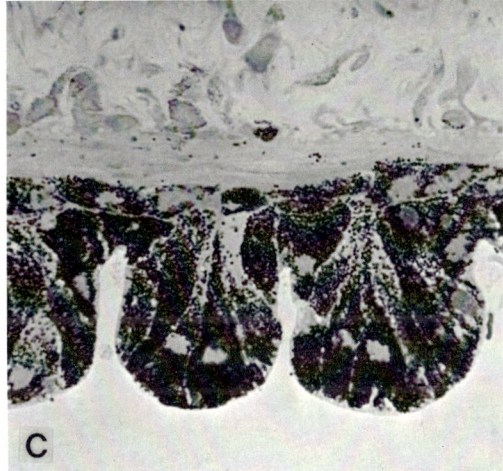

FIG. 10–25. A. Tapering out of one iris epithelial ridge between two others as seen by scanning electron microscopy. (Rhesus, X280.) **B.** The staggered arrangement of cells along each ridge can be better appreciated at higher magnification (×900). The free arrows point to two erythrocytes, one in surface view and one in profile. These may serve as internal measures of size. Rhesus, ×900. **C.** Cross sections of epithelial ridges (human) to show incomplete pie-shaped arrangement of sectioned cells. 1.5μ section. PD, ×440. (AFIP Neg. 71–7389.)

stroma. The uniformity and small size of the latter granules produce the light-brown color characteristic of uveal pigmentation.

The melanin granules of iris stromal cells of the rhesus monkey are wholly unlike those of the human in that they are longer and strap-like in shape.[16,20] Their arrangement can be appreciated easily not only in electron micrographs but also by light microscopy, where they can be seen as long pigment "needles." The pigment epithelial melanin granules, however, as well as those in the remainder of the uveal tract (i.e., choroidal) resemble closely those of the human.

The anterior layer of epithelium remains as a layer of pigmented cuboidal cells whose cytologic bases have expanded and specialized into the overlapping plate-like smooth muscle "cells" that make up the dilator muscle except in the region behind the sphincter muscle where dilator muscle is lacking. In this region a thin basement membrane is present (*inset 2*, Fig. 10–26). Where dilator muscle is present the basement membrane surrounds the basilar cytoplasmic expansions. The anterior layer is continuous with the layer of pig-

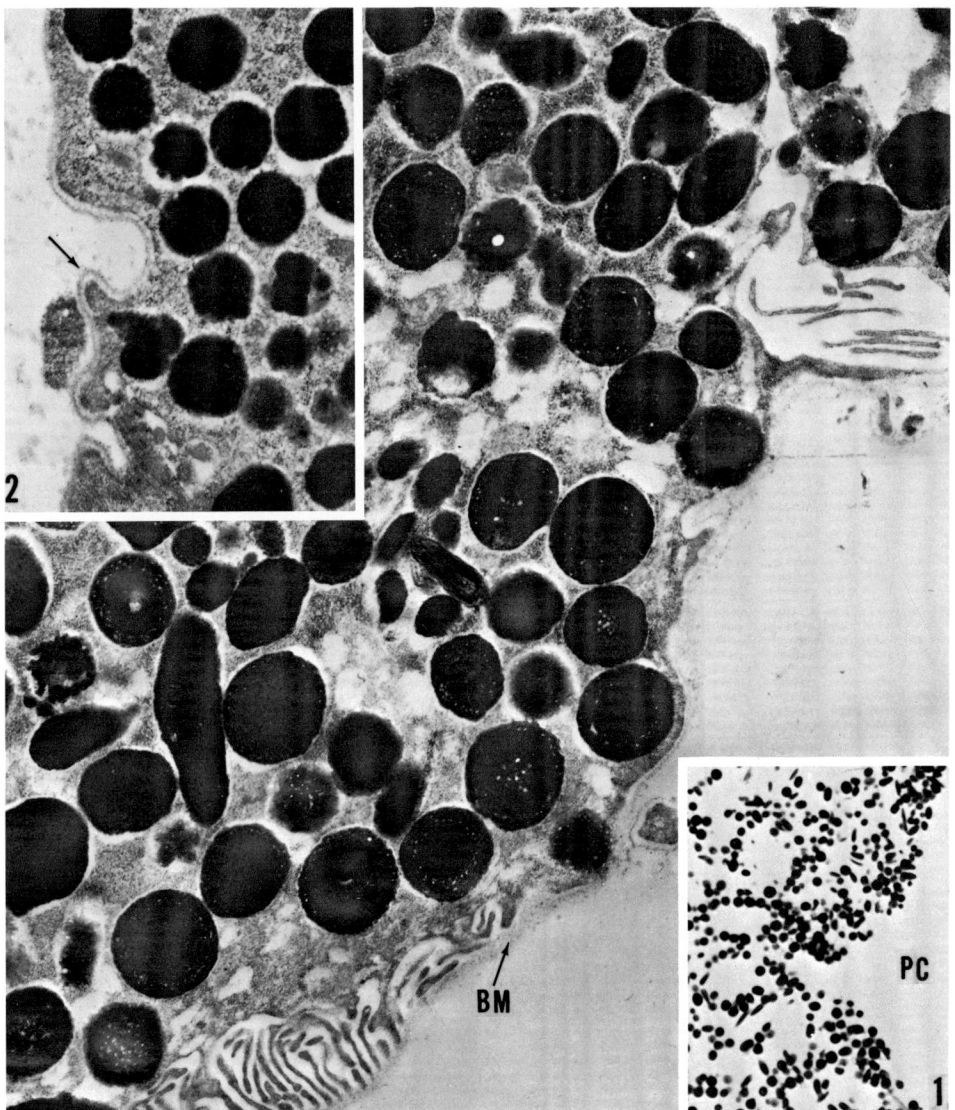

FIG. 10–26. Posterior edge of posterior layer of iris pigment epithelium. **Main figure** shows free edge of cell with many villi (i.e., basilar infoldings), some of which interdigitate with adjacent cells. Surface is covered by delicate, often multilaminar basement membrane (*BM*). **Inset 1** is light micrograph of similar region of pigment epithelial cells. *PC*, posterior chamber. **Inset 2** shows thin basement membrane (*arrow*) lining *anterior* pigment epithelium behind sphincter. **Main figure,** ×14,500; **inset 1,** 1.5-μ section, ×1,000 (AFIP Neg. 68–9949); **inset 2,** ×16,500.

mented epithelium of the ciliary body and the retinal pigment epithelium.

The columnar cells of the posterior layer of pigment epithelium are arranged apex-to-apex with the cells of the anterior layer. Such an arrangement provides a delicate, often multilaminar, basement membrane on the posterior surface[18] and frequent clusters of apical villi on the anterior surface that project into small pockets or spaces between the two layers of epithelium.

Remnants of the lumen of the optic vesicle persist throughout life. That the posterior layer and the anterior layer can be separated easily from each other is evidenced by the cysts that may form spontaneously in the pigment epithelium or be produced at the pupillary margin by the long-term use of strong miotics.

The posterior epithelial cells are attached to one another by very small and poorly formed desmosomes as well as by a terminal-bar-like arrangement. The cell attachments serve to hold the cells together in layers.

The thicker posterior layer of pigment epithelium is of relatively greater importance in maintaining the opacity of the iris.

THE CHOROID

The choroid is that part of the uveal tract extending from the edge of the nerve head to the ora serrata.

SUPRACHOROIDAL "SPACE" AND LAMINA FUSCA

The usually heavily pigmented posterior or choroidal portion of the uveal tract is loosely adherent to the overlying sclera. This plane of loose attachment is a zone of potential separation. When separated it is known as the suprachoroidal ("perichoroidal") space. This "space" is common to both the choroidal and ciliary portions of the uveal tract. Attachment of the longitudinal (meridional) ciliary muscle to the scleral roll limits the space (as in "choroidal detachments") anteriorly. The enlargement is limited posteriorly by attachment of the choroid to the sclera, augmented by the outward passage of the vortex veins, by the perforating short posterior ciliary arteries (Fig. 10-27) and by border tissue at the scleral aperture for the optic nerve.

When the choroid is separated from the sclera (either in vivo or artifactitiously), delicate pigmented bands can be seen traversing the space from the choroidal stroma proper to the innermost few layers of the sclera (Fig. 10-30). Delicate cytoplasmic extensions of

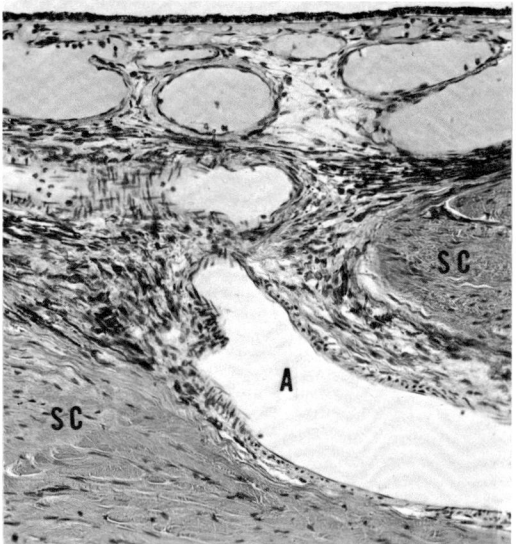

FIG. 10-27. Posterior choroid attached to sclera (SC) through which an artery (A) is passing. H&E, ×115. (AFIP Neg. 62-5291.)

nonpigmented cells parallel the pigmented cells (Fig. 10-31) in the layers. A very small amount of detectable extracellular material accompanies the characteristic pigmented and nonpigmented uveal cells to make up a single choroidal lamella. Deeper within the choroidal stroma, additional quantities of collagen fibrils and other extracellular materials thicken the lamellas further. From behind forward the multilaminar arrangement of loose, flattened, pigmented lamellas is diminished as the meridional ciliary muscle becomes thicker and elastic tissue becomes more noticeable.

The long posterior ciliary arteries and their corresponding long ciliary nerves lie within the suprachoroidal space in the horizontal plane, encased by collagenous tissue. Branches from the nerves to the adjacent choroid form small net-like arrangements where large ganglion cells may be observed.

Smooth muscle cells, at first singly and then in small groupings, appear in the suprachoroid even as far posteriorly as the equator.

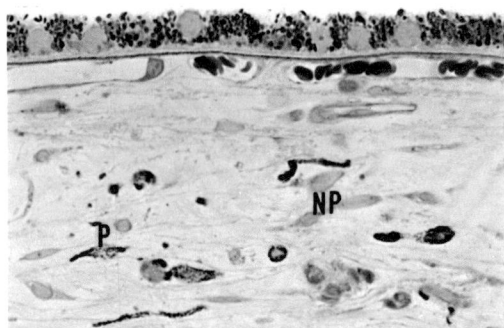

FIG. 10–28. Light micrograph of choroid. Pigmented (*P*) and nonpigmented (*NP*) cells are present in stroma 1.5-μ section, PD, ×440. (AFIP Neg. 66–1298.)

FIG. 10–29. Electron micrograph of choroidal cells. *P*, pigmented cells; *NP*, nonpigmented cell. Rhesus monkey, ×14,500.

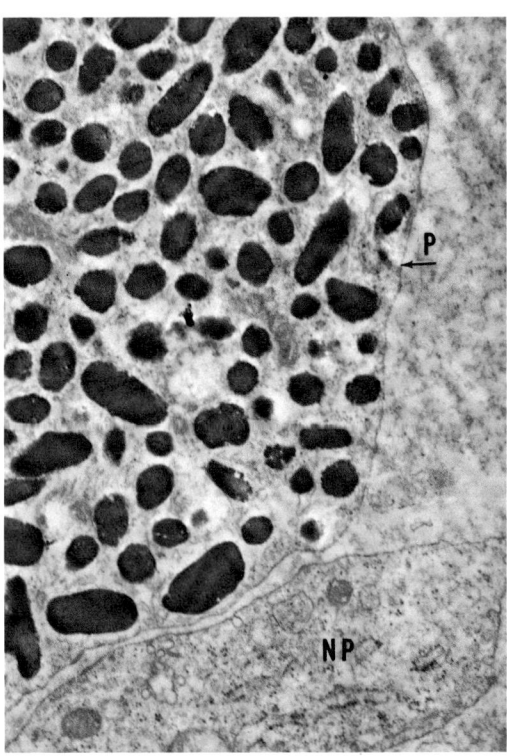

These somewhat stellate cell groupings ("muscle stars") increase in number anteriorly until they form a detectable posterior border to the ciliary muscle.

CHOROIDAL STROMA

Pigmented and Nonpigmented Cells

As in the stroma of the iris, the resident cells are separable morphologically into pigmented and nonpigmented varieties (Figs. 10–28 and 10–29). Because of the flattened or interconnecting lamellar arrangement, the pigmented cells, which are dendritic or branching on surface view, necessarily appear spindle-shaped in meridional section (Figs. 10–30 and 10–31). The pigmented cells or melanocytes are easily observed and produce the deep-brown color of the choroid. Associated with these cells are varying amounts of collagen fibrils and watery mucinous intercellular materials whose composition has not yet been defined.

On surface view the choroid is least pigmented where the larger vessels are located

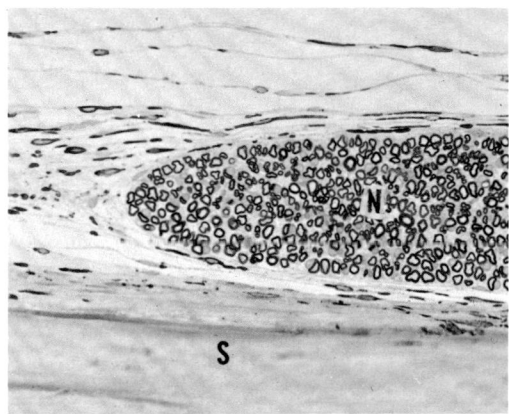

FIG. 10–30. Light micrograph of loose, pigmented suprachoroidal lamellas. Large nerve (*N*) lies in suprachoroidal space. *S*, sclera. 1.5-μ section, PD, ×225. (AFIP Neg. 66–1293.)

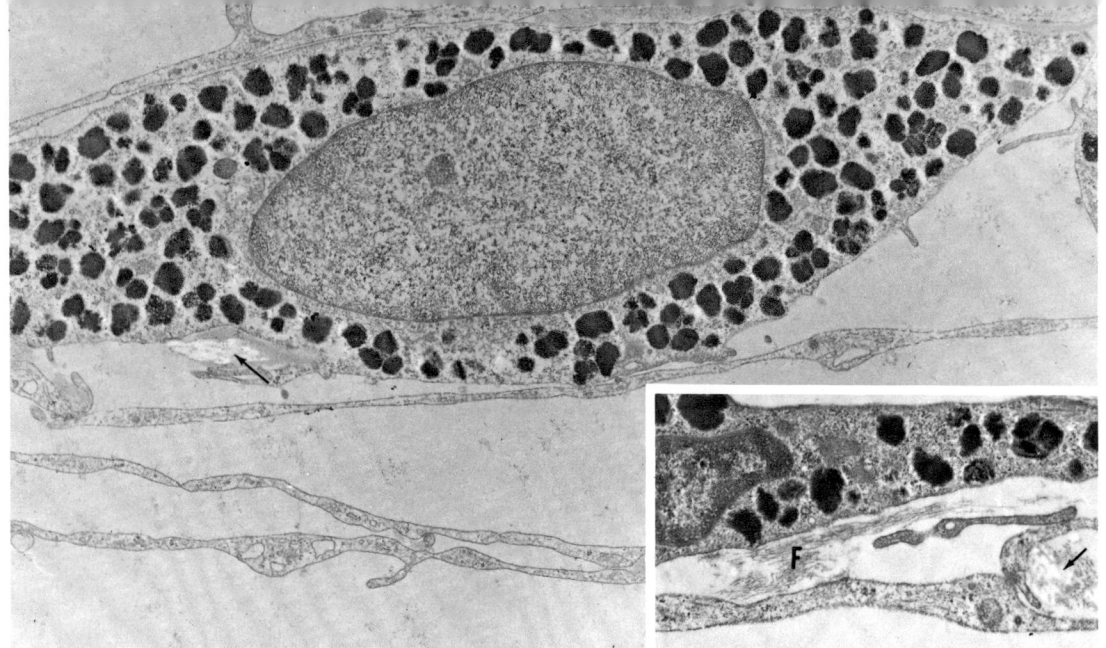

FIG. 10–31. Main figure, electron micrograph of flattened pigmented and nonpigmented cells, that make up loose suprachoroidal lamellas. Collagen is sparse in this plane (*arrow*). ×9,600. **Inset:** Sample of nasal choroid slightly deeper (i.e., internal) to that in the main figure. Filaments (*F*) (presumably collagen or microfibrils of elastic tissue) are present between the melanocyte above and the nonpigmented cell process below. On the right (*free arrow*) a cluster of negatively stained material (elastin) is enveloped by filaments. ×15,000.

and most pigmented in the spaces between the vessels. The contrast between the pigmented zones and the vessels helps to produce the ophthalmoscopic choroidal pattern. In section, the outer layers are more heavily pigmented than the inner.

When the intervascular columns are moderately pigmented, the fundus appears ophthalmoscopically as a uniform red color. The uniformity is due to the layer of retinal pigment epithelium covering the vasculature.

Vessels

The venous drainage system is seen as four vortex systems each located in a posterior quadrant. Each system converges to form a single vestibule, the *ampulla,* which then exits through the sclera by a *vortex vein.*

The larger arteries are found most readily in the outer layers of the posterior choroidal stroma (Fig. 10–27). As elsewhere, the arteries and veins generally are differentiated by size, shape of lumen, and thickness of wall.

The capillary layer of the choroid, the *choriocapillaris* (Fig. 10–32), is very important. It nourishes the pigment epithelium and outer layers of the neural retina to approximately the level of the middle limiting membrane.

An interconnecting network, it lies in a single plane and is formed by typical, fenestrated endothelial cells (Fig. 10–33). The fenestrations can be observed on both inner and outer walls.

A thin basement membrane surrounds the basal plasma membrane of the endothelial cells. Pericytes along the *outer* surface of the choriocapillaris are typically and completely invested by their own thin basement membranes. Similar cells are not normally ob-

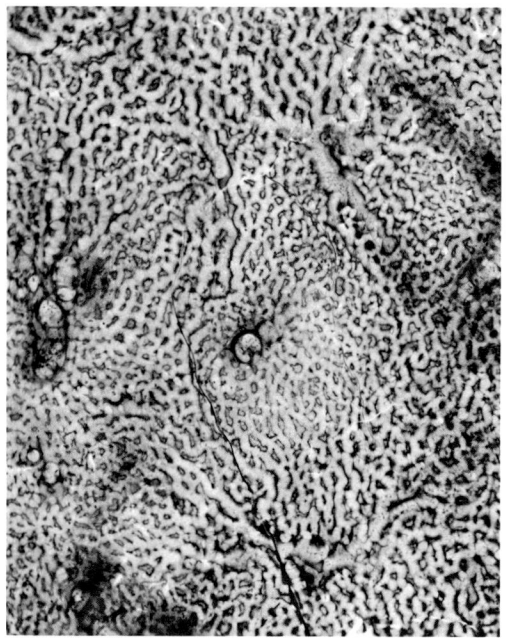

FIG. 10–32. Flat preparation of choriocapillaris (posterior pole). PAS, ×50. (AFIP Neg. 69–5384. Courtesy of Dr. M. O. M. Ts'o.)

served along the *inner* surface of the choriocapillaris.

A layer of acellular collagenous material occupies the plane between the endothelial and pigment epithelial basement membranes. The layer is in continuity with the remainder of the choroidal stroma through the spaces between adjacent segments of the choriocapillaris. The continuities or "columns" are necessarily thin or very delicate in the region of the posterior pole (Fig. 10–32) where the segments of choriocapillaris are more numerous per unit area than they are toward the equator and the retinal periphery.

With aging of tissues, the intervascular columns become dense and thickened (Fig. 10–34), as does the collagenous plane between choriocapillaris and pigment epithelium. The midzone of the latter plane becomes unusually dense with aging, stains positively with orcein, and thus may be identified as "elastic tissue." Crystalline deposits suggestive of calcification may also appear in this zone, a region that is known to become basophilic.[19] The various aging changes are sometimes considered to play a part in the pathogene-

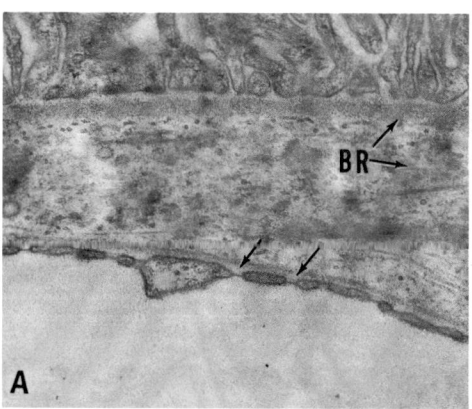 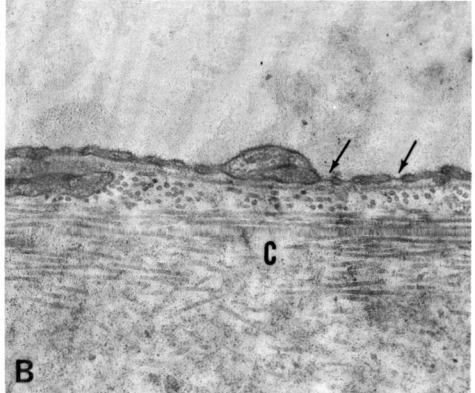

FIG. 10–33. Choriocapillaris. **A.** Fenestrated (*free arrows*) inner wall. BR, Bruch's membrane. **B.** Fenestrated (*arrows*) outer wall. C, collagen fibrils of choroidal stroma. Both rhesus monkey, ×24,000.

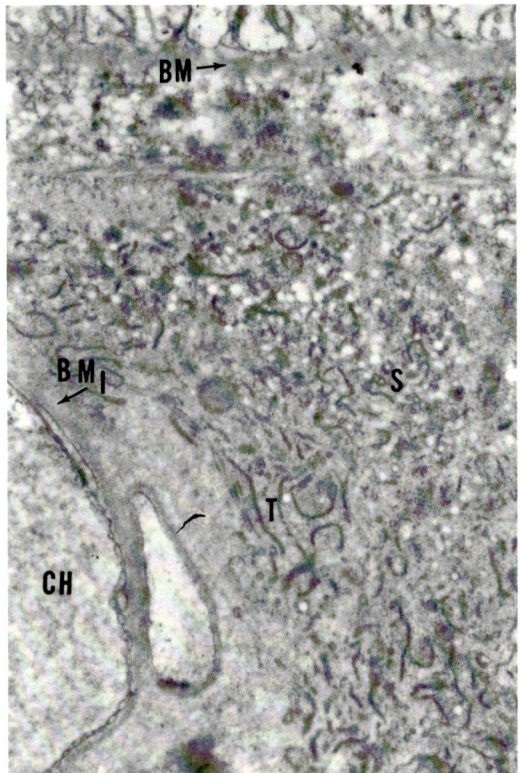

FIG. 10–34. Intervascular columns of choriocapillaris that have become dense with aging. Aggregates of spherical (*S*) and tubular (*T*) forms are present in large number. *BM*, basement membrane of pigment epithelium; BM_1, basement membrane of choriocapillaris; *CH*, choriocapillaris. ×12,000.

sis of *some* forms of macular degeneration, e.g., senile disciform degeneration.

Bruch's Membrane

Bruch's membrane is most usefully considered as a continuous sheet composed of two layers: an inner cuticular layer or lamella (the *lamina vitrea* or basement membrane of the pigment epithelium) and an outer collagenous layer (lamina elastica)[5,9,10,12] (see under Pigment Epithelium, Chap. 6).

The term *lamina elastica* is related to the affinity of the outer plane for such stains as orcein or Weigert's elastic tissue stain. An aging process that occurs focally along the inner plane of this region is the development of small nodules, *drusen*, beneath the pigment epithelium. These nodules are eosinophilic and PAS-positive. By electron microscopy many of the nodules are found to be composed of an aggregate of abnormally formed basement-membrane-like materials (Fig. 10–35).

CILIARY BODY

The triangular ciliary body has its base at the iris root anteriorly and its apex at the ora serrata posteriorly (Fig. 10–36). It is formed by the two neuroepithelial layers internally and the intermediate portion of the uveal tract externally. It measures 6.0 mm* from its apex to its base, where it is thickened by large, smooth muscles.

The anterior surface (the "face") of the ciliary body forms part of the angle of the anterior chamber and continues anteriorly as the uveal trabecular meshwork and the root of the iris. Posteriorly, at the ora serrata, the ciliary body continues as the choroid.

ZONES

The ciliary body is clearly divisible into two parts (Fig. 10–37): an anterior ring, the *pars plicata* (corona ciliaris), and a wider posterior ring, the *pars plana* (orbiculus ciliaris).

The pars plicata consists of a ring of approximately 70 major crests ("processes," Fig. 10–22) meridionally arranged. Each crest measures ~ 2 mm long and is easily recognized by its usually whitish appearance along the free edge. In the valleys between the

---

* The total width of the ciliary body stated here for practical purposes as 6.0 mm is an average figure. In reality, the width is variable, being narrowest in the upper nasal quadrant (4.6 to 5.2 mm) and widest in the lower temporal quadrant (5.6 to 6.3 mm).[12]

FIG. 10–35. Drusen. **Main figure.** Electron micrograph of very small simple druse formed beneath pigment epithelium (*PE*). Basement membrane characteristic of pigment epithelium (*BM*) follows basal contour of cells. Drusen accumulations appear similar to materials accumulating within intervascular columns (Fig. 10–34). *CH,* choriocapillaris. **Inset 1.** light micrograph section of a small druse. **Inset 2.** Light micrograph flat preparation of two small drusen. **Main figure,** ×11,000; **inset 1,** H&E, ×575 (AFIP Neg. 70–3664); **inset 2,** H&E, ×245 (AFIP Neg. 70–1186. Courtesy of Dr. M. O. M. Ts'o.)

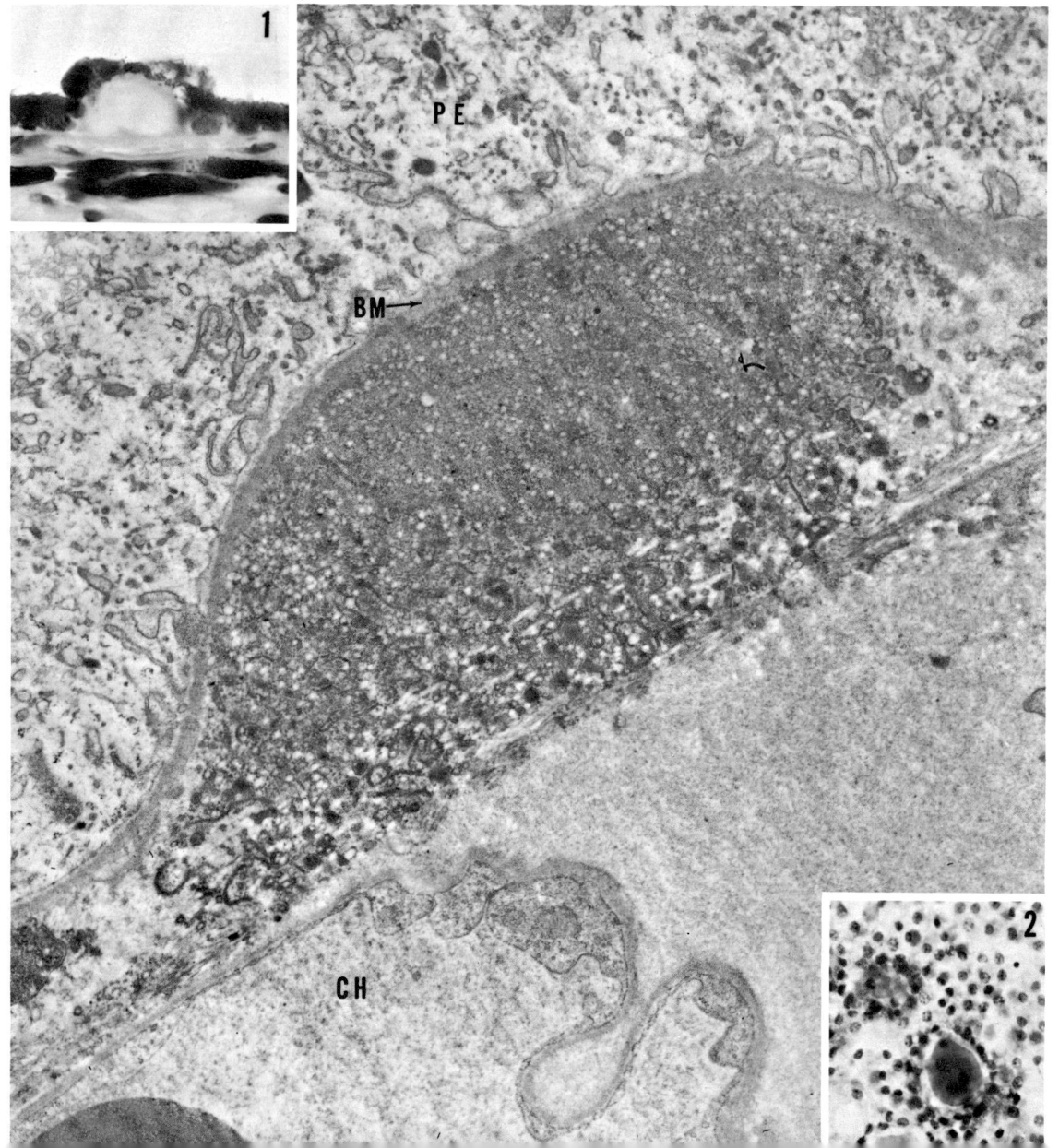

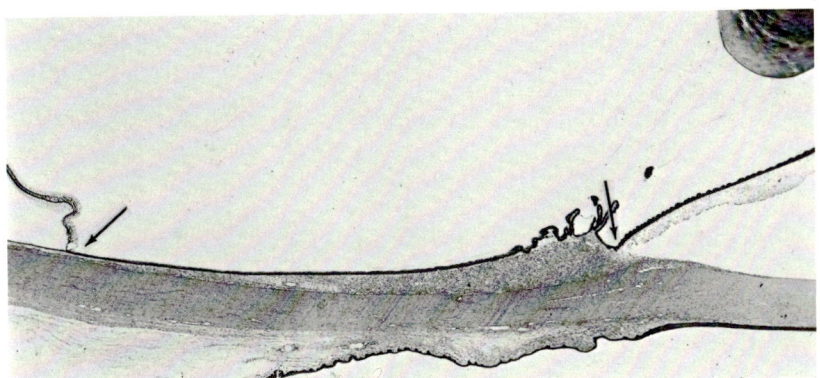

FIG. 10–36. Ciliary body (infant). Structure extends from root of iris to ora serrata (*free arrows*). H&E, ×15. (AFIP Neg. 62–5285. From Fine, B. S., and Zimmerman, L. E. *Invest Ophthal* 2:105, 1963.)

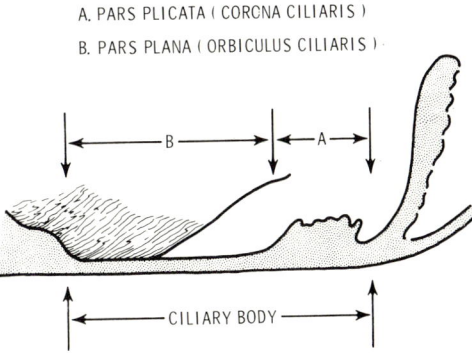

FIG. 10–37. Schematic representation of ciliary body. *A*, pars plicata (corona ciliaris); *B*, pars plana (orbiculus ciliaris).

crests lie various smaller, uniformly pigmented folds which might be called *minor crests* (anteriorly, *plicae ciliares*; posteriorly, *warts*.).

The pars plana is flat, extending from the posterior edges of the ciliary crests to the ora serrata ∼ 4 mm). Thus, the ratio of width of pars plana to pars plicata in meridional section is 2:1.

The pars plana often is not uniformly pigmented. Frequently a dark band (more prominent on the temporal side) can be observed posteriorly paralleling the tooth-like configuration of the ora serrata (see Fig. 6–85). Continuing the direction of the anterior tooth-like projections from the retina at the ora serrata are *pigmented striae* which course meridionally (radially) to blend with the pigmented minor ciliary crests lying in the valleys of the pars plicata. When proper light is reflected obliquely from the inner surface of the pars plana, a very fine meridional surface striation often can be appreciated (Fig. 10–52). The striation is more delicate than the circumferential furrows present on the posterior iris surface.

COMPONENTS

For descriptive purposes the ciliary body may be subdivided in many ways. Separation into anterior (pars plicata) and posterior (pars plana) parts has already been described. Another method is to separate the ciliary body into at least six layers: (1) the suprachoroidal (potential) space, (2) the ciliary muscle, (3) the layer of vessels, (4) the external basement membrane, (5) the epithelium, and (6) the internal basement membrane. On the basis of embryonic development, the ciliary body may be separated into *two layers*: the inner retinal (neuroepithelial) and the outer uveal (mesodermal) portion. This is the classification used in the following discussion.

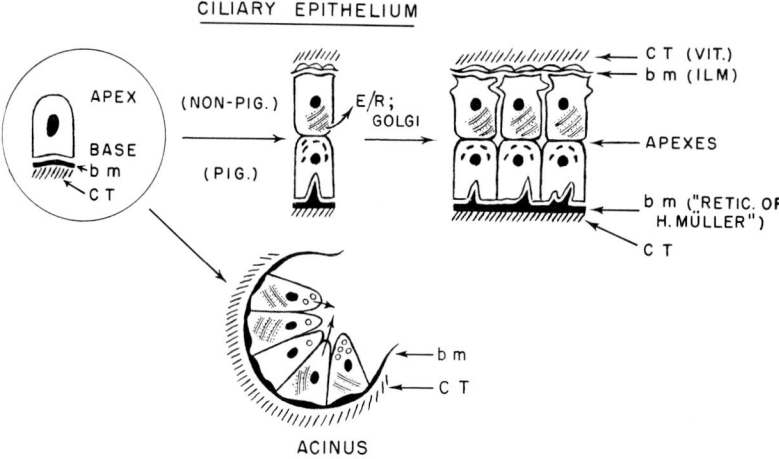

FIG. 10–38. Schematic drawing of arrangement of ciliary epithelium compared with that of an acinus of a more typical secretory tissue. *bm,* basement membrane; *ct,* connective tissue; *E/R,* granular endoplasmic reticulum; *ILM,* internal limiting membrane; *VIT,* vitreous filaments.

Retinal (Neuroepithelial) Portion

The two epithelial layers of the ciliary body derived from the embryonic optic vesicle retain their original orientation and are applied to one another, apex-to-apex (Fig. 10–38). The pigment epithelium of the retina is continuous with the pigment epithelium of the ciliary body without marked change, whereas the multilayered retina ceases abruptly at the ora serrata (see Fig. 6–86) to continue anteriorly as the single-layered *nonpigmented* ciliary epithelium.

Here, the lumen of the optic vesicle becomes essentially obliterated in contrast with persistence of the narrowed lumen posteriorly between the sensory retina and its pigment epithelium and anteriorly between the two layers of iris pigment epithelium. The two epithelial cell layers become strongly attached to each other by a blending of the two systems of terminal bars: the external limiting membrane of the neural retina and the fenestrated membrane of the pigment epithelium (Fig. 10–39). Any attempt at mechanical separation of this attachment invariably tears the adjacent apical cell cytoplasm (Fig. 10–40).

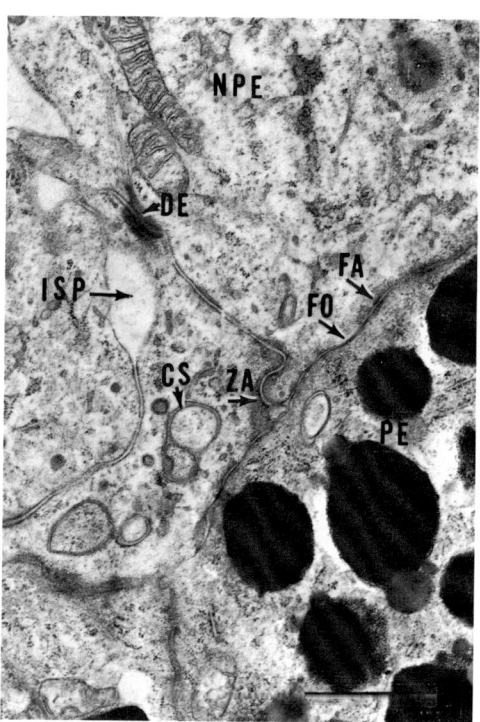

FIG. 10–39. Attachment between nonpigmented epithelium (*NPE*) and pigmented epithelium (*PE*). Attachment consists of three parts: fascia occludens (*FO*), fascia adherens (*FA*), and intercellular cement (*CS*). Zonula adherens (*ZA*) of retinal external limiting membrane persists along apical plane of nonpigmented epithelium. *DE,* desmosome; *ISP,* enlarged intercellular space. ×18,000.

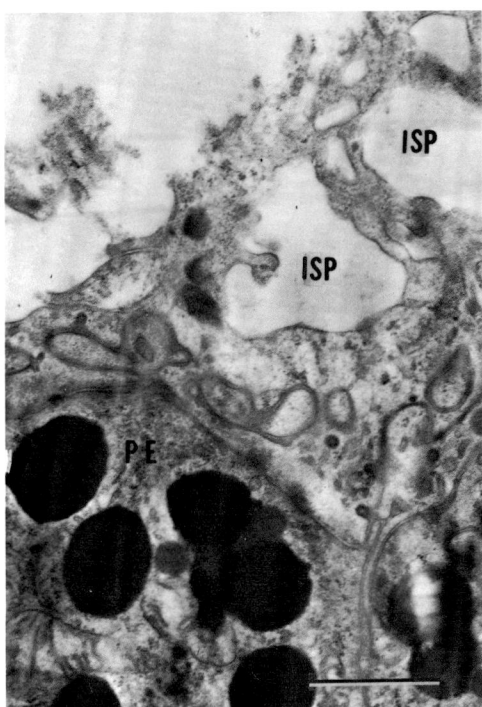

FIG. 10–40. Tearing of apex of nonpigmented cell layer produced by mechanical separation. Enlarged intercellular spaces (*ISP*) and apical attachments remain behind. *PE*, pigment epithelium. ×18,000.

Although the photoreceptor cells disappear at the ora serrata, the apical ends of the nonpigmented ciliary epithelial cells remain united to each other by a series of zonula-adherens-like attachments (Fig. 10–41), continuous with those of the external limiting membrane of the retina. Similarly, except for occasional foci, the apical villous processes of the retinal pigment epithelium disappear, and the terminal bar attachments, *both* occludens *and* adherens portions (Fig. 10–42), lie *between* the now-apposing surfaces of the two cell layers (i.e., uniting pigmented epithelium to nonpigmented epithelium).[4]

Because the attachment zones *between* the two layers are observed segmentally in any single section (Figs. 10–39, 10–40, and 10–42), they presumably are incomplete girdles and therefore should be termed *fascia occludens* and *fascia adherens* (see Chap. 3). The zonulae adherentes of the neural retina (external limiting membrane) are carried over into the apical plane of the nonpigmented ciliary epithelium.

Lying between the two epithelia, in addition to the fascia occludens and fascia adherens, is a third component to the attachment plane, the *intercellular cement substance* (Fig. 10–43).[4] The cement substance lies extracellularly between the apical plasma membranes and occasionally can be observed to be continuous with similar cement material of a nearby desmosome (Fig. 10–44).

The union of nonpigmented epithelium with pigmented epithelium is therefore tripartite: fascia adherens, fascia occludens, and an intercellular cement. Desmosomes (*maculae adherentes*) attach adjacent cells to each other within a single epithelial layer (Fig. 10–45).

Nonpigmented Epithelium and Internal Basement Membrane. At the ora serrata (Fig. 10–46) the delicate, multilaminar basement membrane of the peripheral retina is carried over onto the nonpigmented epithelial cells as the *internal basement membrane* (internal limiting membrane). The basal surfaces of the cells are thrown into folds or processes which intermingle with the multilaminar basement membrane. The intermingling is also continuous with the myriad collagenous filaments of the adjacent vitreous body (known as the "base" of the vitreous body).

Intercellular spaces of varying size between the nonpigmented cells (Fig. 10–47) reach to their apexes.[4] The spaces (easily distinguishable from deep surface invaginations by lack of a basement membrane lining) are occupied by a lucent extracellular mucinous material that stains positively for acid mucopolysaccharides (Fig. 10–48).

In the region of the posterior pars plana the apical cytoplasm of the nonpigmented epithelial cells possess considerable quantities

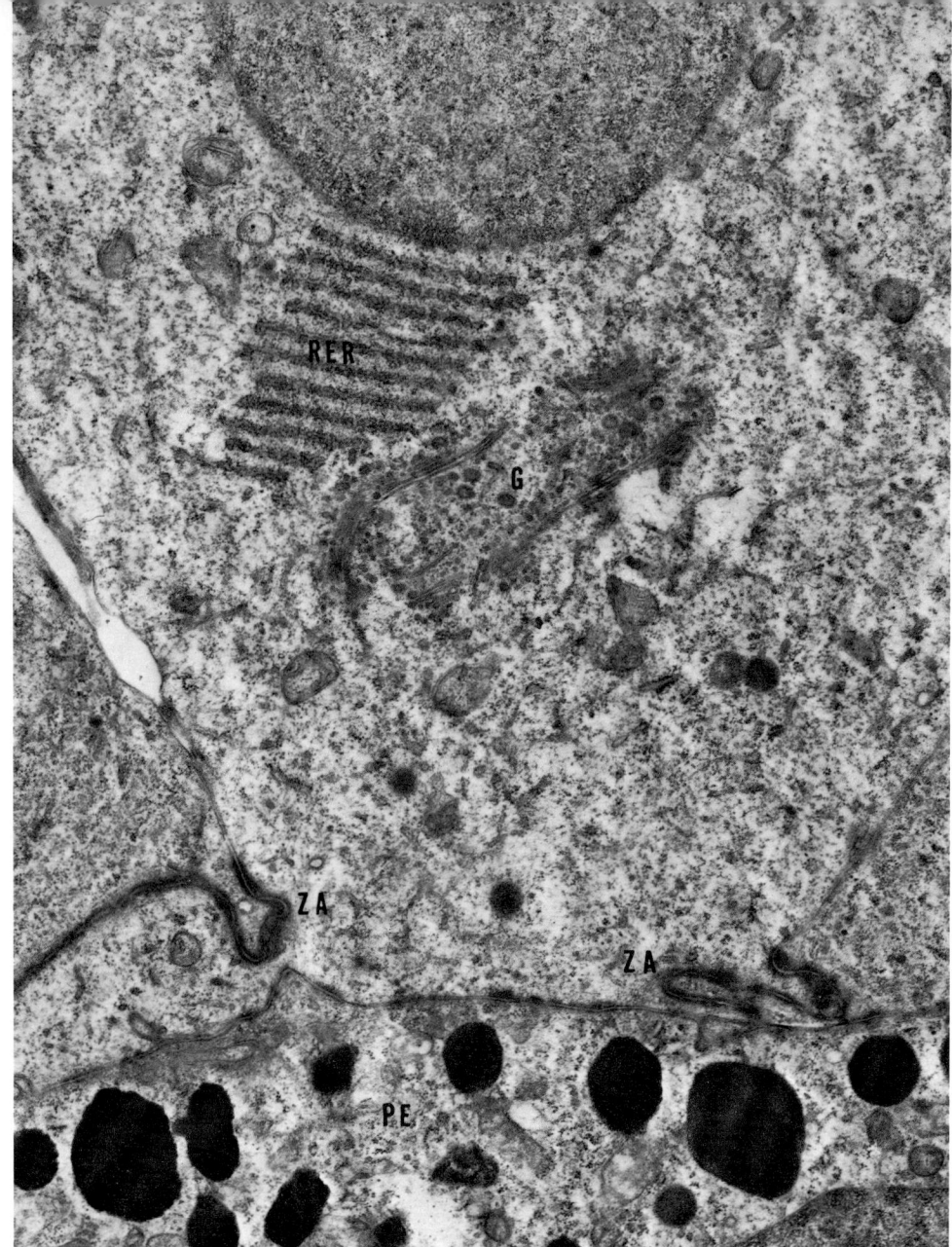

FIG. 10–41. Apical ends of non-pigmented epithelial cells linked by zonulae adherentes (ZA). Segments of granular endoplasmic reticulum (RER) and a Golgi complex (G) lie within apical cytoplasm. PE, pigment epithelium. ×16,800.

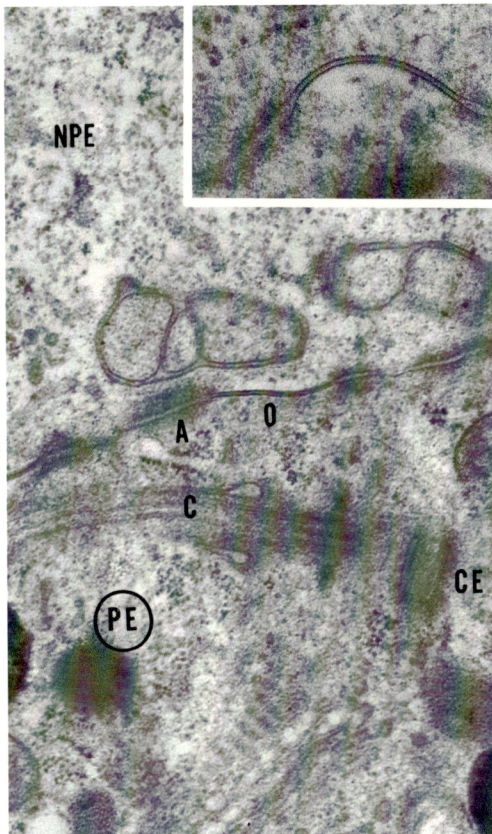

FIG. 10–42. Occludens (*O*) and adherens (*A*) portions of fenestrated membrane of pigment epithelium lying *between* two cell layers (*NPE* and *PE*). Note cilium (*C*), free centriole (*CE*), within apical cytoplasm of pigment cell. **Inset** shows occludens portion at higher magnification. **Main figure,** ×30,000; **inset,** ×46,000.

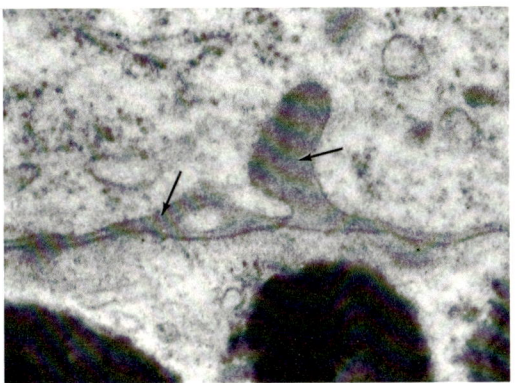

FIG. 10–43. Intercellular cement (*arrows*) lying between adjacent apical plasma membranes of non-pigmented and pigmented epithelial cells. ×46,280. (From Fine, B.S., and Zimmerman, L. E. *Invest Ophthal* 2:105, 1963.)

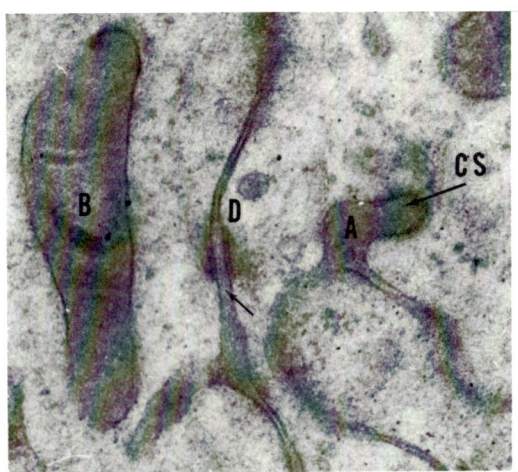

FIG. 10–44. Intercellular cement (*CS*) between apposing plasma membranes at *A*. The cement (*free arrow*) is continuous with that of a nearby desmosome (*D*) or fascia adherens. *B*, finger of intercellular cement protruding from another plane (see Fig. 10–43). ×59,600. (From Fine, B. S., and Zimmerman, L. E. *Invest Ophthal* 2:105, 1963.)

of organized granular endoplasmic reticulum (Fig. 10–49). Such quantities are lacking in all other mature ocular tissues. In addition, the apical Golgi complex often may be observed to be enlarged with a lucent material, probably an acid mucopolysaccharide, that appears morphologically identical with the material present outside the cell in the en-

FIG. 10–45. Typical desmosomes attaching adjacent nonpigmented epithelial cells to each other. ×30,000.

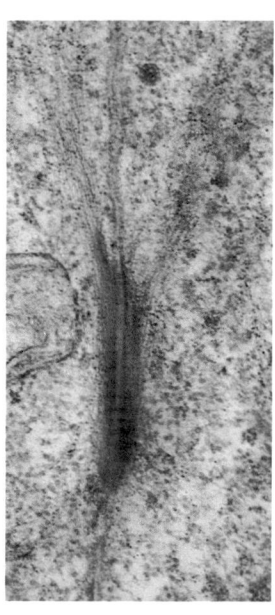

FIG. 10–46. Internal basement membrane of retina near ora serrata. **Inset.** Light micrograph showing density of vitreous base. Electron micrograph shows multilaminar basement membrane (*arrows*). Many empty-appearing intercellular spaces (i.e., not lined with basement membrane) are present in peripheral retinal tissue. **Main figure,** ×18,000; **inset,** Wilder reticulum stain, ×70 (AFIP Neg. 70–1425).

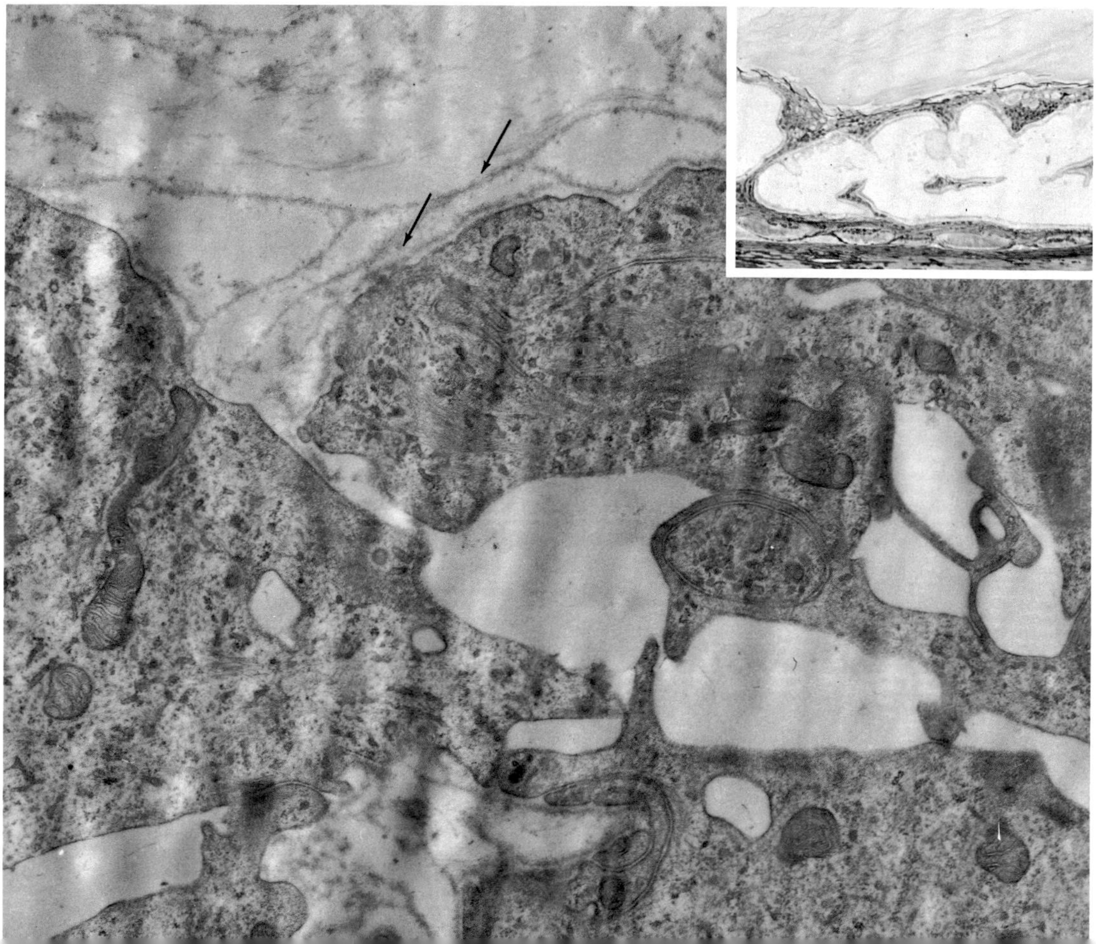

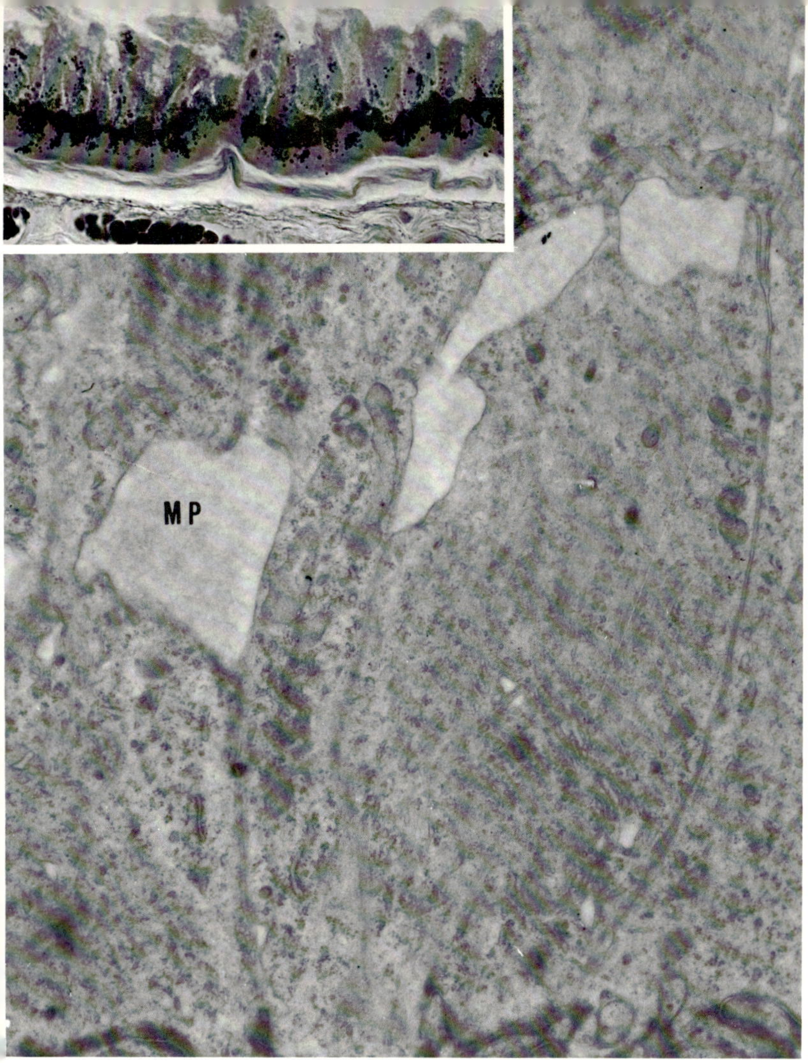

FIG. 10–47. Posterior pars plana. **Inset.** Light micrograph of columnar non-pigmented cells of posterior pars plana. Electron micrograph shows enlarged intercellular spaces filled with mucopolysaccharides (*MP*) between adjacent nonpigmented cells. Apical cell attachments to pigment epithelium are seen below. **Main figure,** ×14,000; **inset,** 1.5-μ section, PD, ×300 (AFIP Neg. 63–9615). (From Fine, B. S., and Zimmerman, L. E. *Invest Ophthal* 2:105, 1963.)

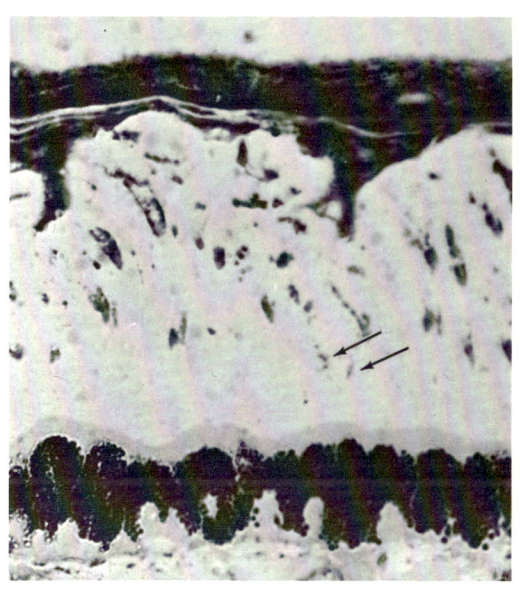

FIG. 10–48. Posterior pars plana of rhesus monkey eye stained for mucopolysaccharides. Blue-staining dots (*arrows*) are seen almost to apexes of nonpigmented cells. Blue-staining dots may be intra- or extracellular. 1.5-μ section, Hale mucopolysaccharide stain, ×840. (From Fine, B. S., and Zimmerman, L. E. *Invest Ophthal* 2:105, 1963.)

larged intercellular spaces (Fig. 10–50).

The acid mucopolysaccharide component of the vitreous body is synthesized mainly by the nonpigmented epithelial cells of the pars plana. It is secreted into the adjacent lateral intercellular spaces (canaliculi) from whence it makes its way *basally* (Fig. 10–51) to enter the vitreous body at its "base" of attachment.[4]

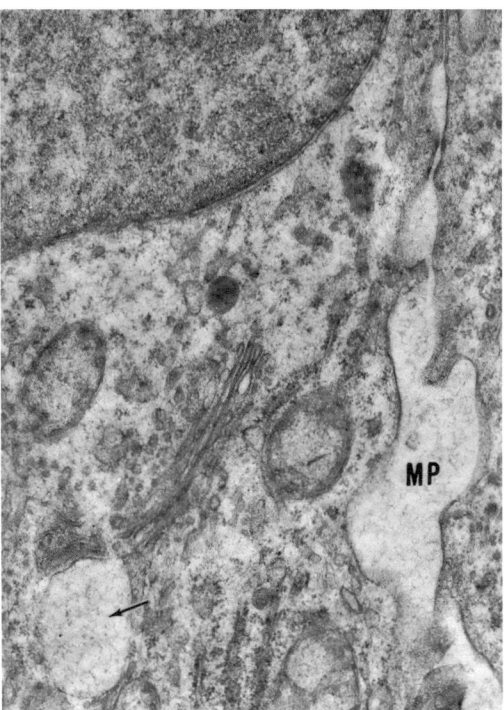

FIG. 10–50. Lucent material (*MP*) within intercellular space, morphologically similar to lucent material (*arrow*) within Golgi complex of adjacent nonpigmented epithelial cell. ×32,000. (From Fine, B. S., and Zimmerman, L. E. *Invest Ophthal* 2:105, 1963.)

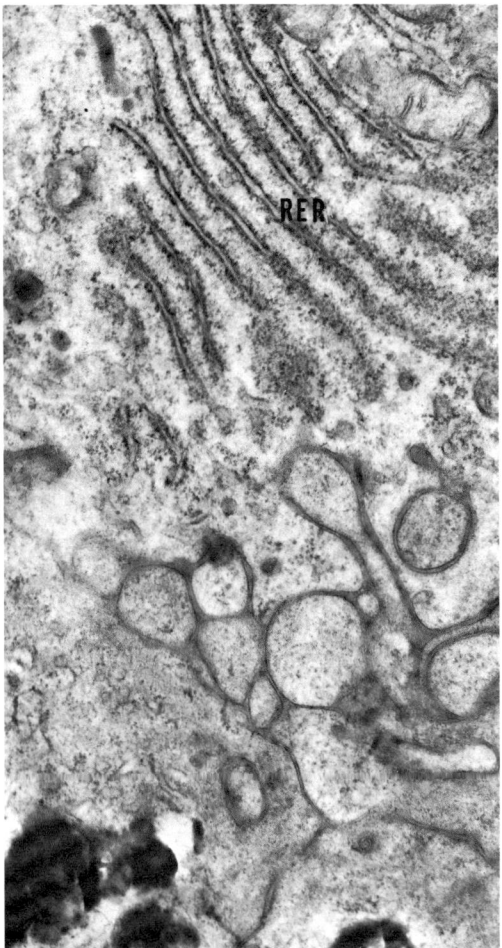

FIG. 10–49. Abundant granular endoplasmic reticulum (*RER*) in apical cytoplasm of posterior nonpigmented epithelium. ×24,900. (From Fine, B. S., and Zimmerman, L. E. *Invest Ophthal* 2:105, 1963.)

Flow of secreted materials toward the pigment epithelium is inhibited by the plane of strong attachment between the two cell layers. Flow toward the vitreous base is not obstructed because only desmosomes join adjacent nonpigmented cells to each other. The vitreous body content of acid mucopolysaccharide is therefore maintained or replaced in the adult eye from the nonpigmented cells of the posterior pars plana, the rate of formation probably being very slow normally.

In the anterior pars plana (Fig. 10–52) the nonpigmented epithelium is more cuboidal, whereas the mid pars plana is a transition zone of cuboidal to columnar shape. This transition region includes the attachment of the anterior face of the vitreous body. The multilaminar basement membrane of the cuboidal cells (Fig. 10–53) occupies the marked infoldings of their basilar surfaces. Issuing from many of the infoldings and attached to a thin basement membrane are the myriad collagenous filaments which in aggregate form the zonular fibers of the lens (Fig. 10–54).

It has long been known that the zonules are so strongly attached to the ciliary epithelium that a pull on them tears the epithelium. The firmness of this attachment is somewhat analogous to the attachment of hair to one's head; pulling out a single hair is relatively easy, whereas pulling out a handful is much more difficult. In addition, the basilar infoldings increase further the number of filament attachments per unit area of basilar surface, thus increasing the strength of adherence.

Zonular fibers in the anterior pars plana are considered vestigial, for few if any originate there. Fibers produced by similar cells found in the valleys of the pars plicata ("interciliary fibers," Figs. 10–55 through 10–58) are also considered vestigial.

That the major zonular fibers do not originate from the ciliary crests ("processes") but pass in the valleys between them (Figs. 10–59 through 10–61) is also supported by clinical observations. When the equator of a subluxated lens is examined biomicroscopically through a peripheral defect in the iris, the zonules can be observed to be taut, whereas the ciliary crests do not appear to be under any tension.[1] Further clinical evidence in support of the anatomic observations is derived from iridocyclectomies in which the zonules frequently remain untouched.

The nonpigmented epithelial cells of the pars plicata are cuboidal (Figs. 10–56 and 10–57).[13] The nonpigmented cells lying on the apexes of the ciliary crests are covered by a minimum of basement membrane (Fig. 10–62) Although the basilar surfaces of these cells on the ciliary crests appear quite smooth

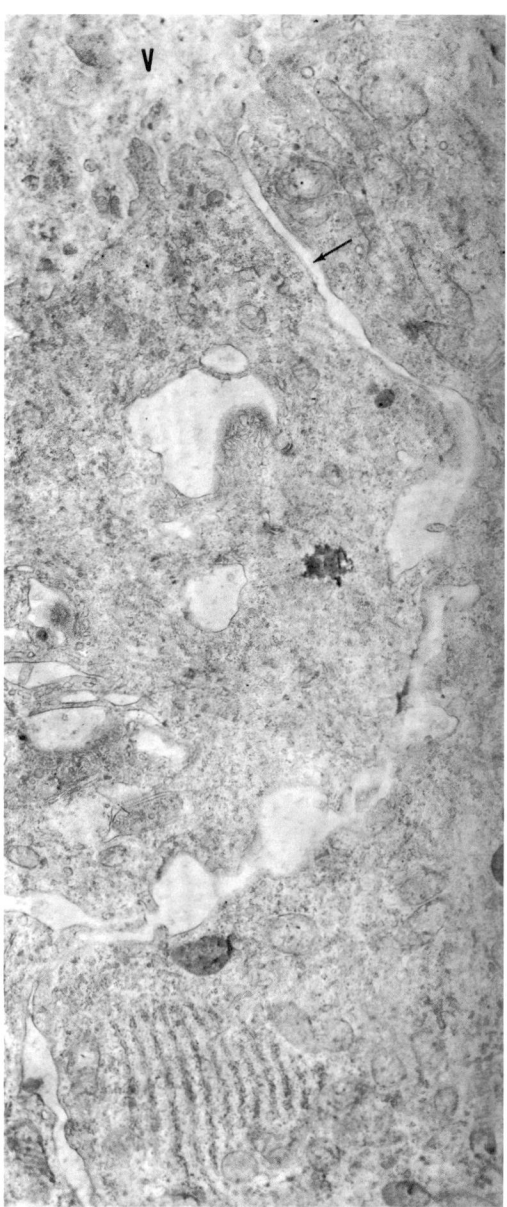

FIG. 10–51. Intercellular canaliculus (*arrow*) carrying mucopolysaccharides synthesized within apical cytoplasm of nonpigmented epithelial cells into base of vitreous body (V). ×12,480. (From Fine, B. S., and Zimmerman, L. E. *Invest Ophthal* 2:105, 1963.)

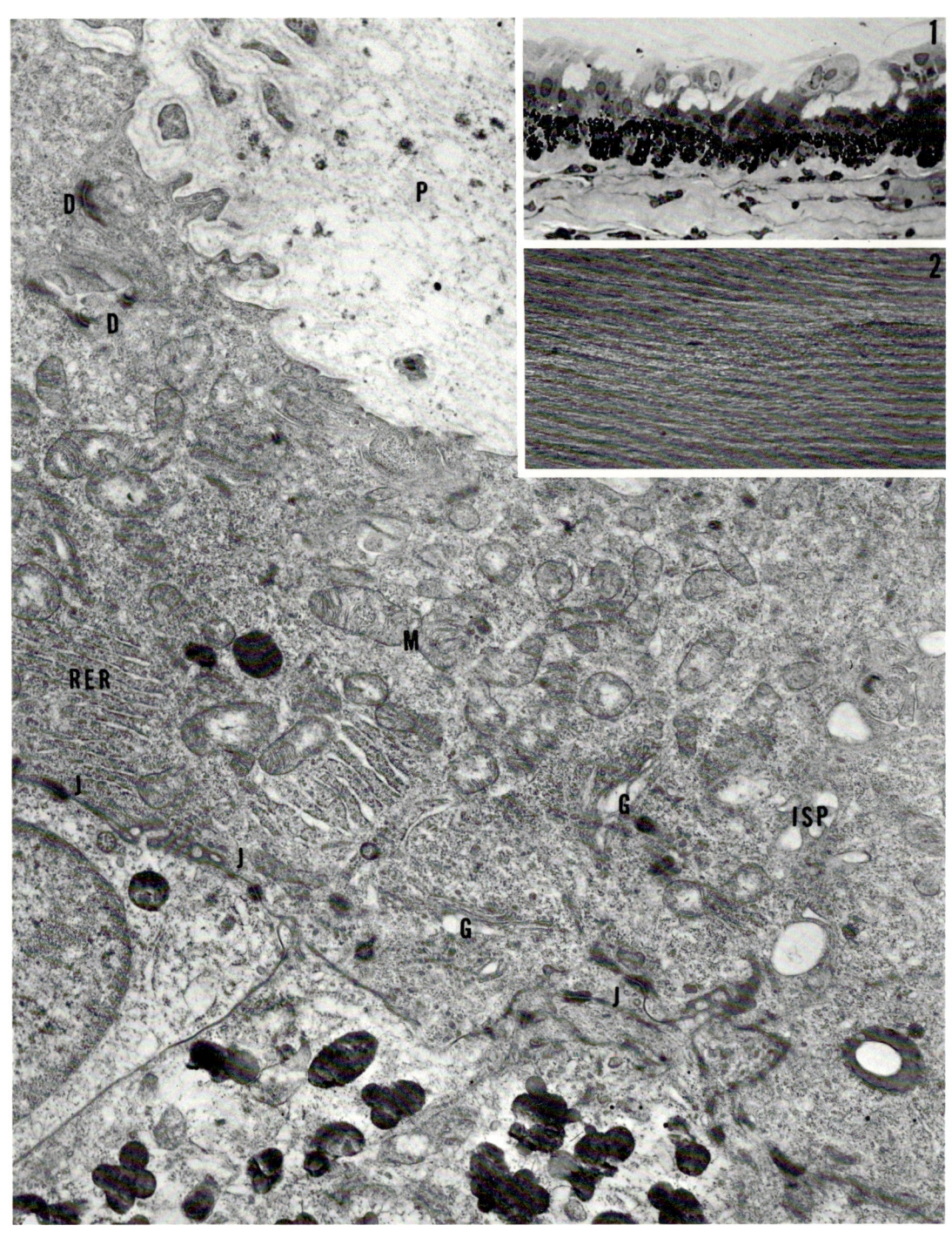

10–52. (*Opposite page.*) ...erior pars plana. **Inset 1.** Light ...ograph showing cuboidal ...earance of epithelium. ...**t 2.** Scanning electron micrograph ..."cleaned" inner surface showing ...cate meridional surface striations ...arently produced by "pocketing" ...ngement. **Main figure.** Electron ...ograph showing large "pocket" ...on basal surface of nonpigmented .... Large spaces are occupied ...xpanded, multilaminar basement ...nbrane, basement membrane ...erials, and vitreous filaments. ...nplex junctional regions (*J*) ...veen nonpigmented and pig-...ted epithelia are seen, as well as ...nosomes (*D*) between adjacent ...oigmented epithelial cells. ...ochondria (*M*) are abundant. ...nular endoplasmic reticulum ...) and a dilated Golgi complex ...are present in apical cytoplasm of ...oigmented epithelium. Dilated ...rcellular spaces (*ISP*) are ...ent. Pigment epithelial cells ...oined to one another by small ...nosomes. No enlarged inter-...lar spaces are present in pigment ...helial cell layer. **Main figure,** ...,960; **inset 1,** 1.5-μ section, ...5 (AFIP Neg. 60–976); **inset 2,** ... (Modified from Fine, B. S., and ...nerman, L. E. *Invest Ophthal* ...5, 1963.)

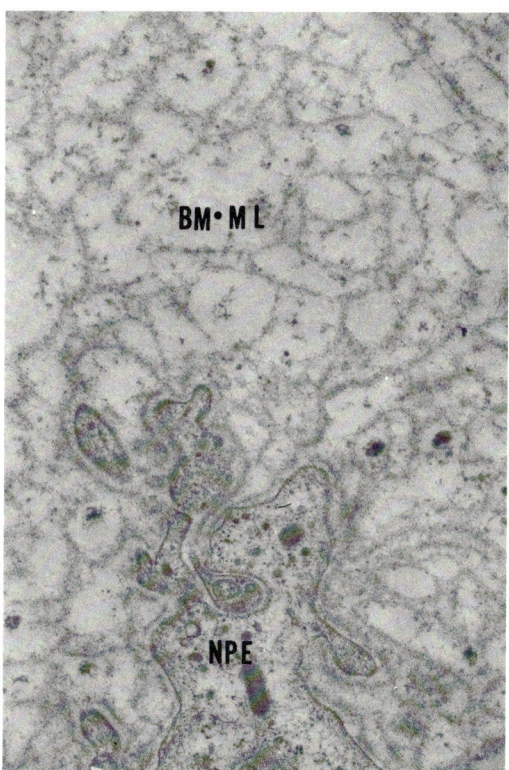

FIG. 10–53. Multilaminar basement membrane (*BM·ML*) characteristic of "internal limiting membrane" of nonpigmented ciliary epithelium (*NPE*). ×19,800.

by light microscopy, by electron microscopy, these surfaces show multiple deep infoldings many of which are formed by interdigitation of myriad villous-like projections from adjacent epithelial cells (Fig.. 10–62).[14,15] The white color of these apical regions (Fig. 10–22) is partly due to the high content of glycogen of the nonpigmented cells as well as to the lesser pigmentation of the underlying pigment cell layer.

Focal enlargement of adjacent intercellular spaces between the nonpigmented cells is also present in the pars plicata, but since the content of the spaces in this location is generally less well visualized than in the pars plana, the former is considered to be more watery.

On the anterior declivity of the ciliary crests (Fig. 10–63) the nonpigmented cells become more and more pigmented until they are completely altered into a layer of pigment epithelium *before* reaching the iris root.

Although the inner layer of ciliary epithelium is called nonpigmented, in the adult eye it may, in fact, contain a considerable amount of pigment, both lipofuscin and melanin. On occasion the granules may be so numerous in the apical cytoplasm that the nonpigmented cells of the pars plana appear markedly pigmented. Presumably, therefore, these cells retain some ability to synthesize melanin even in adult life.

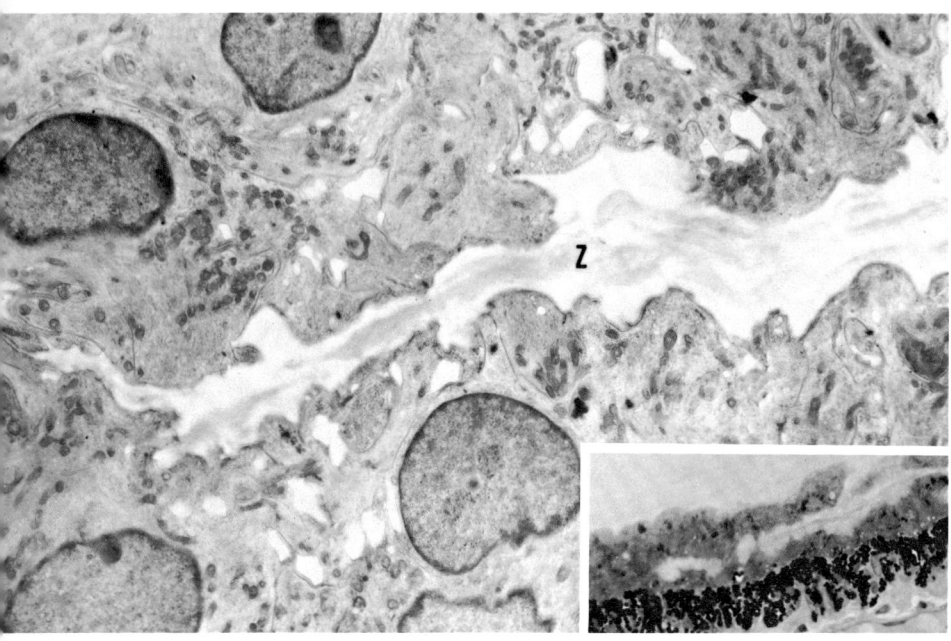

FIG. 10–54. Zonular fibers arising in basilar infoldings of nonpigmented epithelial cells. **Inset.** Light micrograph. **Main figure.** Electron micrograph showing thin basement membrane present on cell surface. Myriad filaments attached to basement membrane aggregate to form zonular fiber (Z). **Main figure,** ×5,500; **inset,** 1.5-μ section, ×305 (AFIP Neg. 60–972). (From Fine, B. S., and Zimmerman, L. E. *Invest Ophthal* 2:105, 1963.)

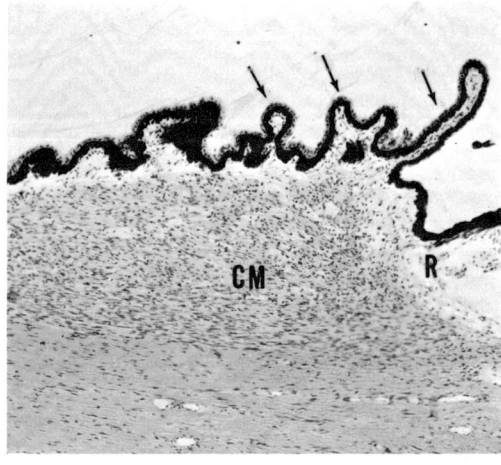

FIG. 10–55. Meridional section of pars plicata of infant showing ciliary "processes" (*arrows*) on a ciliary crest. Large ciliary muscles (*CM*) thicken greatly this portion of uveal tract. Pars plicata extends from root of iris (*R*) to posterior extremity of ciliary crests (see Fig. 10–22). H&E, ×55. (AFIP Neg. 65–234.)

THE UVEAL TRACT   203

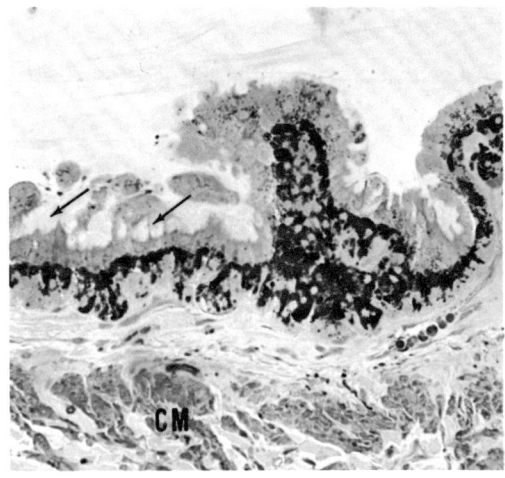

FIG. 10–56. Meridional section showing transition from anterior pars plana onto posterior edge of ciliary crest. "Motheaten" cuboidal cells characteristic of anterior pars plana (*arrows*) diminish in relative numbers, presenting anteriorly mostly in ciliary valleys (see Fig. 10–57). CM, ciliary muscle. 1.5-μ section, PD, ×165. (AFIP Neg. 66–1289.)

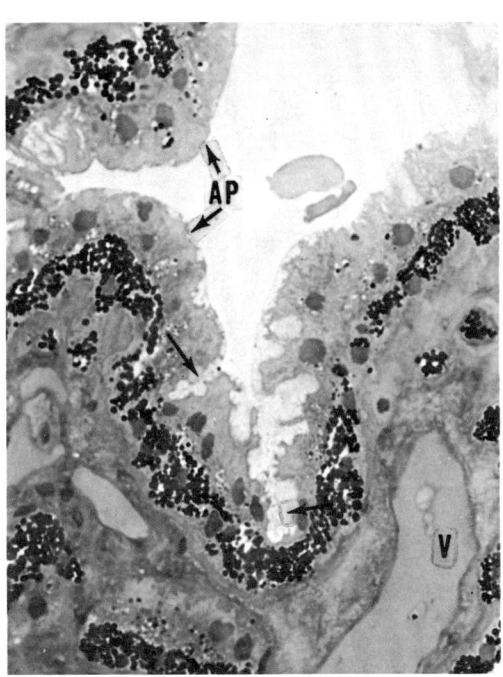

FIG. 10–57. Coronal section of pars plicata. "Motheaten" cells characteristic of anterior pars plana nonpigmented epithelial cells are present in ciliary valleys. Vestigial zonules (*free arrows*) are present. Nonpigmented epithelial cells at apex of crests (AP) possess smoother basal surface. V, blood vessel in connective tissue core. 1.5-μ section, H&E, ×530. (AFIP Neg. 60–5331. From Fine, B. S., and Zimmerman, L. E. *Invest Ophthal* 2:105, 1963.)

Pigment Epithelium and External Basement Membrane. This layer of pigmented cells is continuous with the retinal pigment epithelium at the ora serrata where the cell groups, as seen in section, form evaginations (Fig. 10–64) to give a thickened appearance to the layer. Grossly, this accounts for the heavily pigmented zone that parallels the ora serrata (see Fig. 6–85).

As in other epithelia, the melanin granules are located mostly apically and the basal plasma membranes are lined by a basement membrane (*external basement membrane, cuticular lamella,* or *outer glass membrane*

FIG. 10–58. Ciliary processes. Their vascular collagenous core is covered by a double layer of epithelium, nonpigmented and pigmented. A small cyst (C) of nonpigmented epithelium is present at right and two large granules of unknown composition (*arrows*) are present in nonpigmented epithelial layer at left. 1.5-μ section, PD, ×70. (AFIP Neg. 63–6909.)

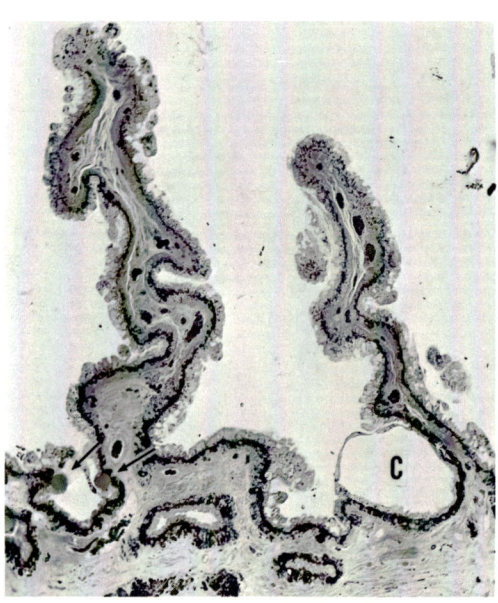

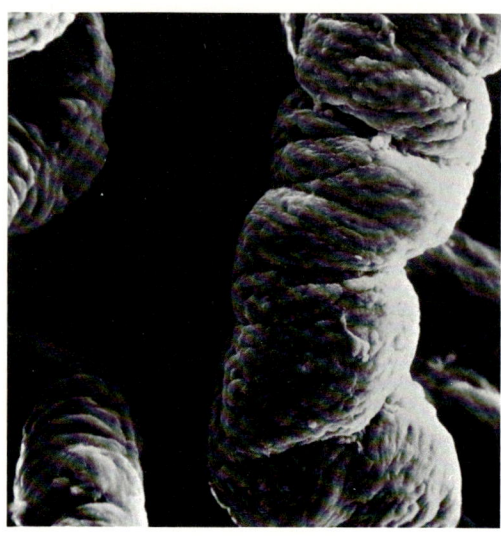

FIG. 10–60. Free surface of single ciliary crest as seen by scanning electron microscopy. Ciliary valley between crests is at right. ×300.

FIG. 10–59. Coronal section showing ciliary crests. Anterior border layer of vitreous body can be seen (*arrows*). H&E, ×40. (AFIP Neg. 70–7287.)

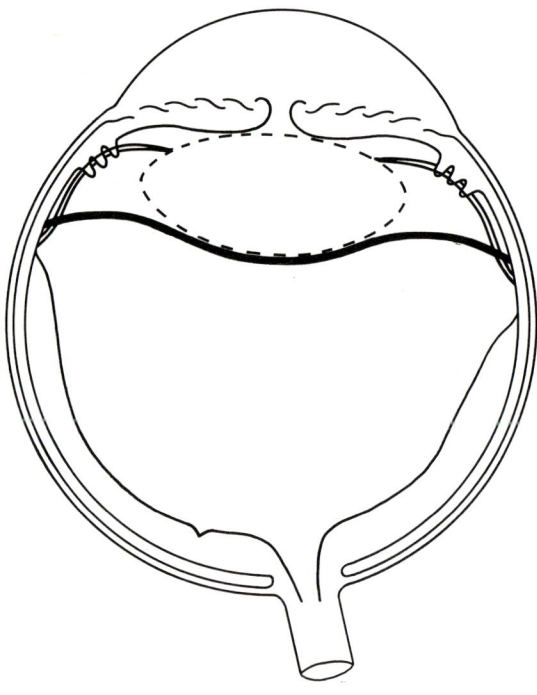

FIG. 10–61. Schematic representation of relation of orbiculoanterior zonular fibers to ciliary crests. Anterior border layer of vitreous body limiting posterior aqueous chamber posteriorly is indicated by *thick black line.*

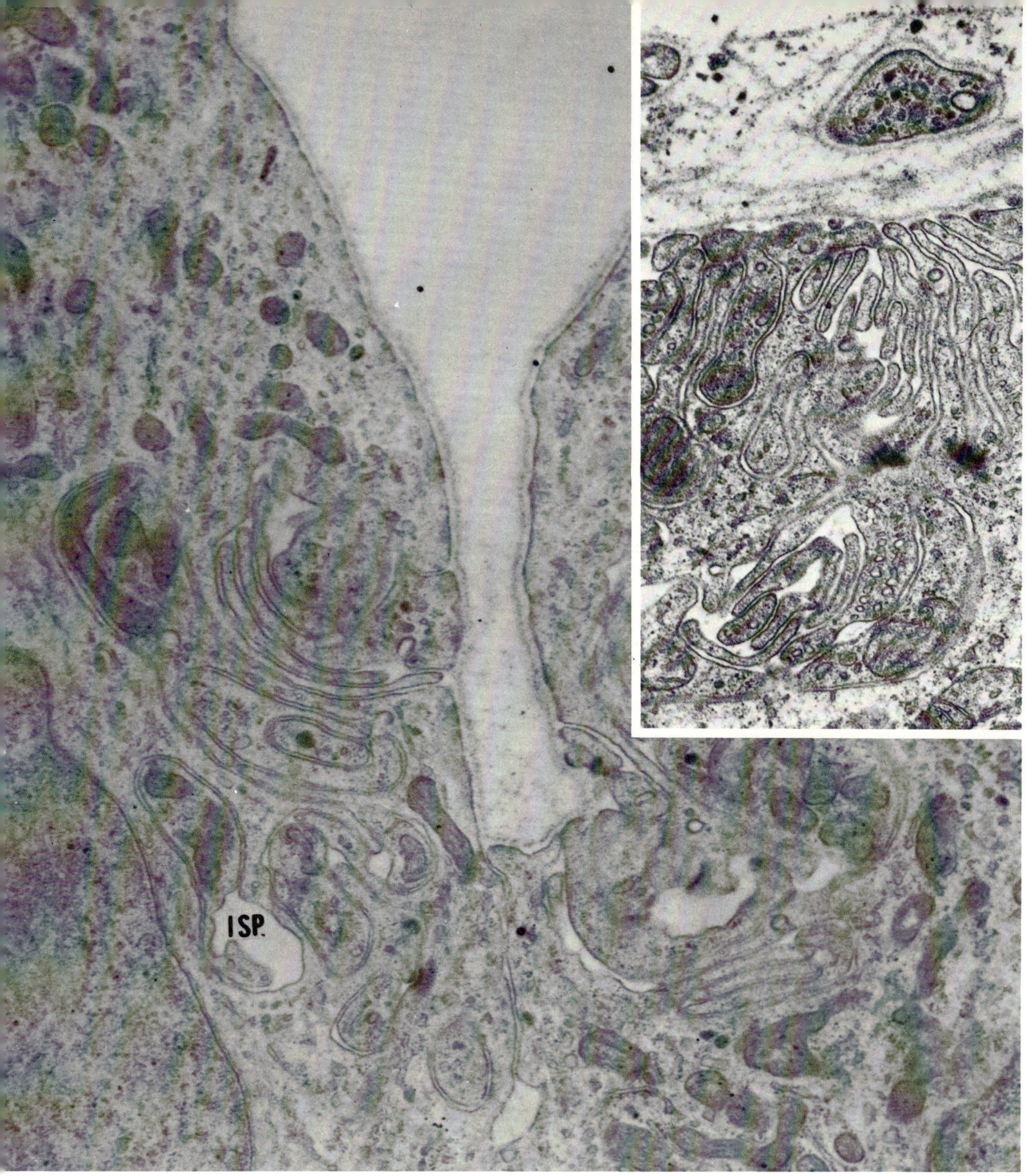

FIG. 10–62. Cuboidal nonpigmented epithelium present on free surface of ciliary crests. Basilar surfaces are lined by thin basement membrane. Infoldings are common along this basilar surface. Enlarged intercellular spaces (*ISP*), apparently empty, are present. **Inset** shows another region in which basement membrane is multilaminar or "expanded." A few villi are embedded in this basement membrane. Tortuous basilar infoldings are in part due to interdigitation of basilar infoldings of adjacent cells. **Main figure,** ×25,000; **inset,** ×16,500. (From Hogan, M., and Zimmerman, L. E. *Ophthalmic Pathology: An Atlas and Textbook.* Philadelphia, Saunders, 1962.)

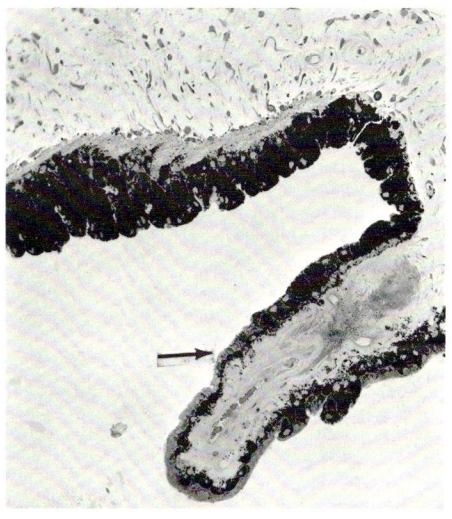

FIG. 10–63. Anterior declivity of ciliary crest. Transition of non-pigmented epithelium to fully pigmented epithelium occurs here (*arrow*). 1.5-μ section, PD, ×115. (AFIP Neg. 65–6684.)

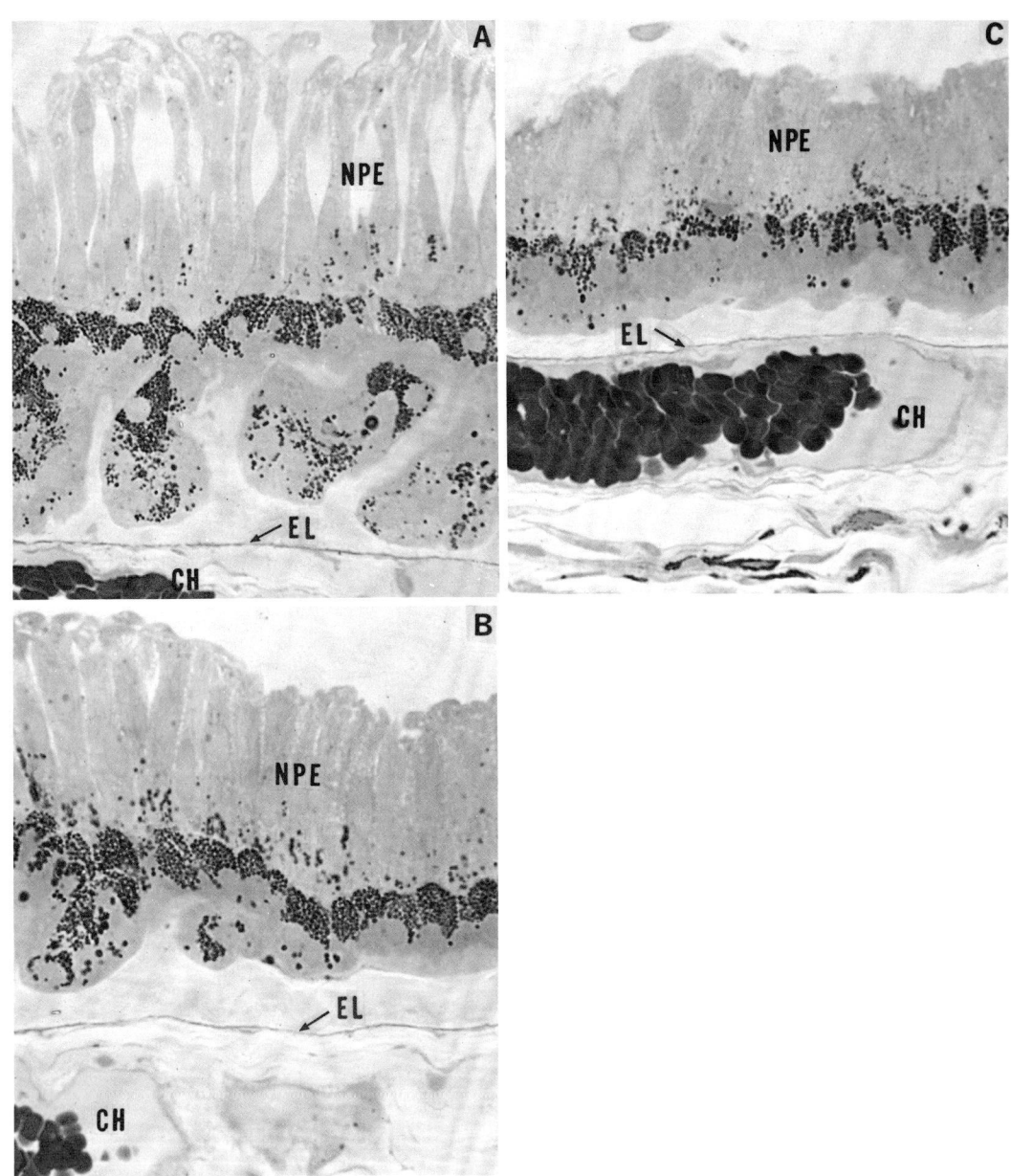

of the pars ciliaris retinae; Fig. 10–65 through 10–68).

Anteriorly in the region of the pars plicata, the pigment epithelial cells become more cuboidal than they are posteriorly. They are less pigmented on the apexes of the ciliary crests where they rest upon the external basement membrane of the ciliary epithelium. This membrane becomes thicker in the anterior pars plana (Fig. 10–66, 67) and on the sides of the ciliary crests (Fig. 10–68A). It becomes so folded that on surface view of a suitable preparation of the pars plana, the appearance is that of a raised network or reticulum (reticulum of Heinrich-Müller, Fig. 10–67).[12]

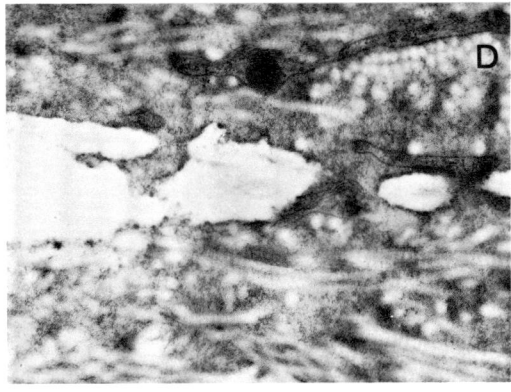
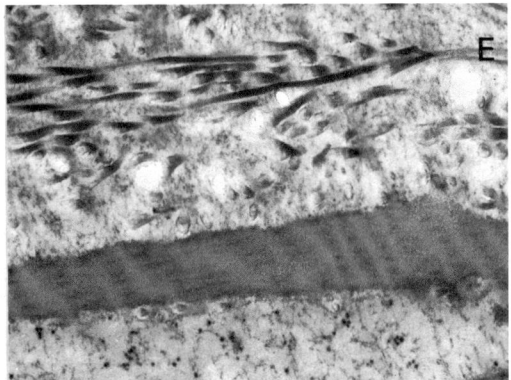
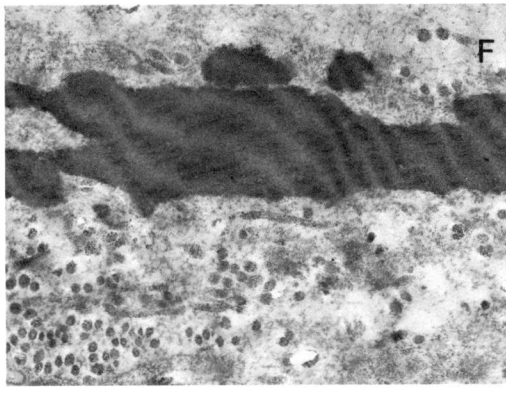

FIG. 10–64. Pigment epithelium in eye of 40-year-old. Near ora serrata pigment epithelium forms evaginations (**A**) which terminate rather abruptly (**B**) in line which roughly parallels ora serrata. In mid and anterior pars plana pigment epithelium is a more uniform single layer (**C**). Note comparable transition in height of nonpigmented epithelium (*NPE*) from tall columnar to short columnar to more cuboidal cells. *CH,* choriocapillaris; *EL,* elastic lamina. Prominent elastic lamina of the pars plana appears "negatively stained" (**D**) when section is treated only with uranyl acetate and lead. With TSC treatment alone both the elastic lamina and adjacent collagen fibrils become dense (**E**). With uranyl acetate, lead and Ag-TPPS (Chap. 1) the elastic tissue is densely stained with greater specificity (**F**). **A,** 1.5-μ section, PD, ×485 (AFIP Neg. 70–8858); **B,** 1.5-μ section, PD, ×485 (AFIP Neg. 70–8859); **C,** 1.5-μ section PD, ×485 (AFIP Neg. 70–8857); **D, E, F,** all ×21,000.

Uveal (Mesodermal) Portion

As elsewhere, the external basement membrane of the pigment epithelium is associated with collagen fibrils. In this region the fibrils are of considerable diameter (i.e., thick). A layer of vessels, continuous in a plane with those of the choriocapillaris, is separated from the pigment epithelium by a basement membrane and collagen fibrils (Fig. 10–65). Between these two aggregates may be found denser collagenous tissue or long patches of homogeneous elastic tissue, the elastic lamina. Anteriorly the vascular tissue is folded, forming the connective tissue cores of the ciliary crests ("processes"). The capillaries of the

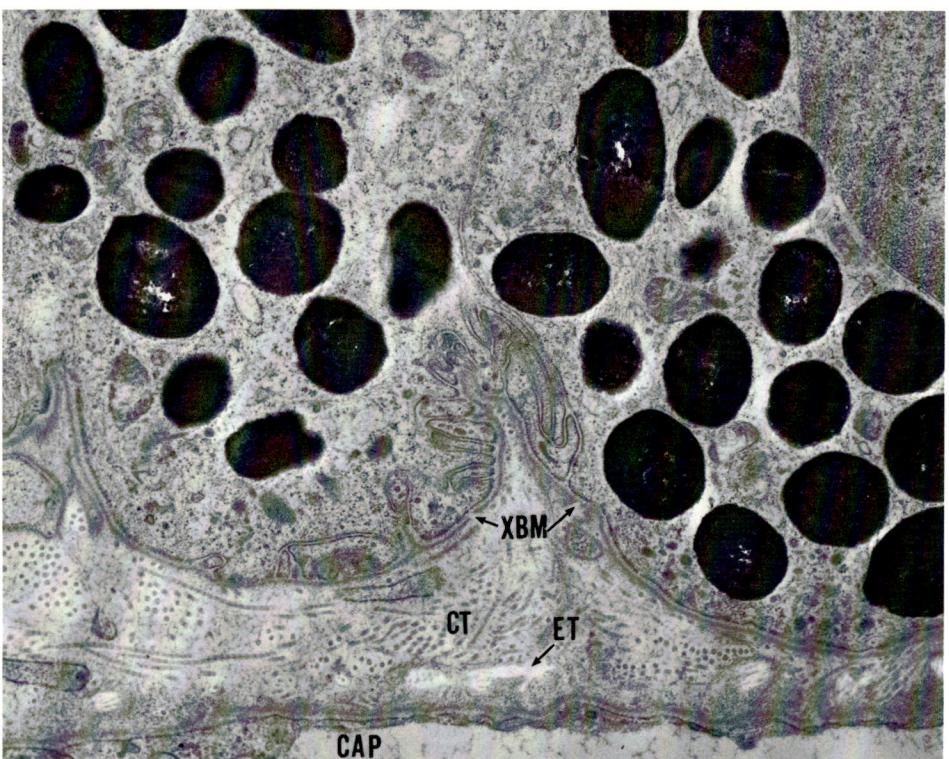

FIG. 10–65. Pigment epithelium of anterior pars plana from eye of 10-month-old infant. External basement membrane (*XBM*) is thin but markedly folded. Thin basement membrane is separated from similar thin basement membrane of adjacent fenestrated capillary layer (*CAP*) by collagenous connective tissue (*CT*) in which small patches of homogeneous-appearing elastic tissue (*ET*) are present. ×16,500.

thelium is a characteristic histologic feature of the eye in many patients with long-standing diabetes mellitus.[21]

Continuous with the choroid proper and outside the plane of the ciliary crests lies the large mass of smooth muscle, the ciliary muscle (Fig. 10–70), that generally is subdivided into three parts: the meridional or longitudinal muscle externally; the circular, round, or radial muscle internally; and a variable oblique muscle in between.

The smooth muscle cells are typical[7] of other smooth muscles (Fig. 10-71), being enveloped by a continuous basement membrane (except for regions of apposition with another muscle cell or nerve ending) and having myriad surface plasma membrane infoldings (micropinocytotic vesicles or caveolae), surface densities, cytoplasmic filaments, and scattered densities among the cytoplasmic filaments.

When the criteria for a cell type are sufficiently numerous, as here for smooth muscle cells, they may be useful in diagnosing a tumor even if the material has been preserved inadequately.[8]

ciliary crests are large, and many possess *fenestrated* endothelia similar to those of the choriocapillaris (Figs. 10–68B and 10–69).

In contrast with the situation in the pars plana (Fig. 10–64), the capillaries of the ciliary crests are not separated by a prominent elastic tissue layer from the very thick, more conspicuously multilaminar basement membrane (Fig. 10–68A) of the adjacent pigmented ciliary epithelium.

Excessive thickening of this normally thick external basement membrane of the ciliary epi-

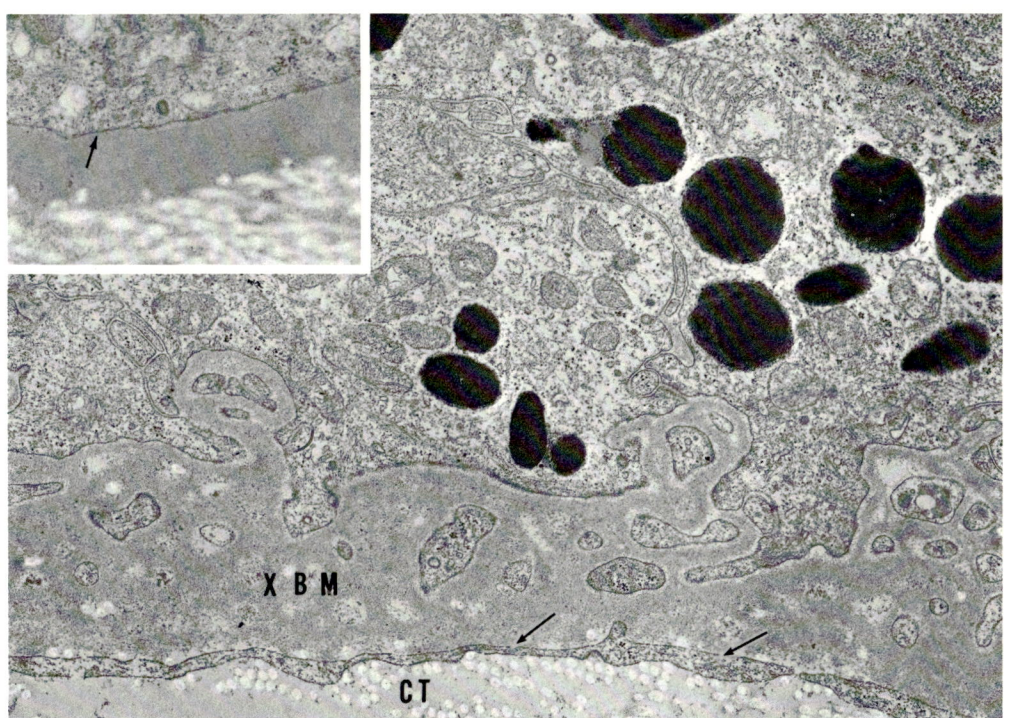

FIG. 10–66. Pigment epithelium from eye of 18-year-old. External basement membrane (*XBM*) is markedly thickened. A layer of elongated nonpigmented uveal cells (*arrows*) lines outer surface of this thick basement membrane. Collagenous connective tissue (*CT*) lies outside layer of nonpigmented uveal cells. **Inset** shows close apposition (*arrow*) of thick basement membrane to cell plasmalemma. **Main figure,** ×11,000; **inset,** ×15,000.

The muscle cells are separated by delicate collagenous fibrils and spaces presumably occupied by watery intercellular mucinous materials. The muscles are liberally provided with arterial supplies from the long posterior ciliary and anterior ciliary arteries, whereas their drainage is mainly posteriorly via the small vessels of the inner uveal layers and externally (i.e., laterally) via the anterior ciliary veins through the sclera. These latter veins also participate in drainage from the canal of Schlemm.

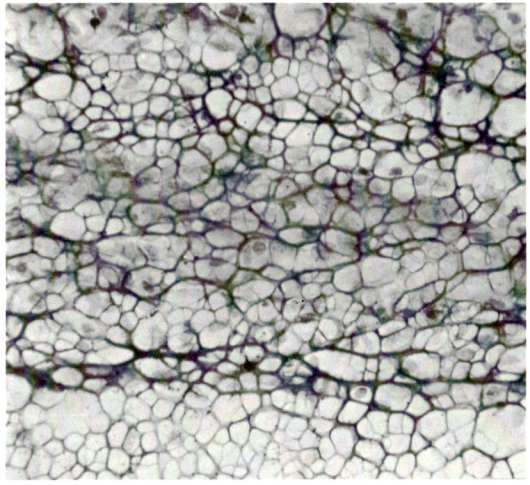

FIG. 10–67. Flat preparation of ciliary body external basement membrane (anterior pars plana) showing its reticular topography (reticulum of Heinrich-Müller). PAS, ×115. (Courtesy of Dr. G. de Venecia.)

FIG. 10–68. A. Prominent multilaminar basement membrane (*M·BM*) of the pigmented ciliary epithelium of the pars plicata in a 40-year-old man. ×12,300. **B.** Fenestrated endothelium (*arrows*) of vessel in collagenous "core" of ciliary crest. *RBC*, red blood cell in lumen. ×18,000.

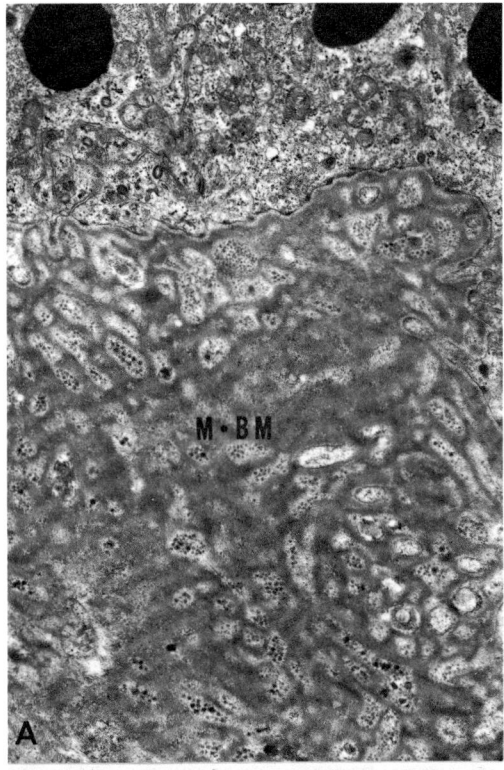

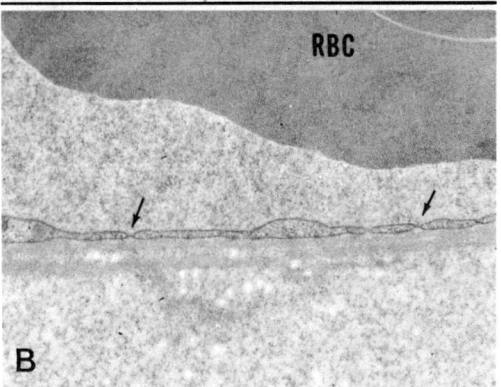

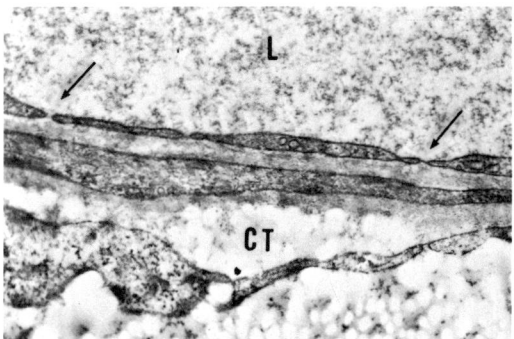

FIG. 10–69. Fenestrated (*arrows*) innermost capillary of pars plana. *CT*, connective tissue; *L*, lumen of vessel. ×16,500.

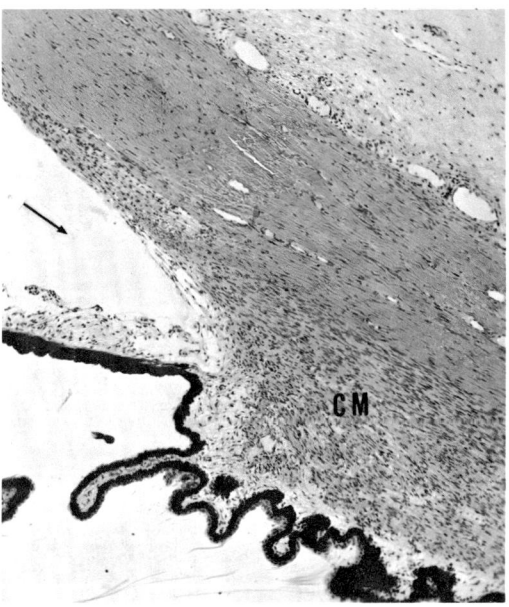

FIG. 10–70. Marked thickening of uveal tract in region of pars plicata due to ciliary muscle (*CM*). *Arrow* indicates drainage angle of anterior chamber. H&E, ×55. (AFIP Neg. 65–233.)

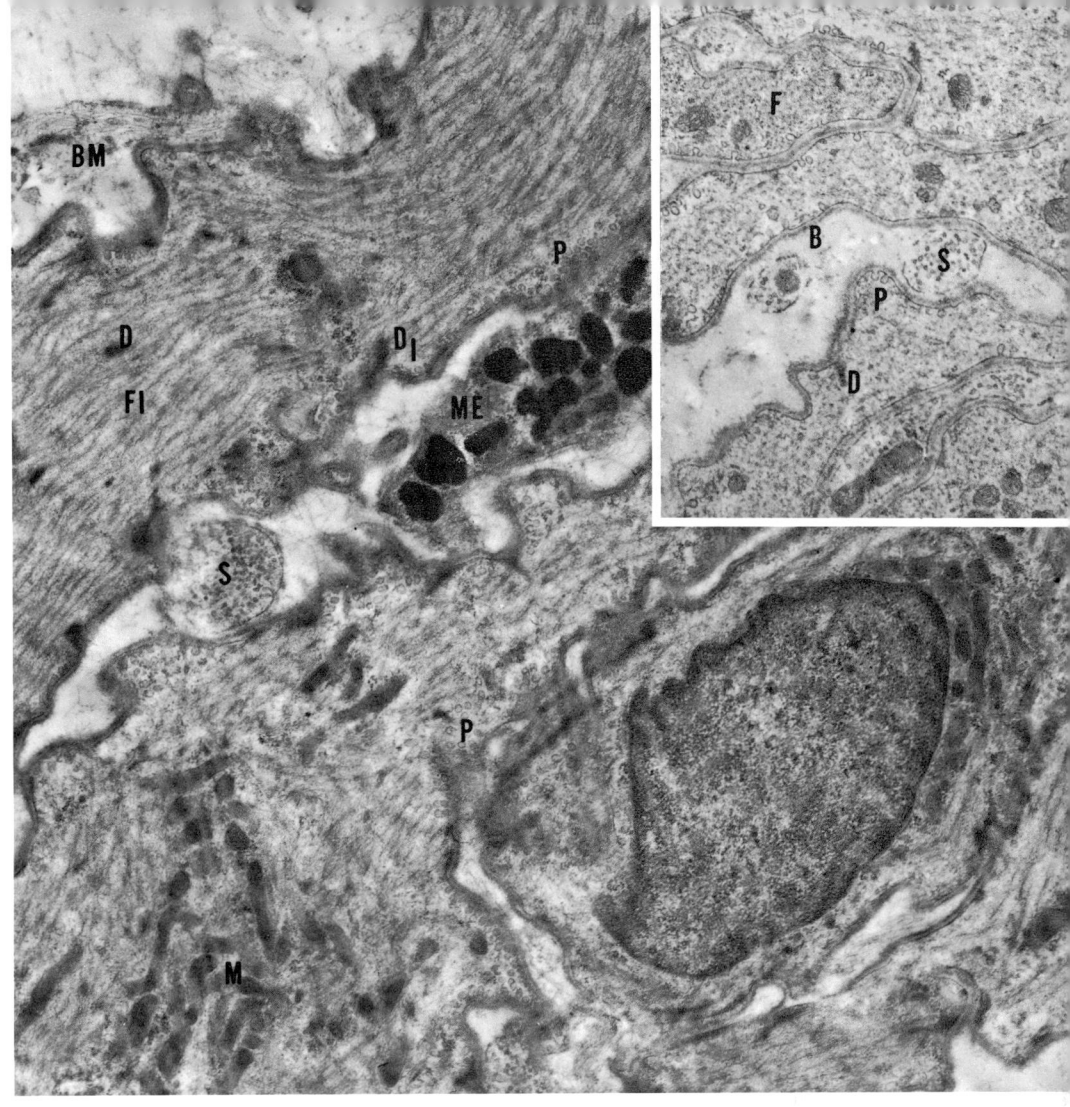

FIG. 10–71. Smooth muscle cells of ciliary muscle. *BM*, basement membrane; *P*, caveolae or micropinocytotic vesicles; *FI*, intracellular filaments; *D*, densities among the filaments or (D_1) attached to cell plasma membranes; *M*, mitochondria. Circular profile containing myriad vesicles represents a nerve near its synaptic ending (*S*). Melanocyte (*ME*) is typical of uveal melanocytes. **Inset** shows smooth muscle cells in cross-section. Their basement membrane (*B*), filaments in cross-section (*F*), densities (*D*), vesicles (*P*), and nearby nerve endings (*S*) are clearly seen. Both ×16,500.

Lying within the loose uveal tissue anterior to the circular or oblique muscles and close to the root of the iris is a circumferential artery, the *major arterial circle*. This circle is formed from the terminal bifurcations of the two long posterior ciliary arteries which also anastomose with branches from the anterior ciliary arteries. The iris and ciliary crests are supplied from this arterial circle.

Venous drainage from the uveal tract is somewhat unusual in that it does not accompany the arterial supply, but rather leaves the globe posteriorly as a separate and distinct system of four vortex veins, each located within one of the quadrants (see Fig. 5–5, on Color Plate I).

REFERENCES

1. Berliner, M. L. *Biomicroscopy of the Eye.* New York, Hoeber, 1949.
2. Butler, T. H. *An Illustrated Guide to the Slit-Lamp.* London, Oxford University Press, 1927.
3. Fine, B. S. Free-floating pigmented cyst in the anterior chamber. *Amer J Ophthal 67:*493, 1969.
4. Fine, B. S., and Zimmerman, L. E. Light and electron microscopic observations on the ciliary epithelium in man and rhesus monkey, with particular reference to the base of the vitreous body. *Invest Opthal 2:*105, 1963.
5. Garron, L. K. The ultrastructure of the retinal pigment epithelium, with observations on the choriocapillaris and Bruch's membrane. *Trans Amer Opthal Soc 61:*545, 1963.
6. Hogan, M., and Zimmerman, L. E. *Ophthalmic Pathology: An Atlas and Textbook.* Philadelphia, Saunders, 1962.
7. Ishikawa, T. Fine structure of the human ciliary muscle. *Invest Ophthal 1:*587, 1962.
8. Meyer, S. L., Fine, B. S., Font, R. L., and Zimmerman, L. E. Leiomyoma of the ciliary body: Electron microscopic verification. *Amer J Ophthal 66:*1061, 1968.
9. Nakaizuma, Y. The ultrastructure of Bruch's membrane. I. Human, monkey, rabbit, guinea pig, and rat eyes. *Arch Ophthal (Chicago) 72:*380, 1964.
10. Nakaizuma, Y., Hogan, M. J., and Feeney, L. The ultrastructure of Bruch's membrane. III. The macular area of the human eye. *Arch Ophthal (Chicago) 72:*395, 1964.
11. Richardson, K. C. The fine structure of the albino rabbit iris, with special reference to the identification of adrenergic and cholinergic nerves and nerve endings in its intrinsic muscles. *Amer J Anat 114:*173, 1964.
12. Salzmann, M. *The Anatomy and Histology of the Human Eyeball in the Normal State.* Tr. by E. V. L. Brown. Chicago, University of Chicago Press, 1912.
13. Smelser, G. K. Electron microscopy of a typical epithelial cell and of the normal human ciliary process. *Trans Amer Acad Ophthal Otolaryng 70:*738, 1966.
14. Tormey, J. McD. Relationship Between the Structure of the Ciliary Epithelium and the Secretion of Aqueous Humor. In Rohen, J. W. (ed.). *The Structure of the Eye.* Stuttgart, Schattauer, 1965, p. 237.
15. Tormey, J. McD. The ciliary epithelium: An attempt to correlate structure and function. *Trans Amer Acad Ophthal Otolaryng 70:*755, 1966.
16. Tousimis, A. J., and Fine, B. S. Ultrastructure of the iris: An electron microscopic study. *Amer J Ophthal 48:*397, 1959.
17. Tousimis, A. J., and Fine, B. S. Ultrastructure of the iris: The intercellular stromal components. *Arch Ophthal (Chicago) 62:*974, 1077, 1959.
18. Tousimis, A. J., and Fine, B. S. Electron Microscopy of the Pigment Epithelium of the Iris. In Smelser, G. K. (ed.). *The Structure of the Eye.* New York, Academic Press, 1961, p. 441.
19. Verhoeff, F. H., and Sisson, R. J. Basophilic staining of Bruch's membrane. *Arch Ophthal (Chicago) 55:*125, 1926.
20. Wobmann, P. R., and Fine, B. S. The Clump Cells of Koganei: A Light and Electron Microscopic Study. Paper presented at meeting of the Eastern Section, Association for Research in Ophthalmology, Philadelphia, 1970.
21. Yamashita, T., and Becker, B. The basement membrane in the human diabetic eye. *Diabetes 10:*167, 1961.
22. Yanoff, M., and Zimmerman, L. E. Pseudomelanoma of anterior chamber caused by implantation of iris pigment epithelium. *Arch Ophthal (Chicago) 74:*30, 1965.

chapter 11

The Anterior Chamber Angle

The Aqueous compartment
The Aqueous
The Limbus
The Drainage angle
The Trabecular meshwork

THE AQUEOUS COMPARTMENT

The opened eye (Fig. 11–1) clearly separates into two unequal *compartments:* The smaller or *aqueous compartment* lies anteriorly in front of the lens and the larger or *vitreous compartment* lies posteriorly behind the lens (Fig. 11–2 on Color Plate V).

The anterior (aqueous) compartment is further subdivided by the iris diaphragm into two *chambers:* the anterior and the posterior (Fig. 11–3). The anterior chamber is bounded by the anterior surface of the iris and the pupil posteriorly, by the corneal endothelium anteriorly, and by the arciform boundary of the drainage angle laterally.

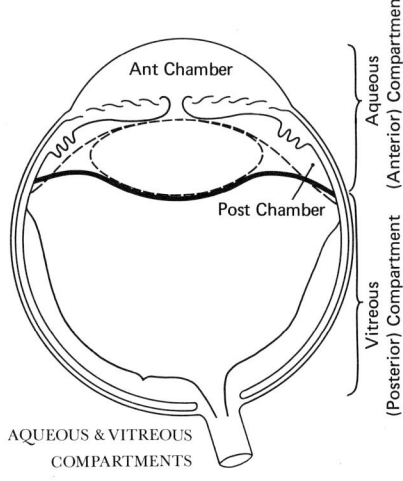

FIG. 11–2. Aqueous and vitreous compartments. Compartments are separated by anterior face of vitreous body (*heavy black line*) which also demarcates posterior boundary of posterior (aqueous) chamber.

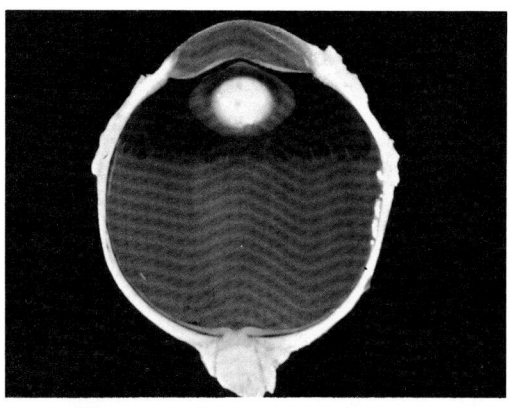

FIG. 11–1. An opened, fixed globe. Temporal part of globe has been removed by cut running through optic nerve and just temporal to pupil.

FIG. 11–3. Chambers of aqueous compartment. Aqueous compartment is subdivided into two *chambers* by iris diaphragm. Anterior chamber in front of diaphragm is indicated in *stippled black;* the posterior chamber behind iris diaphragm is indicated in *solid black.*

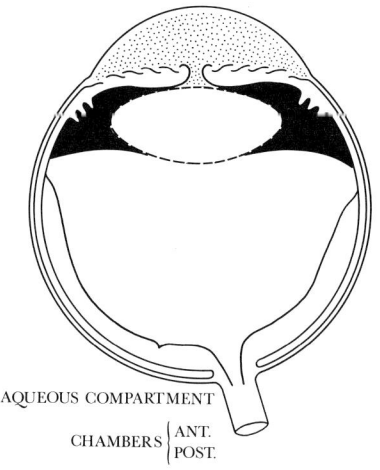

The angle of the anterior chamber, or drainage angle, is formed where the corneoscleral and uveal coats of the eye unite (Fig. 11–4, on Color Plate V). The angle consists of a circumferential furrow or groove in the mesodermal tissue at the periphery of the anterior chamber. In meridional section, therefore, this iridocorneoscleral *angle,* sometimes called the *angular sinus,* appears as a concavity lying on the "face" of the ciliary body

Color Plate V

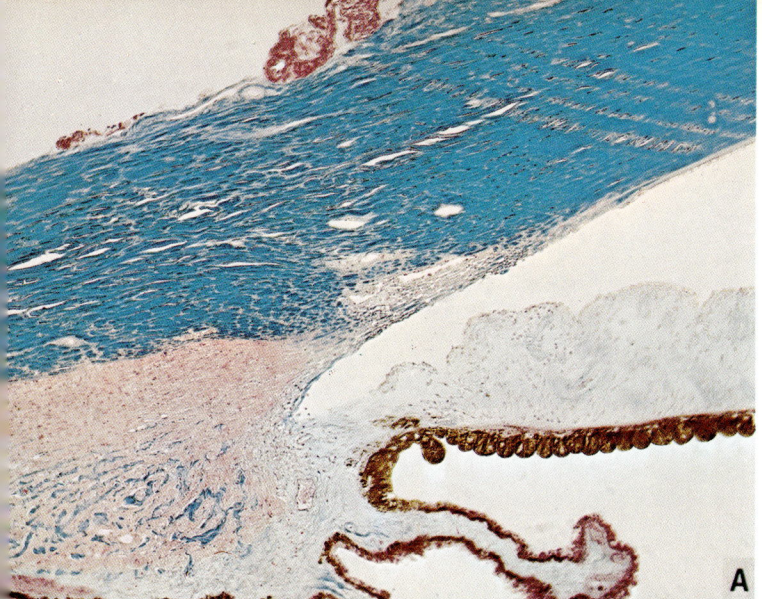

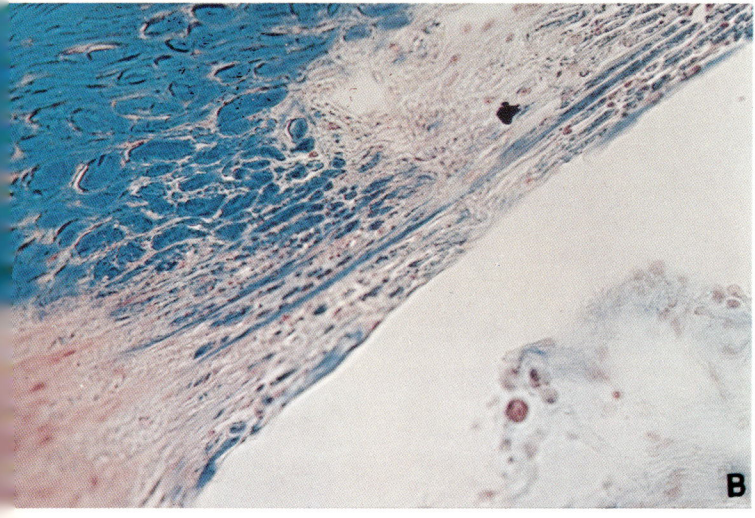

FIG. 11–4. The drainage angle. **A.** Angle is formed by union of corneoscleral and uveal coats at periphery of anterior chamber. Trabecular meshwork fills inner scleral sulcus (Fig. 11–6), posterior boundary of which is scleral roll. **B.** Higher magnification of region of adult scleral *spur* showing its bipartite composition of circumferentially arranged bundles of collagen from scleral roll and more meridionally arranged bundles of collagen from adjacent trabecular (uveal) meshwork. **C.** Rhesus monkey drainage angle. Most of trabecular meshwork extends from deep corneal periphery to become continuous with uveal tract. A small outer portion (corneoscleral) blends with scleral roll. Compare drainage angle of infant, Figure 11–24. All Masson; **A,** ×60 (AFIP Acc. 84029); **B,** ×300; **C,** ×80 (AFIP Neg. 60–4865). (From Fine, B. S. *Invest Ophthal* 3:609, 1964.)

whose blunt end faces the anterior chamber. Histologically, this concavity or arc begins where Descemet's membrane (Schwalbe's line or ring clinically) ends, and it terminates at the iris root where the anterior border layer of the iris begins abruptly.

THE AQUEOUS

The aqueous appears to be actively secreted mostly by the nonpigmented ciliary epithelium that lines the posterior chamber. It flows continuously from the posterior to the anterior aqueous chamber via the pupil, and within the anterior chamber it flows in at least two directions (Fig. 11–5): first, from the pupillary entrance toward the drainage angle in the chamber periphery; and second, from where it rises as it becomes warmed by the iris posteriorly to where it falls anteriorly as it is cooled by the cornea.

These currents of flow can be best appreciated clinically when small amounts of material such as the products of inflammation are present in the aqueous. Excessive accumulation of such materials may even halt the currents of flow. In such a case resumption of motion by these materials in the anterior chamber may be the first sign of improvement.

deep inner circumferential groove, the *inner scleral sulcus* or *furrow*, and a shallow outer one, the *outer scleral sulcus* or *furrow* (Fig. 11–6) can be seen on gross examination of a dissected preparation.[14] On meridional section, the transition line from cornea to sclera extends from the end of Bowman's membrane to the end of Descemet's membrane and appears as an arcuate line with its convexity directed laterally and posteriorly.

This "convexity" of peripheral transparent corneal tissue within the "concavity" of the adjacent opaque sclera can also be well appreciated clinically by slit-lamp examination (gonioscopy). The radius of curvature of this union is shorter superiorly and inferiorly than laterally.

THE DRAINAGE ANGLE

The posterior boundary of the internal scleral sulcus is a collagenous ridge, composed mainly of circumferentially oriented bundles of collagen fibrils, the *scleral roll* (Figs. 11–7 and 11–8)[4,14] or Schwalbe's posterior border ring. After continuing for a short distance posteriorly, the ridge tapers and finally blends with the more predominant,

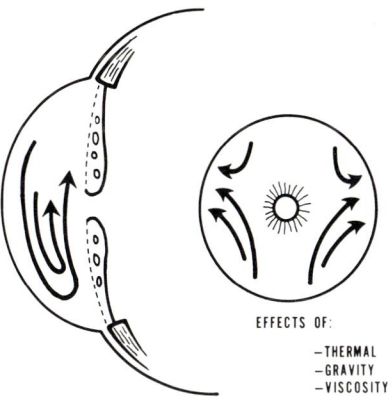

FIG. 11–5. Aqueous currents in anterior chamber.

THE LIMBUS

In the region of the limbus where clear cornea is transformed into opaque sclera, a

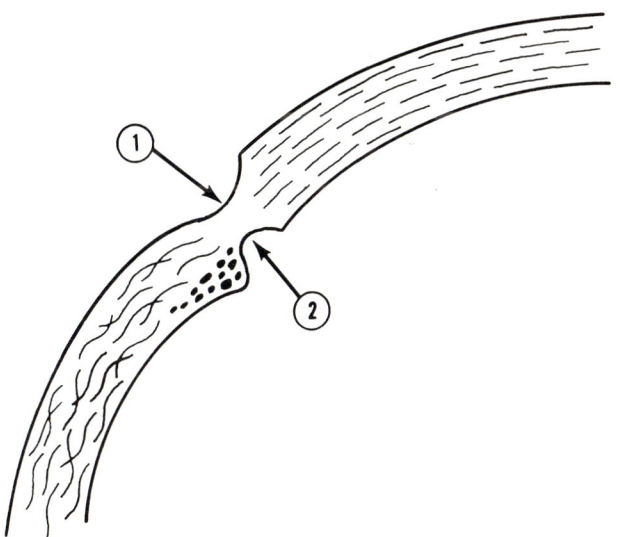

FIG. 11–6. Schematic representation of meridional section of corneoscleral coat. Circumferential shallow outer sulcus (*1*) and deeper inner sulcus (*2*) are present in region of union of cornea with sclera. Posterior boundary of inner sulcus is thickened by scleral roll.

216 OCULAR HISTOLOGY

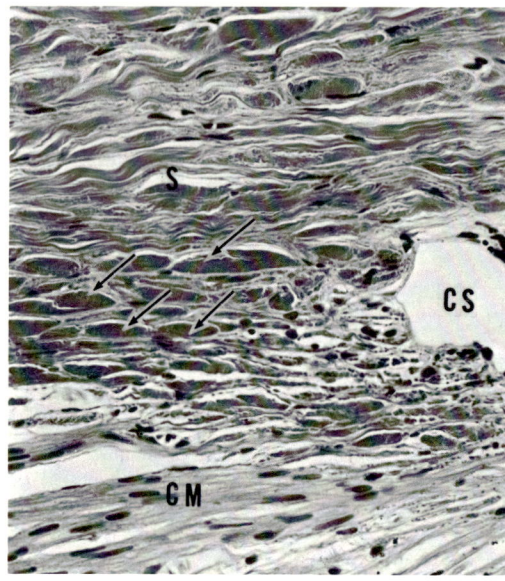

FIG. 11–7. Light micrograph of region of scleral roll showing cross-sectioned bundles of circumferentially arranged collagen (*arrows*) characteristic of this region. Contrast these collagen bundles with those of adjacent sclera (*S*) and with nearby ciliary muscles (*CM*). *CS*, canal of Schlemm. Meridional section, H&E, ×265. (AFIP Neg. 68–9954.)

FIG. 11–8. Electron micrograph of scleral roll. Collagen fibrils, though varying in diameter, are all circumferentially arranged within each collagen bundle. A number of cells are present, as well as patches of aging collagen or elastic tissue. Meridional section, ×16,000.

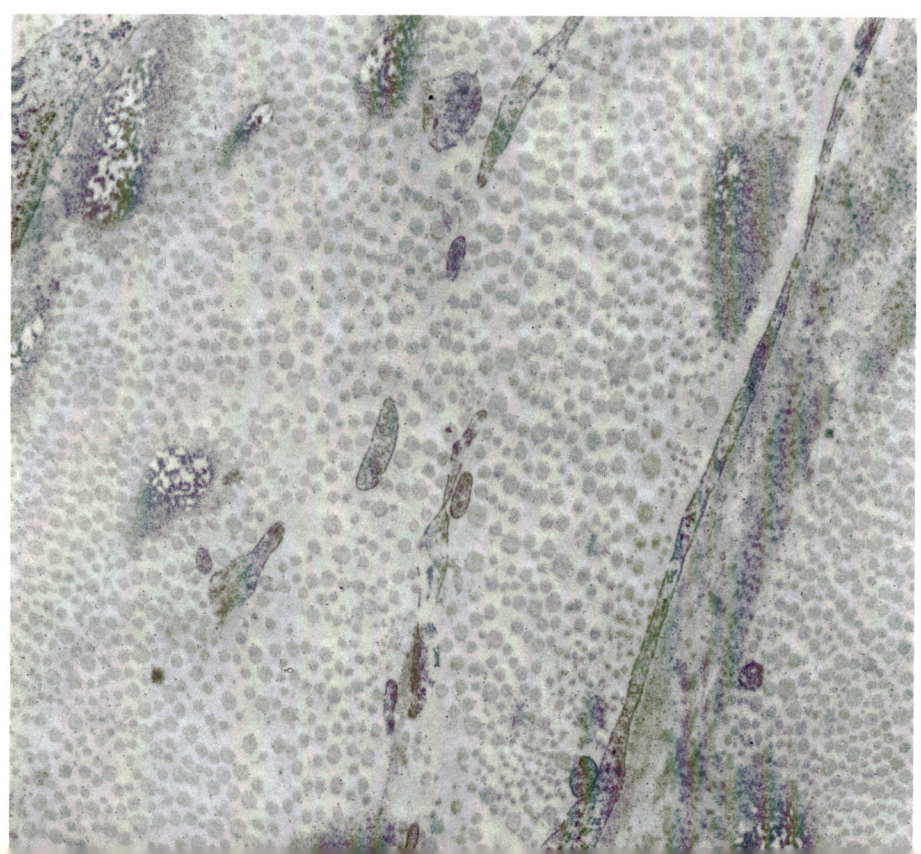

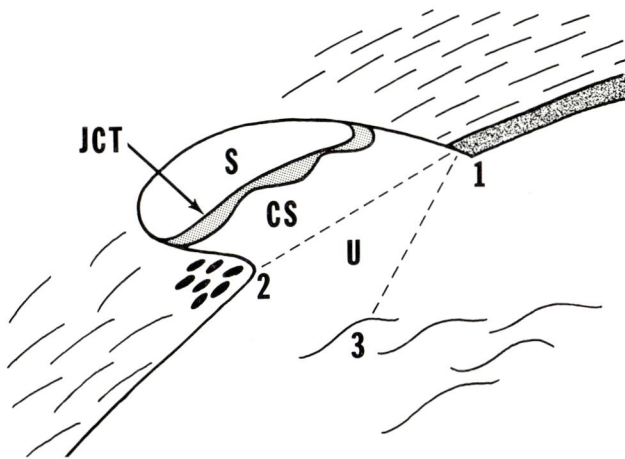

FIG. 11–9. Schematic drawing of drainage angle. *1*, end of Descemet's membrane; *2*, tip of scleral roll; *3*, end of anterior border layer of iris; *CS*, corneoscleral meshwork; *JCT*, juxtacanalicular connective tissue layer; *S*, canal of Schlemm; *U*, uveal meshwork. (From Fine, B. S. *Trans Amer Acad Ophthal Otolaryng* 70: 777, 1966.)

FIG. 11–10. Light micrograph of drainage angle of adult. Scleral roll (*SR*) is outlined by *dotted line*. Adjacent longitudinal ciliary muscle (*LCM₁*) is partially "hyalinized," more markedly in anterior and outer parts of longitudinal ciliary muscle (*LCM*). *TR* indicates imaginary line extending from scleral roll to Schwalbe's line. Meshwork is separable into inner or uveal (*U*) portion and outer or corneoscleral (*C*) portion. A special region, the juxtacanalicular connective tissue (*JCT*), variable in thickness, lies adjacent to endothelium-lined (*EN*) canal of Schlemm (*CS*). *AC*, anterior chamber; *CL*, canaliculus of canal of Schlemm; *SC*, sclera. 1.5-μ meridional section, hematoxylin and phloxin, ×420. (AFIP Neg. 63–4155. From Fine, B. S. *Invest Ophthal* 3:609, 1964.)

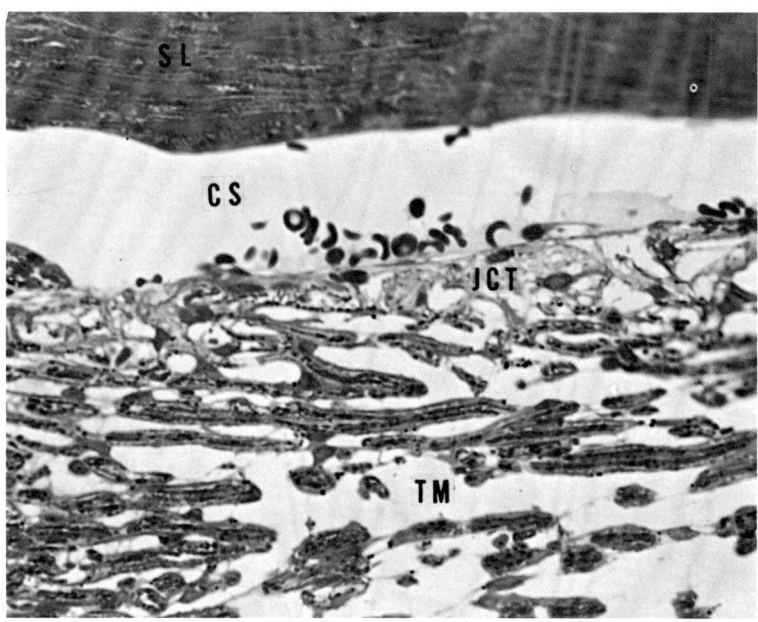

FIG. 11–11. Light micrograph of region of canal of Schlemm (CS). Density of adjacent scleral lamellas (SL) is apparent. Trabecular meshwork (TM) blends with irregular juxtacanalicular connective tissue (JCT). Red blood cells are present in lumen of canal of Schlemm. 1.5-μ section, hematoxylin-phloxin, ×530. (AFIP Neg. 63–4154. From Fine, B. S. *Trans Amer Acad Ophthal Otolaryng* 70:777, 1966.)

obliquely arranged collagenous lamellas of the sclera.

Deep within this sulcus (Fig. 11–4, on Color Plate V, and Fig. 11–9) and applied closely to the collagenous tissue of the corneosclera lies the large vessel known as the *canal of Schlemm*. This circumferentially arranged vessel (or complex of vessels, for it branches frequently,* Fig. 11–24) is formed by a continuous layer of endothelial cells (Figs. 11–10 through 11–13) with a rather patchy or diffuse basement membrane (Figs. 11–14 and 11–20). The structure of the canal of Schlemm closely resembles the structure of a lymphatic. It is

---

* The projections of tissue intervening between segments of the canal are called *septa*. Each septum contains a dense collagenous core resembling closely those of the adjacent corneoscleral tissue.

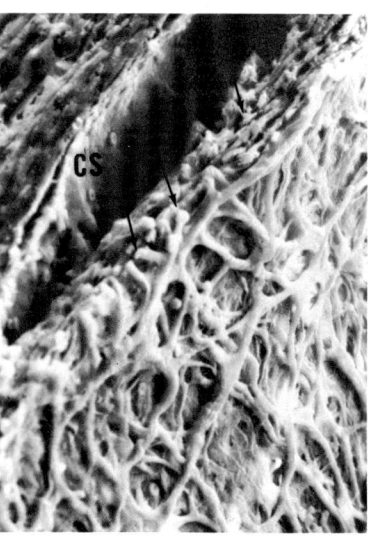

FIG. 11–12. Scanning electron micrograph of cut open canal of Schlemm (CS) and anterior chamber surface of adjacent uveal meshwork of young adult. A few red blood cells lie on smooth outer endothelial lining of canal, and cut edges of full thickness meshwork can be seen at *arrows*. ×310.

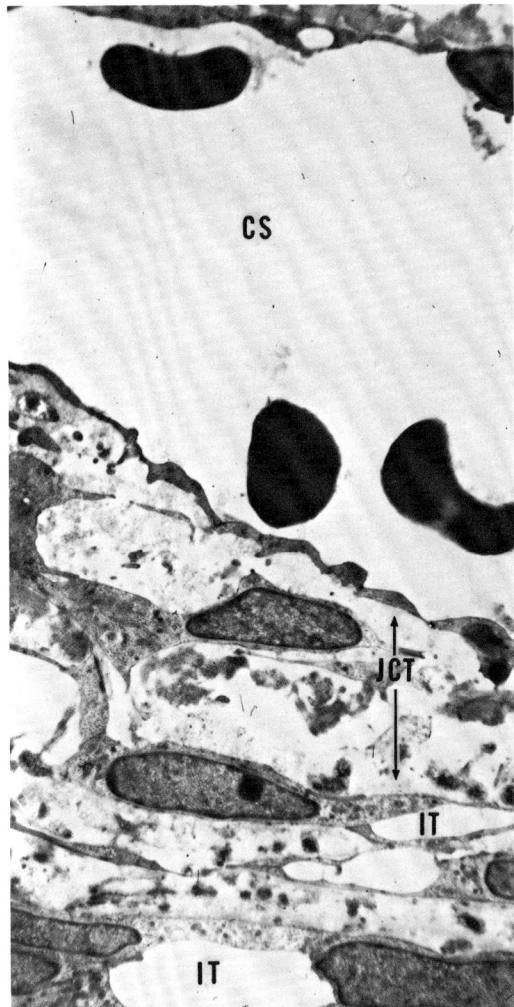

FIG. 11–13. Electron micrograph showing canal of Schlemm (CS), adjacent juxtacanalicular connective tissue (JCT), and corneoscleral meshwork with intertrabecular spaces (IT). ×4,000.

called an aqueous vessel because in vivo it contains aqueous fluid alone.

The canal of Schlemm is lined by endothelial cells whose bodies protrude into the lumen and whose basal cytoplasm is attenuated into large area-covering sheets. The endothelial cells abut one against the other in a number of forms which vary from single end-to-end apposition to more complex arrangements resembling interlocking mortise-and-tenon joints (Fig. 11–14). The cell-to-cell appositions form an attachment zone resembling a zonula adherens (Fig. 11–15) where the cytoplasm often is thrown into large villi or folds. The folds project either into the lumen of the canal or into the adjacent *juxtacanalicular connective tissue* (Fig. 11–16)[4] and often assume bizarre configurations. There is some evidence that they may even function as a form of *macropinocytotic* transport mechanism across the endothelial cell barrier.[4,16,17,18]

The large villi or flaps may explain in part the presence of large "cytoplasmic vacuoles" (Fig. 11–17) that are frequently seen in the endothelial layer.[10,11,16] Variability in fixation may also be responsible, and some of these "vacuoles" may be due to extreme dilation of the sensitive agranular endoplasmic reticulum. "Holes," "apertures," or "pores" have been reported to occur on both the trabecular and scleral sides of the large cytoplasmic vacuoles. It is not yet clear whether these latter spaces are *inter*cellular, *intra*cellular, or the products of fixation, processing or misinterpretation.

All the endothelial cells lining the canal of Schlemm possess small invaginations of the plasmalemma. Their appearance and their pinocytotic-like activity have given them the term *micropinocytotic* vesicles.[3-5] They are present at both the apical and basal surfaces of the endothelial cells. Free forms of these vesicles also are present and functional within the cell cytoplasm.

Pinocytotic vesicles often may be *induced* in a cell by contact of its plasmalemma with a foreign substance such as a tracer material. Vesicles observed in normal tissues otherwise untouched except by fixation procedures generally are considered to be present in vivo and presumably carrying out some normal function. Demonstration of uptake of a small amount of tracer material by these preexistent, surface-connected vesicles on one side of the cytoplasmic barrier, and their presence in "free" vesicles within the cytoplasm as well as in lumen-connected vesicles suggests transport of the tracer material together

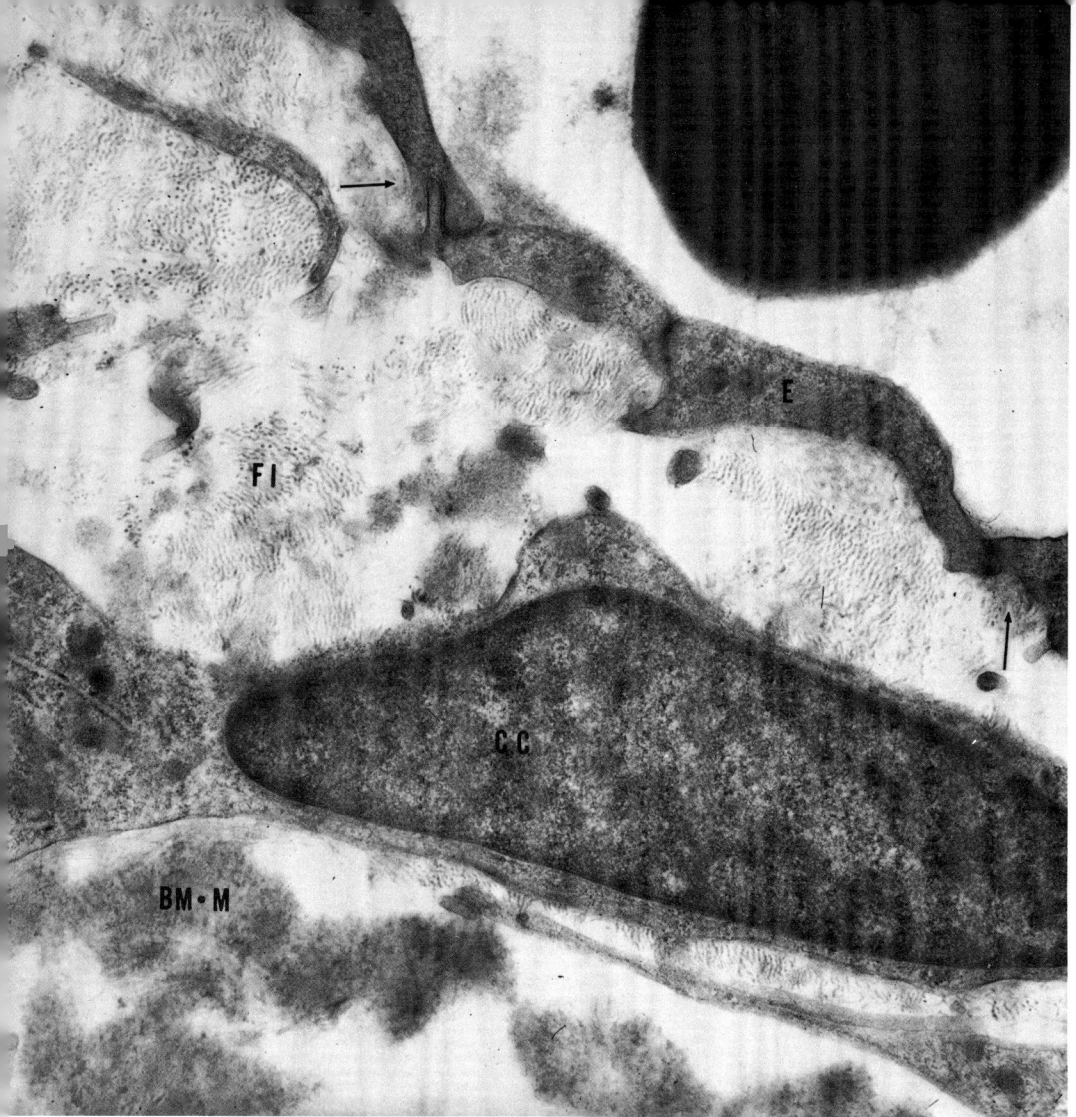

FIG. 11–14. Electron micrograph of endothelium (*E*) lining inner wall of canal of Schlemm. Endothelial basement membrane is diffuse and patchy (*arrows*). Endothelial cell-to-cell abutments are present. Juxtacanalicular connective tissue contains cells surrounded by extracellular materials varying from delicate collagenous filaments (*FI*) to patches of basement-membrane-like material (*BM·M*). *CC*, connective tissue cell. ×21,500. (Modified from Fine, B. S. *Invest Ophthal* 3:609, 1964.)

with its supporting milieu (Fig. 11–18).[13] Larger quantities of tracer material may be transported by the *macropinocytotic* mechanism presumably similar to the recently proposed mechanism of "endothelial vacuolation."[18] In the normal eye, the milieu alone (i.e., the aqueous) would be transported.

Not infrequently, red blood cells are present in the lumen of the canal of Schlemm in histologic material, because blood can reflux easily into this vessel. The red blood cells often separate endothelial cells and thus gain access to the underlying loose connective tissue and even to the anterior chamber. Such histologic observations indicate that the endothelial cell-to-cell attachments are not excessively strong and that they can be separated with comparative ease. These

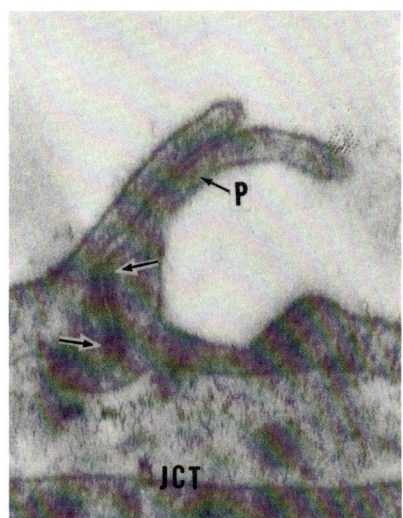

FIG. 11-15. Zonula adherens type of attachment (*free arrows*) between endothelial cells of inner wall of canal of Schlemm. *JCT*, juxtacanalicular connective tissue, *P*, villous projections or flaps into lumen. Rhesus monkey, ×41,600. (From Fine, B. S. *Trans Amer Acad Ophthal Otolaryng* 70:777, 1966.)

any single meridional section; it is more delicate in the younger eye and more prominent in the adult eye.

The cells embedded within this tissue are mesodermal and presumably retain the capacities of fibroblasts. Excessive increase in density of this connective tissue layer in the aging eye may produce disease.

Large endothelium-lined channels (collector channels; Figs. 11-21 and 11-22) connect the canal of Schlemm either anteriorly

FIG. 11-16. Endothelial cell villi or folds projecting into lumen of canal of Schlemm (*L*) or into adjacent juxtacanalicular connective tissue. ×40,500. (From Fine, B. S. *Invest Ophthal* 3:609, 1964.)

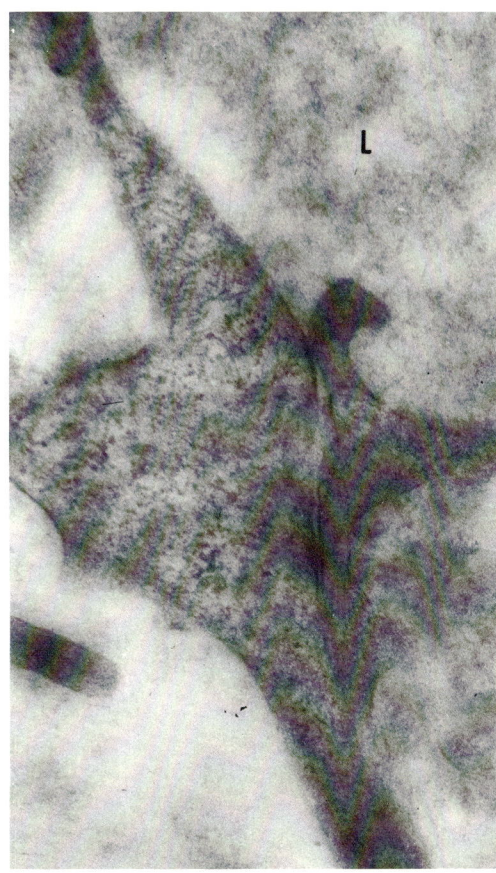

observations appear compatible with the cytologic observations of a zonula-adherens-type of attachment.

The outer wall of the canal rests on a basement membrane that is separated from the dense collagenous lamellas of the cornea and the sclera by a few loose cells (Fig. 11-19). The "looser" arrangement is more prominent posteriorly than anteriorly.

The inner wall rests on a thinner or patchy basement membrane that is associated with a zone of delicate connective tissue, the *juxtacanalicular connective tissue* (Figs. 11-13, 11-14, and 11-20).[4,5] This is a special zone of the corneoscleral trabecular meshwork and consists of cells surrounded by a variety of fibrous and mucinous extracellular materials. The juxtacanalicular connective tissue is irregular in thickness from front to back in

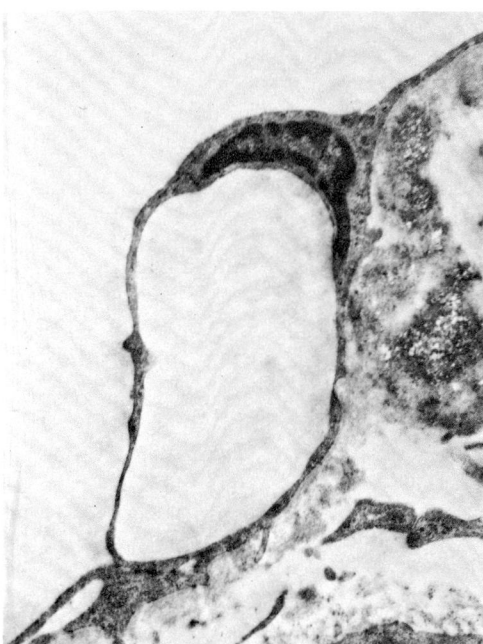

FIG. 11–17. Large vacuole in endothelial layer of canal of Schlemm. ×20,000.

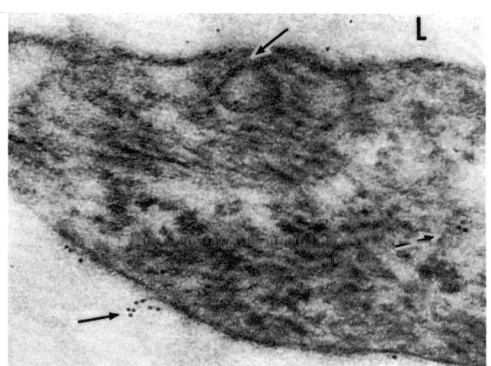

FIG. 11–18. Ferritin particles being transported across endothelial cell by micropinocytotic vesicles (*arrows*). L, lumen of canal of Schlemm. Rhesus monkey, ×86,500. (From Fine, B. S. *Invest Ophthal* 3:609, 1964.)

FIG. 11–19. Endothelium (*E*) lining outer wall of canal of Schlemm. Patches of basement membrane are present as well as a few loose connective tissue cells (*CC*). L, lumen of canal. ×11,000.

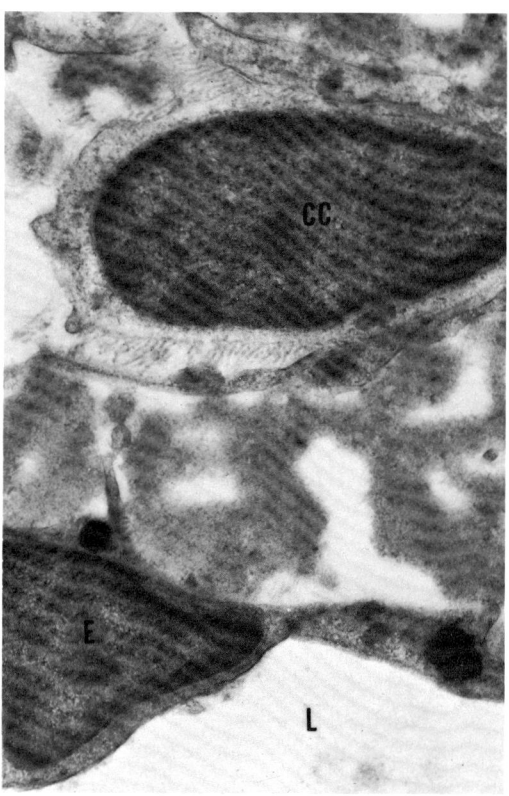

or, more commonly, posteriorly to the intrascleral venous system. If they reach the surface of the sclera unconnected, they can be observed in vivo as the clear *aqueous veins* described by Ascher (Fig. 11–23).[1]

THE TRABECULAR MESHWORK

In meridional section a loose collagenous meshwork can be seen to fill the internal sulcus and extend as an open fan to the root of the iris (Fig. 11–24). The "handle" of this fan is located just anterior to the end of Descemet's membrane, where a few layers

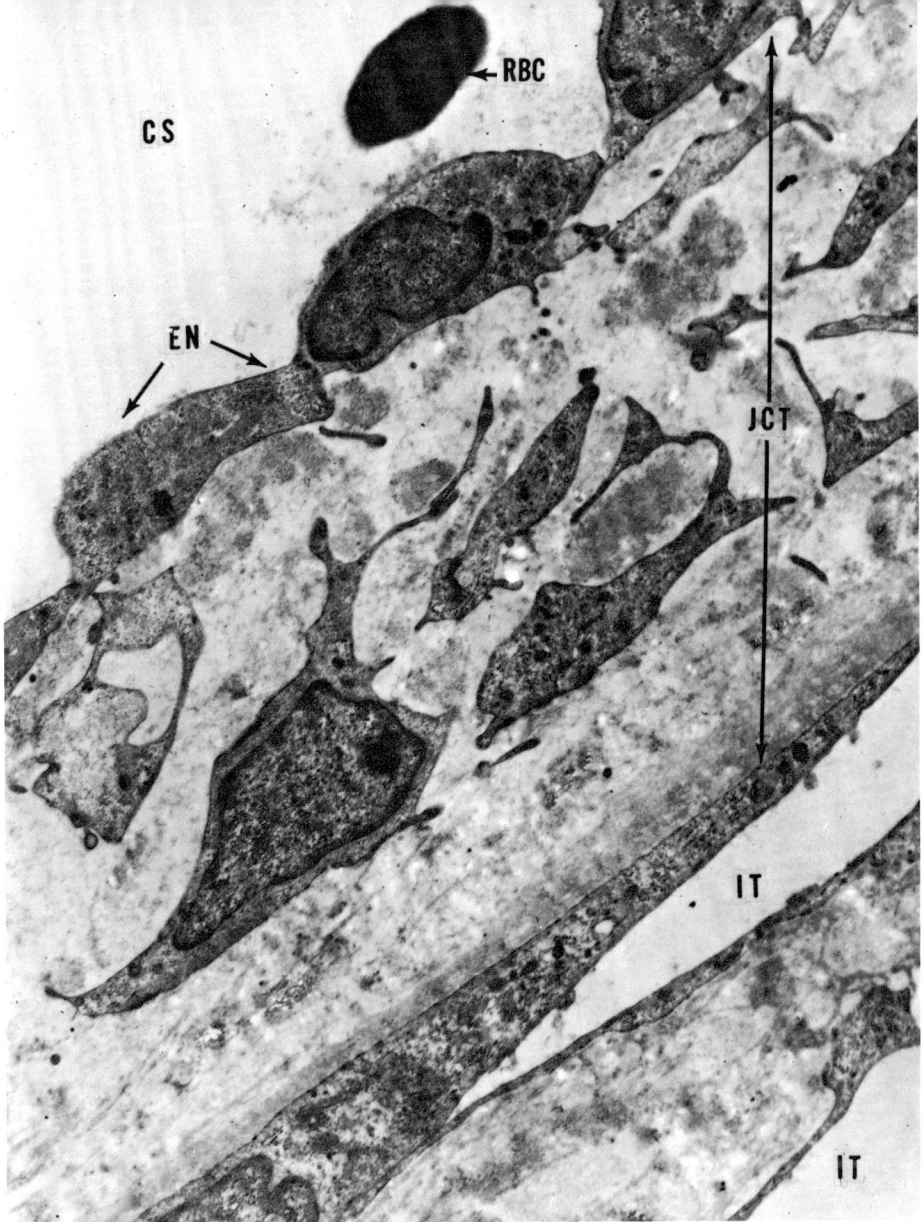

FIG. 11–20. Region of juxtacanalicular connective tissue (*JCT*). Tissue separates endothelial cell lining (*EN*) of canal of Schlemm (*CS*) from terminal intertrabecular spaces (*IT*) in this plane of section. Intertrabecular spaces represent distalmost extensions of anterior chamber space *RBC*, red blood cell in lumen of canal of Schlemm. ×12,000. (From Fine, B. S. *Trans Amer Acad Ophthal Otolaryng* 70:777, 1966.)

of meshwork enter into and blend with the deep peripheral corneal stroma (Fig. 11–25). The meshwork may be easily and usefully separated into two parts by an imaginary line extending from the scleral roll to the end of Descemet's membrane (Figs. 11–9 and 11–10).

The meshwork lying external to the line and extending from cornea to sclera is known as the *corneoscleral meshwork;* the meshwork lying internal to the line and in con-

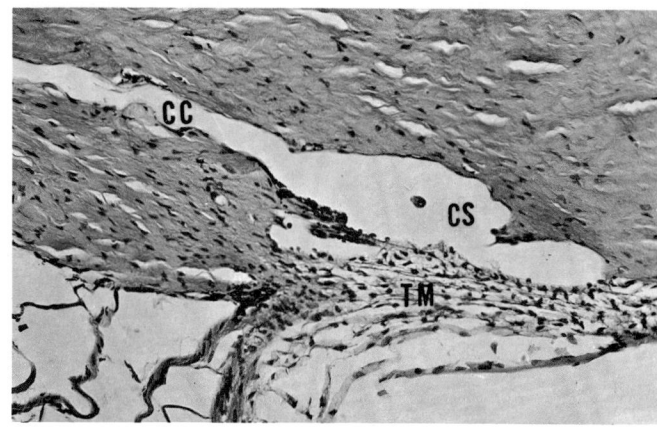

FIG. 11–21. Large collector channel (*CC*) continuous with canal of Schlemm (*CS*). *TM,* trabecular meshwork. Rhesus monkey, alcian blue, ×115. (AFIP Neg. 60–4872. From Fine, B. S. *Trans Amer Acad Ophthal Otolaryng* 70:777, 1966.)

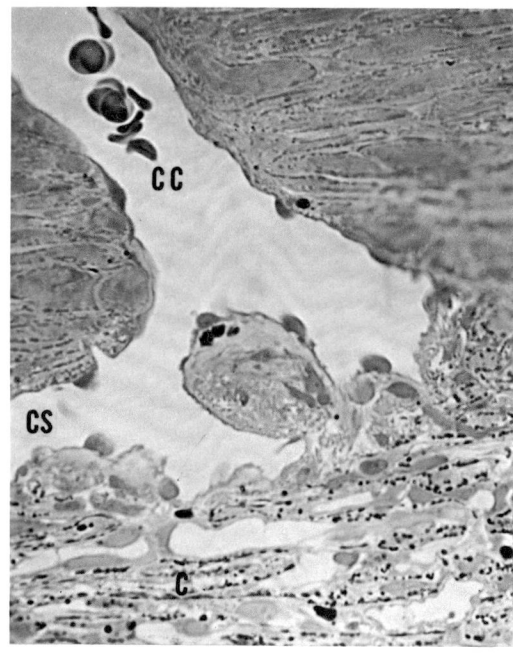

FIG. 11–22. Large collector channel (*CC*) draining posterior extremity of canal of Schlemm (*CS*). *C,* adjacent corneoscleral meshwork. Cornea is to left. 1.5-μ section, PD, ×530. (AFIP Neg. 60–4569.)

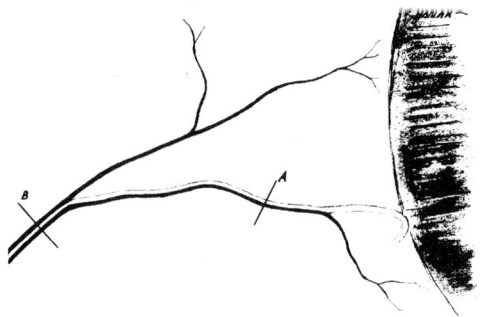

FIG. 11–23. Aqueous veins of Ascher observed biomicroscopically as clear layer uniting with and running parallel to blood-filled layer (or vein) as at *A,* and more noticeably further downstream as central clear or faintly pink band within larger vein formed by union of two adjacent vessels as at *B.* (From Ascher, K. W. *Amer J Ophthal* 25:31, 1942.)

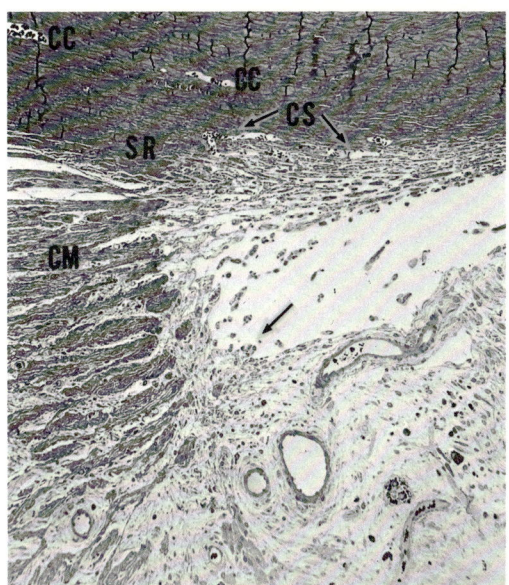

FIG. 11–24. "Open fan" arrangement of trabecular meshwork in infant. *CC,* collector channels; *CM,* ciliary muscle; *CS,* branches of canal of Schlemm; *SR,* scleral roll; *free arrow,* root of iris. 1.5-μ meridional section, PD, ×90. (AFIP Neg. 63–4363.) (From Fine, B. S. *Invest Ophthal* 3:609, 1964.)

FIG. 11–25. Trabecular meshwork (*arrows*) lying external to periphery of Descemet's membrane (*DM*). Meshwork ultimately blends centrally with corneal stroma (*C*). 1.5-μ section, PD, ×395. (AFIP Neg. 65–12339.)

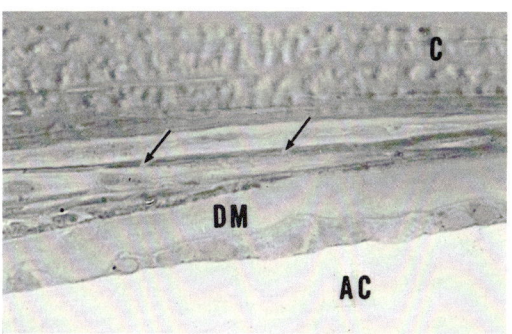

tinuity with the uveal tract posteriorly is known as the *uveal meshwork.*

A single trabecula of uveal meshwork (Fig. 11–26) consists essentially of a collagenous core surrounded by a single layer of cells ("endothelium," in reality a mesothelium). A basement membrane separates the endothelial cells from the underlying collagenous core, and not infrequently, patches of this basement membrane present a periodic structure that measures 1,000 A (*inset 1,* Fig. 11–26).[4,8,9]

The periodic-structured material closely resembles the anteriormost portion of Descemet's membrane (or may be observed in excrescences on Descemet's membrane in the *pathologic* central corneal entity known as cornea guttata). The core of a trabecula is formed of packed collagen fibrils frequently twisted in spiraling fashion. Thin sections of such spiraling fibrils often produce another form of 1,000 A periodicity measured from fibril to fibril.

The 1,000-A periodicity should not be confused with the 2,000-A to 3,000-A periodic collagen fibril which can be produced in vitro by reconstitution of separated collagen molecules (i.e., tropocollagen).[15] Measurements of periodicity similar to those of in vitro or manufactured collagen fibrils have not yet been observed in normal ocular tissues. In vitro collagen fibrils are formed by dissolution of normal or "native" collagen into basic tropocollagen molecules (dimension ~ 14 by 2,800 A). Their subsequent reconstitution in vitro is achieved by manipulating their milieu into a true native type of periodic collagen *fiber* (i.e., with ~ 640-A axial periodicity), a nonnative fibrous long-spacing *fiber* (i.e., with ~ 2,800-A *axial periodicity*), or nonnative *segments* called segment long-spacing (i.e., ~ 2,800 A in *length*) structures.

Lying within the tightly packed collagenous cores of the trabeculae are many aggregates of filamentous material whose density increases with age (*insets 2 and 3,* Fig. 11–26). The aggregates take the stains that demonstrate elastic tissue.[12,19] As in other con-

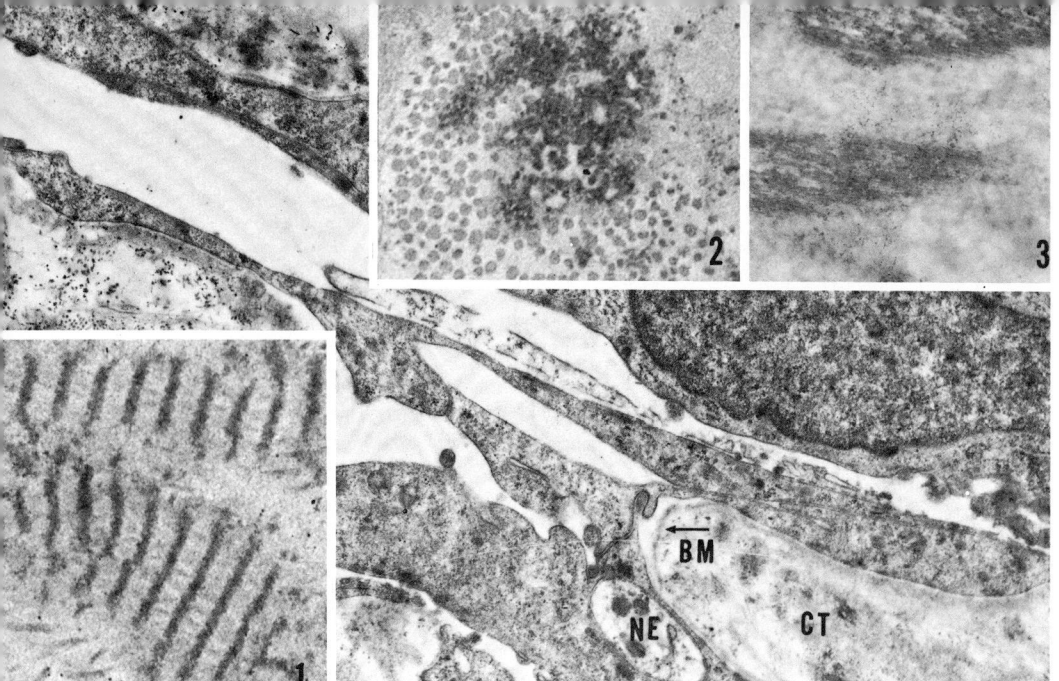

FIG. 11–26. Uveal trabecular meshwork. **Main figure** shows "endothelial" cells with basement membranes (*BM*) and associated connective tissues (*CT*). Apical surfaces line intertrabecular spaces which are prolongations of anterior chamber space. *NE*, neurite enveloped by trabecular endothelial cell. **Inset 1** shows basement membrane material with periodicity. Density of filamentous material within collagen cores is shown in cross-section (**inset 2**) and in oblique section (**inset 3**). **Main figure,** ×13,500; **inset 1,** ×39,500; **inset 2,** ×39,000; **inset 3,** ×23,400. (Modified from Fine, B. S. *Invest Ophthal* 3:609, 1964.)

nective tissues, additional ground substance materials are probably present, but their identification and quantitation remain obscure.

The endothelial cells covering the connective tissue core have apical surfaces, line intertrabecular spaces, and are therefore bathed by aqueous.

The trabeculae of the meshwork are roughly arranged into circumferential sheets lying superimposed one upon the other.[6] They can be fairly easily separated from one another mechanically, especially in the uveal meshwork. Large oval apertures traverse each trabecular sheet and are termed intertrabecular, or more properly "*trans*trabecular," spaces (Figs. 11–27 and 11–28). The apertures are not superimposed and decrease in size in the direction of the corneoscleral meshwork.

The corneoscleral sheets differ only slightly from the uveal in having somewhat flatter trabeculae as observed in cross-section and in lacking the staining characteristics for elastic fibers. The *transtrabecular* apertures are more circular and smaller than those of the uveal meshwork.

Spaces *between* individual sheets are well seen on proper meridional section (Figs. 11–10 and 11–21 through 11–24) and here are termed the *intertrabecular* spaces. In the uveal meshwork, these intertrabecular spaces are observed frequently to pass the scleral roll to be continuous with the tissue spaces lying between the smooth muscle cells of the ciliary muscles (especially those of the meridional ciliary muscle). If serially sectioned in a frontal or coronal plane, these spaces can be

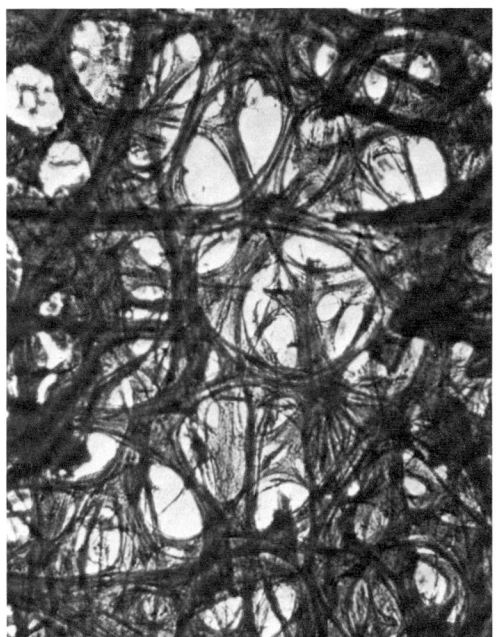

FIG. 11–27. Light micrograph of flat preparation of trabecular meshwork viewed from anterior chamber side. Transtrabecular apertures are not superimposed and decrease in size away from anterior chamber. Rhesus monkey, silver carbonate, ×275. (AFIP Neg. 60–6100.)

scopic evidence in support of this belief has not yet been obtained.

Not infrequently, various trabecular sheets have some interconnection from one level to another, an arrangement that occurs sporadically around the circumference of the globe. Such regions of interconnection appear most prominent within the corneoscleral meshwork, and therefore the transition of corneoscleral trabecular connective tissue "cores" into the looser juxtacanalicular connective tissue may be observed.

All transtrabecular and intertrabecular spaces may thus be considered extensions of the anterior

FIG. 11–28. Scanning electron micrograph of uveal meshwork from anterior chamber side in young adult human. Heavy beams of trabecular meshwork are clearly seen, as are large transtrabecular spaces. Cut edges of layers of trabecular meshwork are seen in upper left. Spaces between adjacent sheets (i.e., intertrabecular) are visible. ×780.

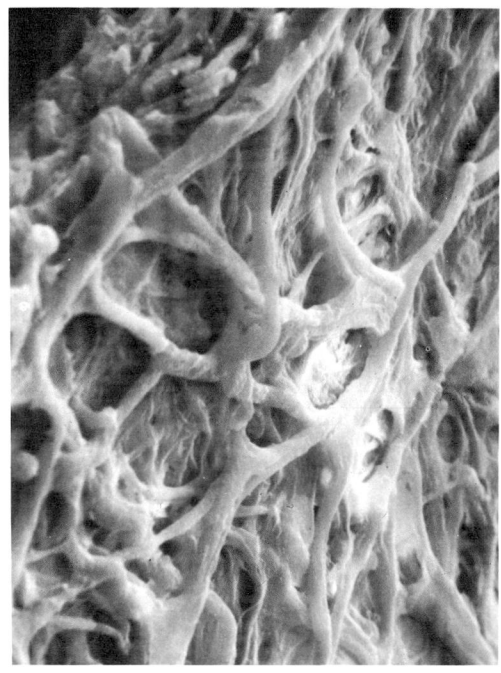

seen as large-apertured, relatively straight short tubes (Fig. 11–29). Such a grouping of tubes with apertured walls might be termed a system of compound *aqueous tubes*.[4,7] In the corneoscleral meshwork, which blends posteriorly with the region of the scleral roll, the intertrabecular spaces (tubes) abut upon the canalicular extensions of the canal of Schlemm (Fig. 11–30). Such extensions are frequent in this region.

The blind inpouchings of the canal of Schlemm, here termed canaliculi, are endothelium-lined and do not appear to be in continuity with the intertrabecular spaces. On the basis of light microscopy, however, they were once thought to be in such continuity, but acceptable electron micro-

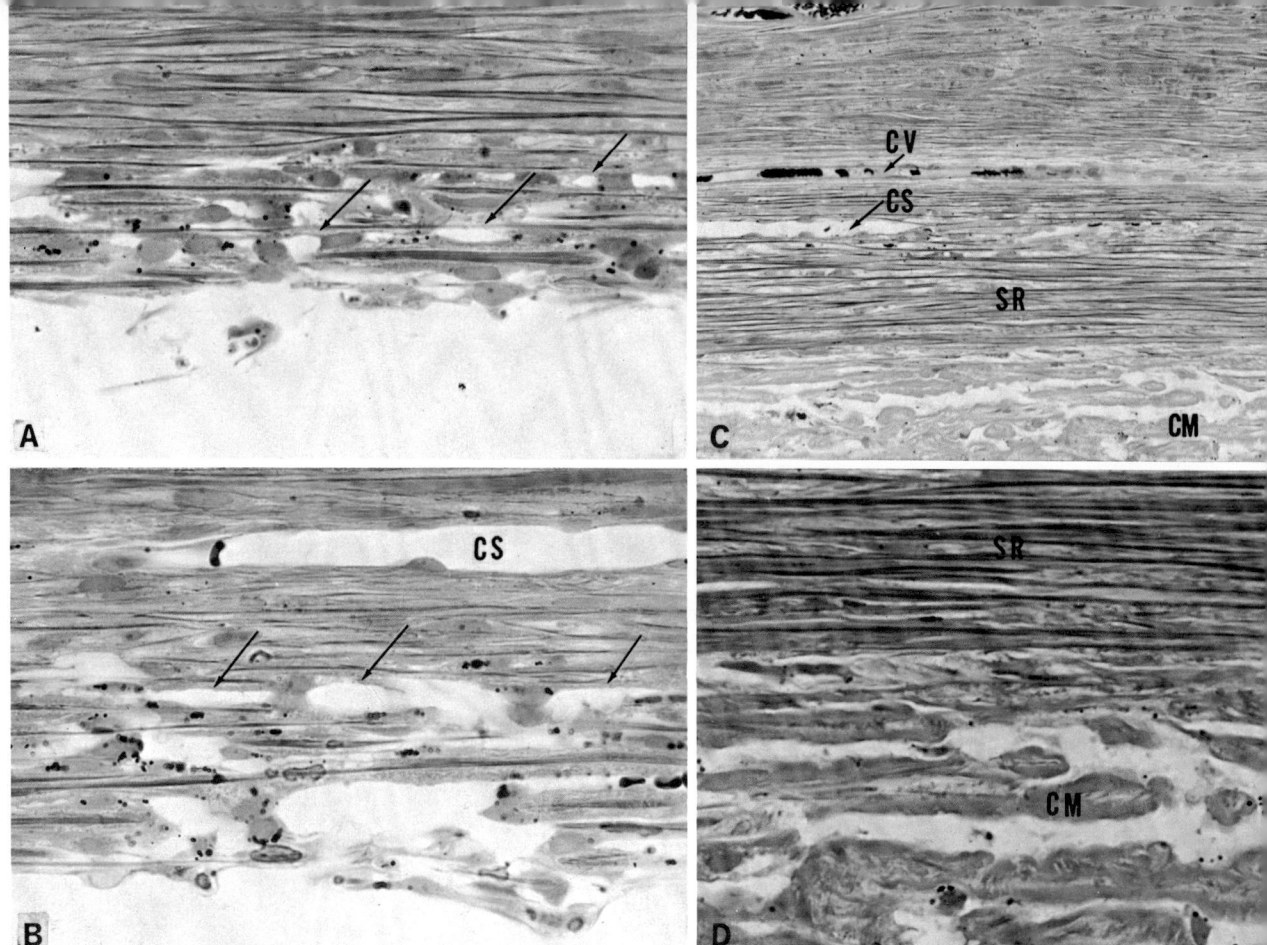

chamber space and can be recognized in any single micrograph by locating the apex (i.e., non-basement-membrane- or non-connective-tissue-related surface) of the endothelial cell. Rarely, "endothelium" of the canal of Schlemm seems capable of producing collagen (Fig. 11–31).

Where the intertrabecular extensions of the anterior chamber space terminate in any single plane of section therefore is defined as the place where the lining cells abruptly depolarize and become completely surrounded by extracellular materials (i.e., the juxtacanalicular connective tissue). The quality and quantity of the nonfibrous extracellular materials present normally in this juxtacanalicular connective tissue layer are not clear. In the normal primate eye, the region is highly permeable to such tracer materials as ferritin.

The collagenous "cores" of the uveal meshwork blend with the looser collagen of the intercellular spaces posteriorly between the smooth muscle cells of the ciliary body.

FIG. 11–29. Coronal sections of trabecular meshwork. **A.** Anterior meshwork. *Arrows* point to spaces between trabecular sheets (intertrabecular spaces). **B.** Spaces (*arrows*) between adjacent trabecular sheets posterior to **A**. Spaces are large and circular to oval in cross-section. Anterior extremity of canal of Schlemm (CS) has appeared in this section. **C.** Low magnification of posterior region of trabecular meshwork showing posterior extremity of canal of Schlemm (CS), circumferentially oriented collagen bundles of the scleral roll (SR), and the cross-sectioned layers of longitudinal ciliary muscle (CM); CV, collector vessel. **D.** Higher magnification showing region of scleral roll (SR) and its relation to adjacent longitudinal ciliary muscle (CM). All 1.5-μ sections, PD; **A, B,** and **D,** ×530; **C,** ×195. (AFIP Negs. 64–563, 64–562, 64–558, 64–559.) (From Fine, B. S. *Invest Ophthal* 3:609, 1964.)

They are also continued somewhat axially as the collagenous bundles or columns within the iris stroma.

It should be understood that anatomic subdivisions of the drainage angle region are somewhat arbitrary and are used to simplify and improve our understanding of the arrangements in this region. All the tissues and layers involved are *mesodermal* (including the so-called endothelia). The anterior chamber represents essentially an exaggerated cleft within mesoderm lying between corneal stroma in front and iris stroma behind. The two mesodermal layers reunite at the periphery of the anterior chamber to form the loose tissues of the "drainage angle."

Functionally, the drainage angle is quantitatively the most significant region for aqueous outflow in the normal primate eye. This is well supported by the clinical fact that removal of the

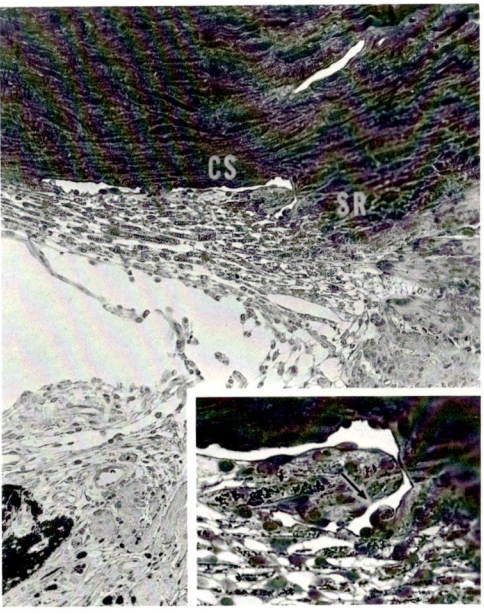

FIG. 11–31. Endothelial "loop" (*E*) along inner wall of canal of Schlemm enveloping a cluster of filaments. A basement membrane is only questionably present. *JCT*, juxtacanalicular connective tissue; *RBC*, red blood cell in lumen of canal of Schlemm. ×17,400.

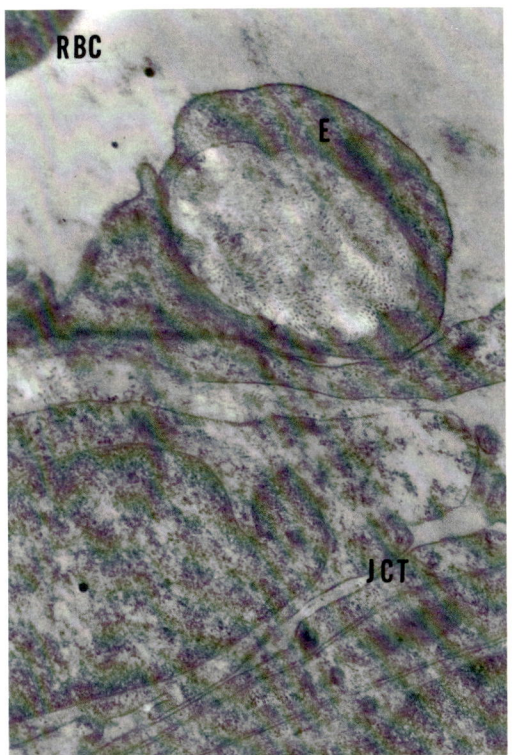

FIG. 11–30. Canaliculi from canal of Schlemm (*CS*) projecting into corneoscleral meshwork anterior to scleral roll (*SR*) in infant. **Inset** shows canal and its canalicular extension (*arrow*) at higher magnification. **Main figure,** 1.5-μ section, PD, ×100 (AFIP Neg. 68–5146); **inset** ×265.

entire iris diaphragm (as has occurred as a surgical accident on occasion) does not immediately or significantly produce an elevation in the intraocular pressure. Because *quantitation* of outflow in these three anatomic regions of the drainage angle (iris root, uvea, and canal of Schlemm) has not yet been determined, it is difficult to assign such quantitation unless it be arbitrary.*

* Attempts have been made[2] to quantitate aqueous outflow from the primate eye and to correlate this with structures in the drainage angle. Only "conventional" outflow and direct "uveoscleral" outflow were considered. No consideration was given to possible *vascular* drainage either from the root of the iris or via the

FIG. 11–32. Iris processes (*arrow*), sporadic continuations of anterior border layer beyond iris root. Processes are most easily appreciated when pigmented, as here. H&E, ×110. (AFIP Neg. 70–1415.)

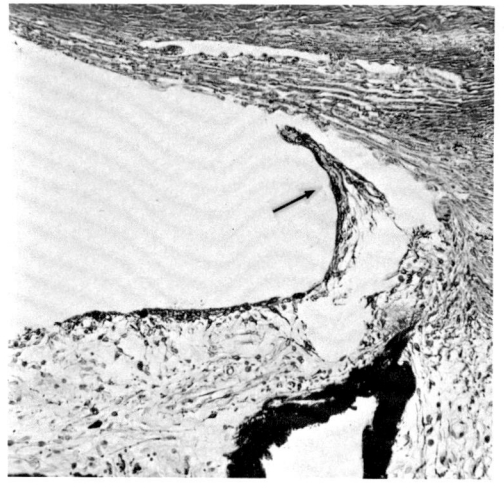

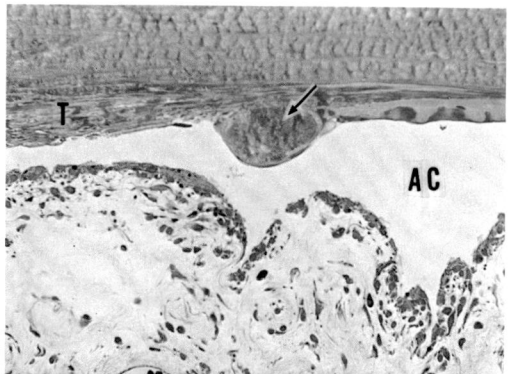

FIG. 11–34. Termination of Descemet's membrane as thickened band which appears as a large nodule (*arrow*). AC, anterior chamber; T, trabecular meshwork. Anterior border layer of iris and adjacent stroma lie below in photograph. 1.5-μ meridional section, PD, ×130. (AFIP Neg. 67–11208.)

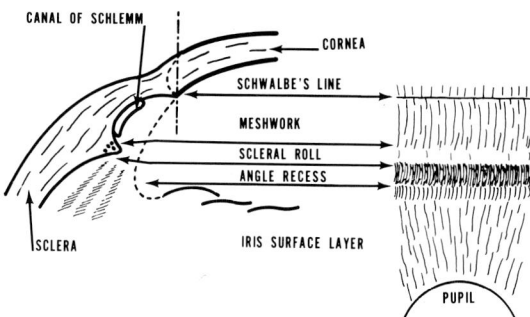

FIG. 11–33. Schematic drawing to illustrate histologic landmarks in drainage angle (left) and comparable zones that may be seen from anterior chamber side by gonioscopy (right).

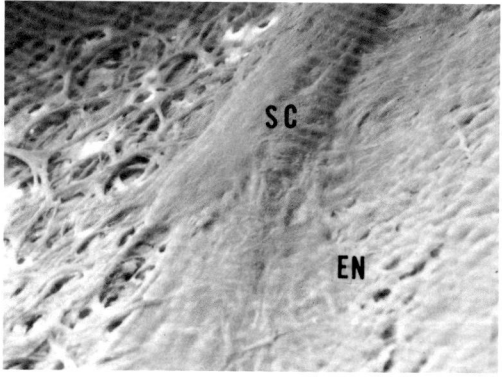

FIG. 11–35. Scanning electron micrograph of region of Schwalbe's "anterior border ring" (SC) in young adult. Uveal meshwork on left disappears anterior to ridge (see Fig. 11–34), while corneal endothelium (EN) continues over free surface of roll. ×320.

anterior ciliary veins draining the region of the longitudinal ciliary muscle. The latter pathways may also be of importance in assignment of quantitation to anatomic structure within the drainage angle, especially when parts of the drainage angle become compromised in pathologic changes.

The anterior border layer of the iris stroma (with or without pigmentation) may occasionally fail to end abruptly at the iris root and continue as short bands or processes along the surface of the uveal meshwork (Fig. 11-32), reaching even to attach to the end of Descemet's membrane (line of Schwalbe). Such anatomic variations intermittently distributed around the circumference of the drainage angle are known as *iris processes* and are best seen when pigmented and when observed clinically by gonioscopy (Fig. 11-33).

In histologic sections, the end of Descemet's membrane may vary from a delicate tapering out beyond visibility by light microscopy to an enlarged nodule (Fig. 11-34) that may be clinically visible as a thickened circumferential cord, Schwalbe's anterior border ring or line (Fig. 11-35). The thickened line generally is composed of a mixture of basement membrane and collagen fibrils, with the contribution of each component being variable.

REFERENCES

1. Ascher, K. W. Aqueous veins: Preliminary notes. *Amer J Ophthal* 25:31, 1942.
2. Bill, A. Conventional and uveo-scleral drainage of aqueous humour in the cynomologus monkey (*Macaca irus*) at normal and high intraocular pressures. *Exp Eye Res* 5:45, 1966.
3. Feeney, L., and Wissig, S. Outflow studies using an electron-dense tracer. *Trans Amer Acad Ophthal Otolaryng* 70:791, 1966.
4. Fine, B. S. Observations on the drainage angle in man and rhesus monkey: A concept of the pathogenesis of chronic simple glaucoma: A light and electron microscopic study. *Invest Ophthal* 3:609, 1964.
5. Fine, B. S. Structure of the trabecular meshwork and the canal of Schlemm. *Trans Amer Acad Ophthal Otolaryng* 70:777, 1966.
6. Flocks, M. The anatomy of the trabecular meshwork as seen in tangential sections. *Arch Ophthal (Chicago)* 56:708, 1956.
7. Frey-Wyssling, A. *Submicroscopic Morphology of Protoplasm.* Amsterdam, Elsevier, 1953.
8. Garron, L. K., Feeney, M. L., Hogan, M. J., and McEwen, W. K. Electron microscopic studies of the human eye. I. Preliminary investigations of the trabeculae. *Amer J Ophthal* 46:27, 1958.
9. Garron, L. K., and Feeney, M. L. Electron microscopic studies of the human eye. II. Study of the trabeculae by light and electron microscopy. *Arch Ophthal (Chicago)* 62:966, 1067, 1959.
10. Holmberg, A. The fine structure of the inner wall of Schlemm's canal. *Arch Ophthal (Chicago)* 62:956, 1959.
11. Holmberg, A. Schlemm's canal and the trabecular meshwork: An electron microscopic study of the normal structure in man and monkey (*Cercopithecus ethiops*). *Docum Ophthal* 19:339, 1965.
12. Iwamoto, T. Light and electron microscopy of the presumed elastic components of the trabeculae and scleral spur of the human eye. *Invest Ophthal* 3:144, 1964.
13. Palade, G. E., and Bruns, R. R. Structural modulations of plasmalemmal vesicles. *J Cell Biol* 37:633, 1968.
14. Salzmann, M. *The Anatomy and Histology of the Human Eyeball in the Normal State.* Tr. by E. V. L. Brown. Chicago, University of Chicago Press, 1912.
15. Schmitt, F. O. Interaction Properties of Elongate Protein Macromolecules, with Particular Reference to Collagen (Tropocollagen). In Oncley, J. L. (ed.). *Biophysical Science: A Study Program.* New York, Wiley, 1959.
16. Tripathi, R. C. Ultrastructure of Schlemm's canal in relation to aqueous outflow. *Exp Eye Res* 7:335, 1968.
17. Tripathi, R. C. Ultrastructure of the trabecular wall of Schlemm's canal. *Trans Ophthal Soc* 89:449, 1969.
18. Tripathi, R. C. Mechanism of the aqueous outflow across the trabecular wall of Schlemm's canal. *Exp Eye Res* 11:116, 1971.
19. Yamashita, T., and Rosen, D. A. The elastic tissue of primate trabecular meshwork: A histologic and electron microscopic study. *Invest Ophthal* 3:85, 1964.

chapter 12

The Optic Nerve

Intraocular (bulbar) portion
    Retinal layer
    Choroidal layer
    Scleral layer
Orbital (retrobulbar) portion
    Sheaths and their spaces
    Intraneural components
Vascular supply

The optic nerve is unlike a peripheral nerve in that it is actually a nerve fiber tract of the central nervous system, a tract formed by axons of the retinal ganglion cells. Although once considered to differ greatly from peripheral nerves, which have a close relation to a layer of Schwann cells (the neurilemmal sheath), the structural differences between peripheral and central nerves appear to be relative. A peripheral nerve is enveloped by a continuous layer of Schwann cells (the neurilemmal sheath) thick enough to be observed by light microscopy. Spiral wrapping of the axon by the doubled plasmalemma of Schwann cells produces the myelin sheath, which therefore is produced in segments by the individual cells (Fig. 12–1A). The doubled plasmalemma which connects the layers of myelin with the rest of the Schwann cell cytoplasm is termed the mesoaxon (outer and inner).

In the optic nerve, as in the white matter of the brain,[11] the ganglion cell axons also are enveloped by a sheath of doubled plasmalemma, to form the myelin. The outer layer of enveloping cytoplasm is derived in this location from oligodendrocytes (Fig. 12–1B). The cytoplasmic sheath is exceedingly thin and cannot be appreciated by light microscopic examination. In addition to this *quantitative* difference in neurilemmal thickness

FIG. 12–1. Formation of myelin sheath. **A.** In peripheral nerve spiral envelopment of a neurite by Schwann cell cytoplasm produces myelin sheath. Schwann cell has a basement membrane. **B.** In central nervous system myelin sheath is formed by oligodendrocyte spirally enveloping a neurite. Oligodendrocyte lacks basement membrane. (Courtesy of Dr. Peter W. Lampert.)

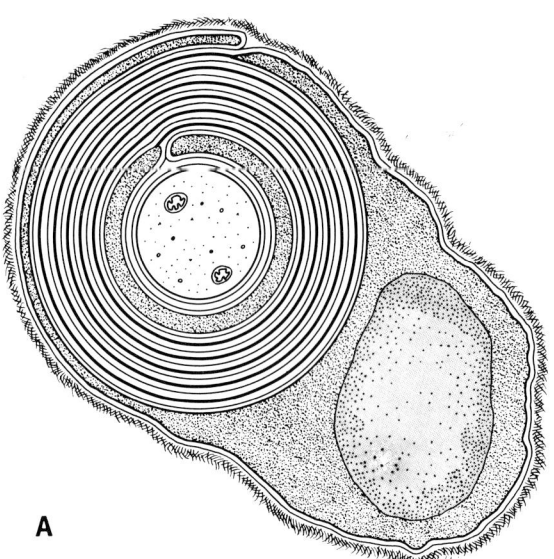
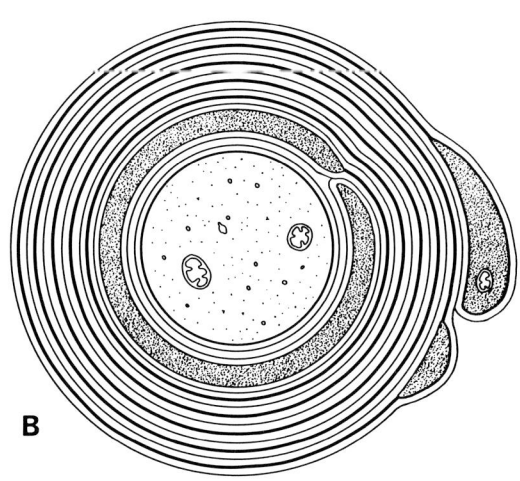

between a peripheral nerve and the optic nerve, there is further difference in that the Schwann cell is enveloped by a thin basement membrane, a structure which is lacking in the oligodendrocyte.

Since the optic nerve is a tract of the central nervous system, it is subject to many of the same diseases that affect other such tracts and reacts similarly to disease processes.

Foci of attenuated myelin, the nodes of Ranvier,[4,11] are present in *both* central and peripheral nerves and represent the endings of adjacent segments of myelin.

Adjacent segments of myelin abut in peripheral nerves but fall short of each other in the central nervous system, leaving short segments of axon exposed (Fig. 12–2).

INTRAOCULAR (BULBAR) PORTION

The intraocular portion of the optic nerve can be conveniently divided into three parts or layers: the inner retinal, the middle choroidal, and the outer scleral, each continuing the plane of an ocular coat across the opening of the scleral canal (Fig. 12–3).

Retinal Layer

The aggregated nerve fibers (ganglion cell axons) of the retinal nerve fiber layer exit from the eye through a circular to slightly oval aperture measuring ~ 1.5 mm in diameter. Clinically and grossly, it is represented by the *optic disc*. (An older term, optic papilla, is presumably histologic in origin and relates to the slight elevations present in this region in the mature eye, especially to the prominence of the nerve fibers on the nasal side.)

A smaller, disc-shaped depression, the *physiologic cup*, whose edge parallels that of the optic disc, lies slightly temporal to the center of the optic disc. The depression (Fig. 12–3) or excavation is lined by a glial plaque, the *central supporting tissue meniscus* (of Kuhnt) (*inset 1*, Fig. 12–4),[13] which may remain developmentally exaggerated as a *Bergmeister's papilla* or may appear relatively empty. In the latter situation the ophthalmologist can see deeply into the cup and may recognize the sieve-like perforations (see *inset 1*, Fig. 12–3) of the scleral lamina cribrosa, the apertures through which the nonmyelinated nerve fiber bundles pass out of the globe. The glial cells of the central supporting tissue meniscus, mostly fibrous astrocytes (Fig. 12–4), are covered on their anterior surface by a thin basement membrane.

Where the periphery of the physiologic cup joins with the remainder of the optic disc, the thin basement membrane of the fibrous astrocytes lining the inner surface of the central supporting tissue meniscus changes over into the thicker basement membrane of the neural retina (*inset 2*, Fig. 12–4). This ring of transition apparently outlines a ring of strong attachment of the (secondary) vitreous body to the nerve head and also the area (Martegiani) exposed to the lumen of the *canal of Cloquet* (primary vitreous).

The innermost retinal layers are occupied by the basal ends of the large atypical specialized fibrous astrocytes known as the Müller cells. About the footplates of these cells are interspersed a number of more typical astrocytes which apparently produce only short segments of thin basement membrane, in contrast to the thick basement membrane produced by the Müller cells. Another difference between these two varieties of fibrous astrocytes is that probably only the typical small astrocytes can undergo proliferation, whereas the Müller cells apparently can at best only undergo swelling, condensation, or thickening. Glial proliferations therefore are observed more frequently in the nerve fiber layer of the retina and in the region of the nerve head, where the small astrocytes are most abundant, than elsewhere in the retina. Detachments of the vitreous from the internal limiting membrane of the retina usually terminate at the strong ring-like attachment at the optic disc. A new surface or posterior hyaloid comes into continuity with

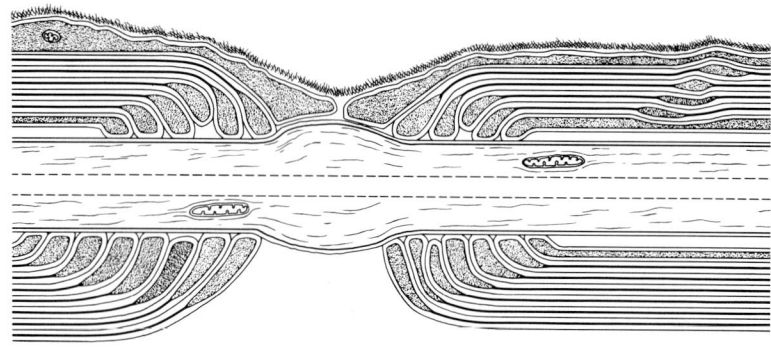

FIG. 12–2. Node of Ranvier from peripheral nerve (**above**) and node of Ranvier in central nervous system (**below**). Ends of adjacent myelin segments fall short of touching one another, and a basement membrane is lacking in node of central nervous system. (Courtesy of Dr. Peter W. Lampert.)

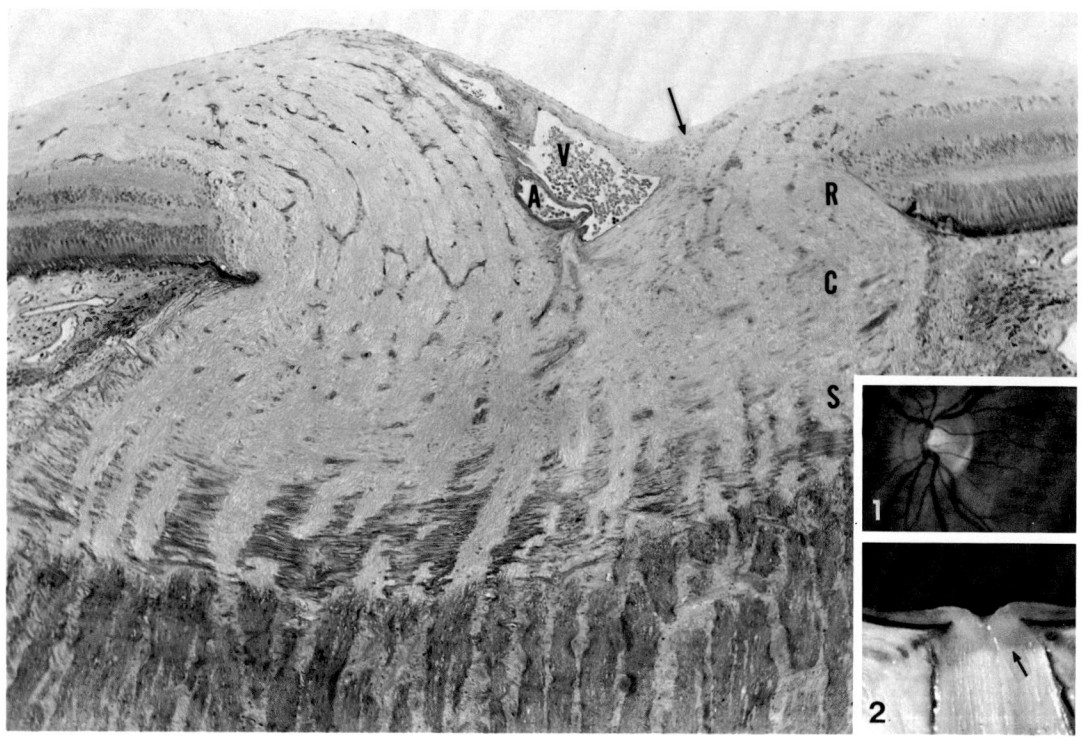

FIG. 12–3. Components of intraocular portion of optic nerve: retinal layer (R), choroidal layer (C), scleral layer (S). Plaque of glial tissue (arrow) lies in floor of physiologic cup adjacent to central retinal vessels. Artery (A) lies nasal to vein (V). Note continuity of some of *supporting tissue meniscus* with that around adjacent vein. Latter tissue, called *intercalary tissue*, separates collagenous vascular adventitia from adjacent nerve fibers. **Inset 1** shows clinical appearance of nerve head or optic disc. **Inset 2** shows nerve head on gross section. Abrupt termination of myelinated part of optic nerve at scleral lamina cribrosa is clearly seen (arrow). **Main figure,** 1.5-μ section, PD, ×56 (AFIP Neg. 62–10521).

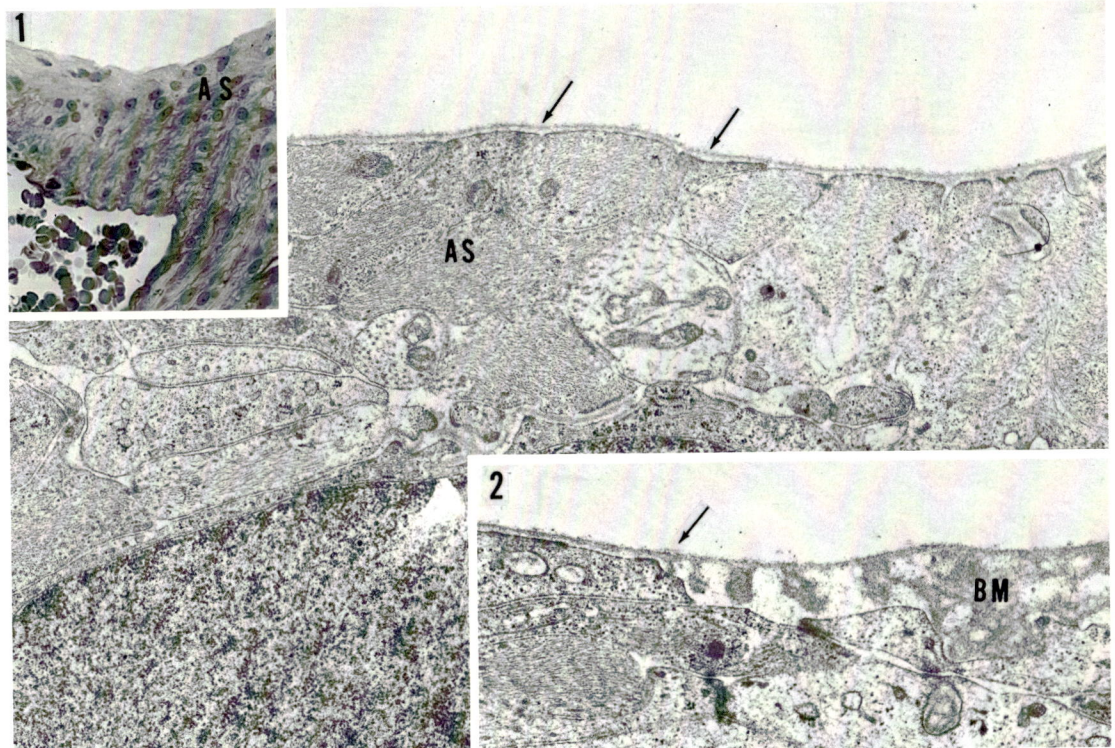

FIG. 12–4. Central supporting tissue meniscus in floor of optic cup. **Inset 1.** Light micrograph showing astrocytes (*AS*) that make up tissue. **Main figure.** Electron micrograph of astrocytes (*AS*) whose thin basement membranes line floor. **Inset 2** shows transition (*arrow*) from thin basement membrane of cup to thicker basement membrane (*BM*) near periphery of optic disc. **Main figure,** glutaraldehyde fixation, ×15,000; **inset 1,** 1.5-μ section, PD, ×220 (AFIP Neg. 69–10520); **inset 2,** ×18,000.

the nerve head. Glial proliferations or neovascularizations can readily gain access to the vitreous compartment by proliferating from the nerve head (Fig. 12–5) along the adjacent retinal surface or along this posterior hyaloid. Such new growths can produce a variety of secondary complications.

When the optic nerve head is viewed in meridional section, the nerve fiber bundles can be seen to be much thicker on the nasal side than on the temporal. The situation is often exaggerated by the slight obliqueness of the nerve head toward the temporal side. At the periphery of the nerve head, the greatly thickened nerve fiber layers of the retina are separated into large bundles by columns of fibrous astrocytes (Fig. 12–3). The posterior retinal layers are separated from the adjacent nerve fiber bundles by a thin delicate sheet of glial tissue, the *intermediary tissue* (of Kuhnt, Fig. 12–6).[13] In this plane of the nerve head the bundles of nerve fibers (axons) are nonmyelinated and closely approximated to one another.

Histologically the earliest sign of pathologic edema (i.e., papilledema) of the nerve head is a swelling of the nerve fibers in this region near the intermediary tissue. The swelling not only alters the linear appearance of the axons, but also displaces adjacent retinal tissue laterally from its usual correspondence with its pigment epithelium (Fig. 12–6).

The axonal bundles are separated by columns of fibrous astrocytes (Fig. 12–7)[2,4,8] recognized by their high content of cytoplasmic filaments.[8,9,12]

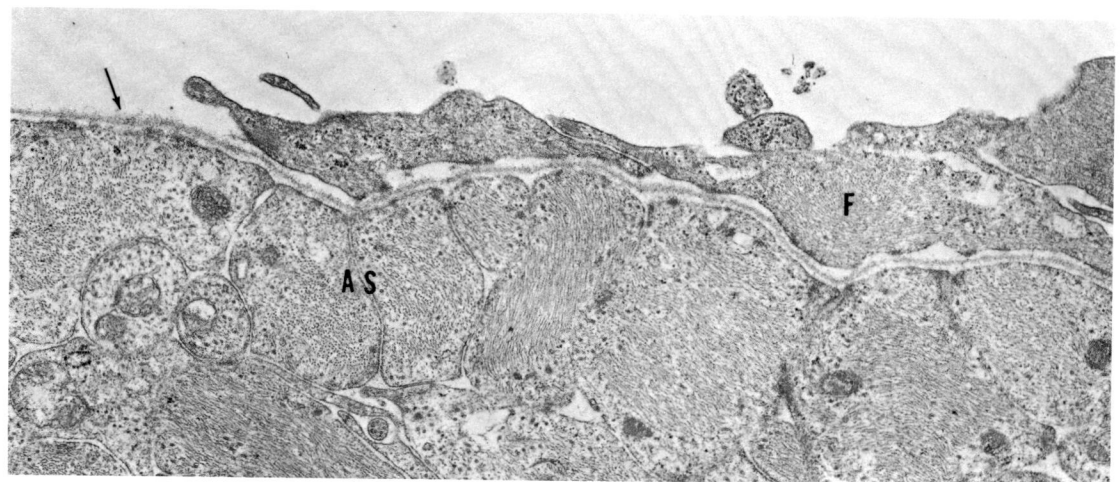

FIG. 12–5. Thin basement membrane (*arrow*) of astrocytes (*AS*) lining optic cup overgrown by layer of free cells. Filamentous (*F*) cytoplasm of these free cells indicates that they are fibrous astrocytes. Glutaraldehyde fixation, ×18,000.

Although the intracytoplasmic appearance of astrocytes varies little with the method used for initial fixation, being mostly filamentous under all conditions, the axonal cytoplasm varies widely in appearance. Initial fixation of axons in osmium tetroxide produces a high percentage of intracytoplasmic filaments (Fig. 12–8), while initial fixation in glutaraldehyde produces a high percentage of intracytoplasmic tubules (Fig. 12–9). This fixation-related change reaches its maximum in well-fixed retinas where all the filaments and tubules of the ganglion cell axons appear to be interconvertible.[5]

Centrally within the physiologic cup the retinal artery and vein enter the eye along the nasal side, with the artery lying nasal to the vein. Both vessels lie closely apposed to the surface of the adjacent thick layer of nasal nerve fibers but separated from them by their collagenous adventitia and by fibrous astrocytes. Occasionally, a thin glial sheet extends from the central meniscus deeper along the vessel bundle and is sometimes called the *intercalary* tissue (of Elschnig). The vessels then divide into their major upper and lower, nasal and temporal branches that lie within and nourish the inner retinal layers.

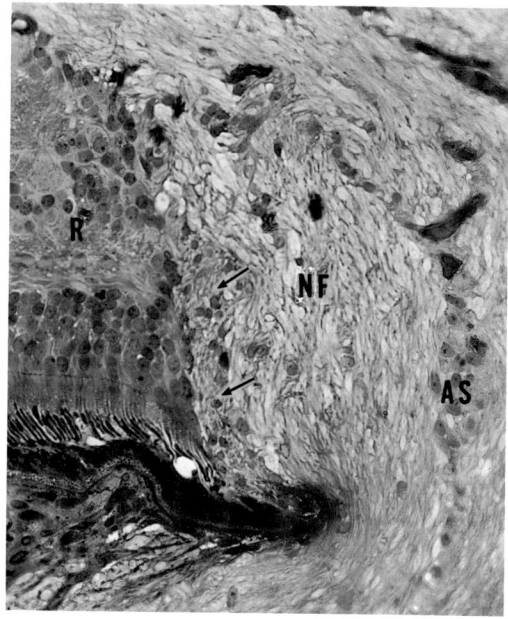

FIG. 12–6. Intermediary tissue (*arrows*) separating retina (*R*) from adjacent nerve fiber bundles (*NF*). Columns of nuclei represent astrocytes (*AS*) which separate nerve fiber bundles in the retinal layer. Note swelling of nerve fibers and slight displacement of both retina and intermediary tissue laterally. 1.5-μ section, PD, ×220. (AFIP Neg. 69-10517.)

(see below) is composed of astrocytes, collagenous connective tissue, and small blood vessels.[1,13]

In meridional section the entire lamina cribrosa shows a slight bowing outward. The concavity may be related to a normal intraocular pressure, but chronic increases in intraocular pressure, as in the chronic glaucomas, usually exaggerate it. Separation of the axonal bundles is here maintained by heavy lines of fibrous astrocytes whose filament-laden processes pass at right angles to the axons and so parallel to the direction of the choroidal plane

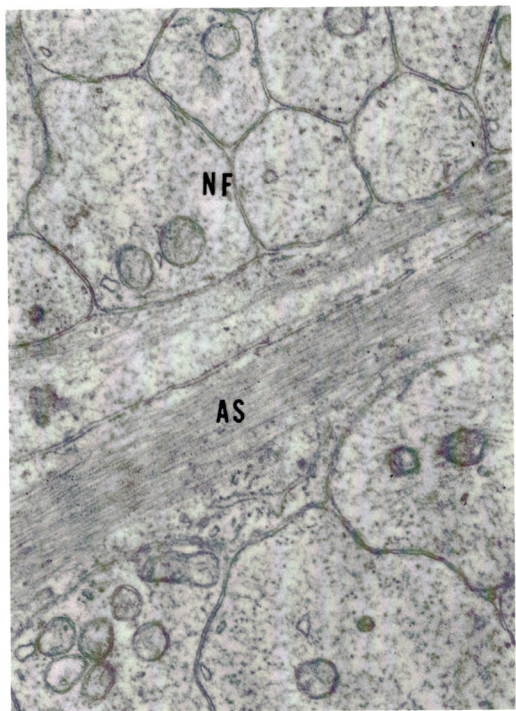

FIG. 12–7. Astrocytes (*AS*) separating bundles of nonmyelinated nerve fibers (*NF*) or axons in region of choroidal layer. Astrocyte is characterized by its cytoplasmic content of filaments. Axons do not possess microtubules in this preparation. Dalton's chrome-osmium fixation, ×18,000.

Choroidal Layer

Approximately in line with the choroidal coat, the nonmedullated nerve fiber *bundles* are even more clearly recognizable as bundles or columns. It is in this region that the lamina cribrosa begins.

Lamina cribrosa (Fig. 12–10), a term generally applied to the striated portions of the bulbar optic nerve, is separable into two parts: the choroidal lamina (lamina choroidalis), present in the choroidal layer is composed mostly of astrocytes, and the scleral lamina (lamina scleralis), present in the scleral layer

FIG. 12–8. In osmium-fixed tissue intracellular filaments (*FIL*) of astrocyte are clearly seen. *BM*, astrocytic basement membranes; *DE*, obliquely sectioned desmosome. Adjacent axons appear to contain mostly filaments. Dalton's fixation, ×24,000.

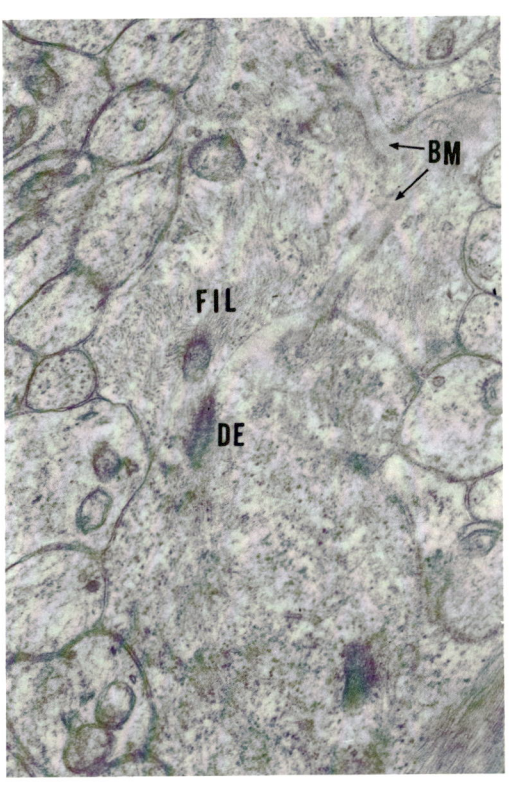

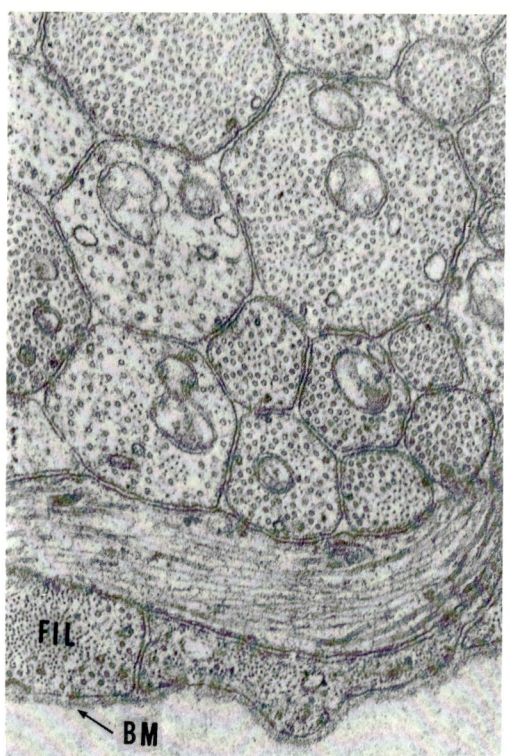

FIG. 12–9. In tissue initially fixed in glutaraldehyde, astrocytes contain mostly filaments (*FIL*), while adjacent axons contain mostly tubules. *BM,* basement membrane of astrocytes. Rhesus monkey, glutaraldehyde perfusion, ×30,000.

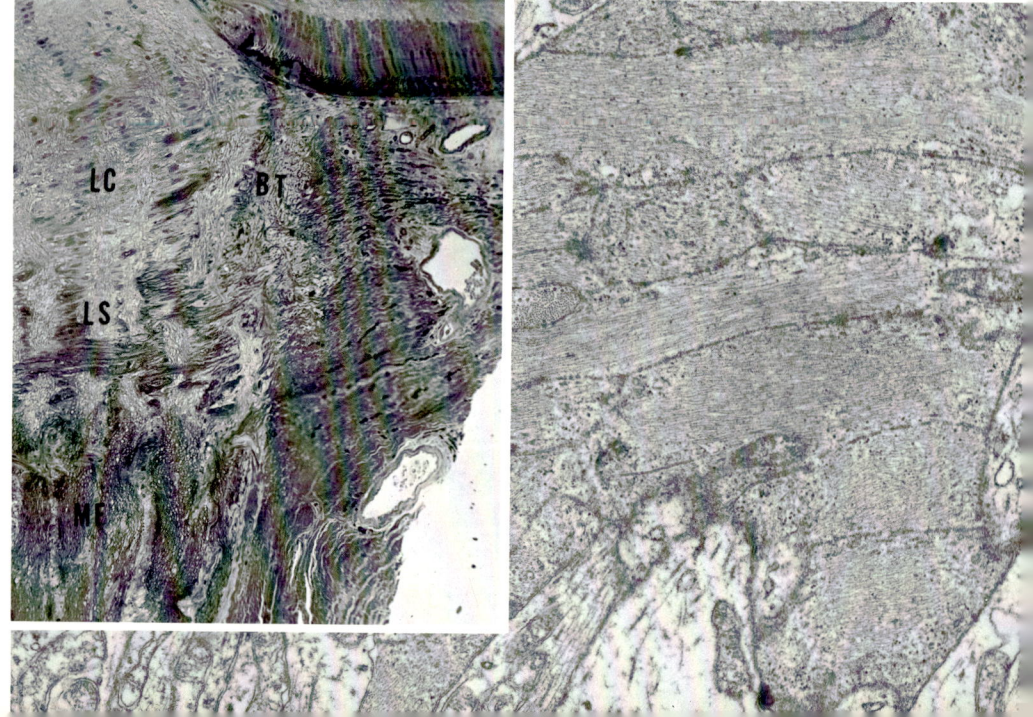

FIG. 12–10. Lamina cribrosa. **Inset.** Light micrograph showing division of lamina cribrosa into lamina choroidalis (*LC*), composed mostly of astrocytes, and lamina scleralis (*LS*), composed of astrocytes, collagen, and small blood vessels. Border tissue (*BT*) separates choroid proper from lamina choroidalis. *MF,* myelinated nerve fibers (retrobulbar). **Main figure.** Electron micrograph showing dense packing of fibrous astrocytes that make up lamina choroidalis. Axons contain microtubules in this preparation. **Main figure,** glutaraldehyde-osmium fixation, ×18,000; **inset,** 1.5-μ section, PD, ×80 (AFIP Neg. 69–10518).

(Fig. 12–10). The astrocytes are identified in conventional microscopic sections by long columns of nuclei, the nuclear columns (Figs. 12–6 and 12–10). Collagenous connective tissue is absent or sparse in the choroidal region except in the area around the *central* vessels (vessel adventitia). Peripherally, a thick looser tissue, the *border* tissue (*inset*, Fig. 12–10) separates the neural elements from the choroid. The border tissue may also contain a number of pigmented cells.

Scleral Layer

In continuity with the scleral coat, the bundles of nonmyelinated nerve fibers are separated by clearly recognizable heavier columns formed by glial cells and collagenous tissue (*inset*, Fig. 12–10) the lamina scleralis. The glial cells, mostly fibrous astrocytes, separate the nerve fibers from the collagenous septa (Fig. 12–11). The fibrous astrocytes commonly attach to one another by desmosome-like attachments (Fig. 12–12). The septa contain the small vascular branches derived from the adjacent scleral vascular complex known as the *circle of Zinn-Haller*, which in turn is derived from the adjacent penetrating branches of the short posterior ciliary arteries.

ORBITAL (RETROBULBAR) PORTION

The orbital portion of the optic nerve extends from the lamina cribrosa to the apex of the orbit in a sinuous manner, permitting freedom of movement of the eye.

Myelination of the retinal ganglion cell axons appears rather abruptly (Fig. 12–13), immediately behind the scleral lamina cribrosa and therefore, actually slightly within

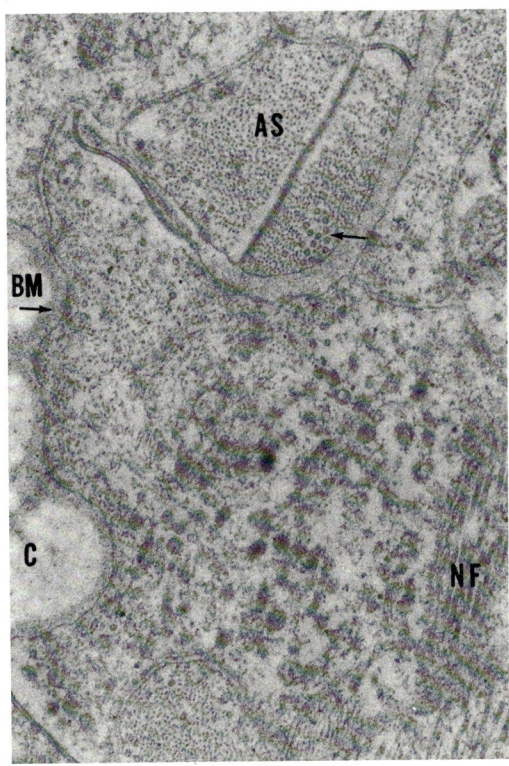

FIG. 12–11. Astrocytes (*AS*) separating nonmyelinated nerve fibers (*NF*), recognized here by their microtubular content, from adjacent collagen (*C*) in region of lamina scleralis. *BM*, basement membrane of cells whose cytoplasmic content is mostly filamentous. A few microtubules (*free arrow*) are present sporadically among filaments. Rhesus monkey, glutaraldehyde fixation, ×35,000.

FIG. 12–12. Fibrous astrocytes (*AS*) adherent to each other by desmosome-like attachments (*arrow*) at free (inner) surface of optic cup. ×18,000.

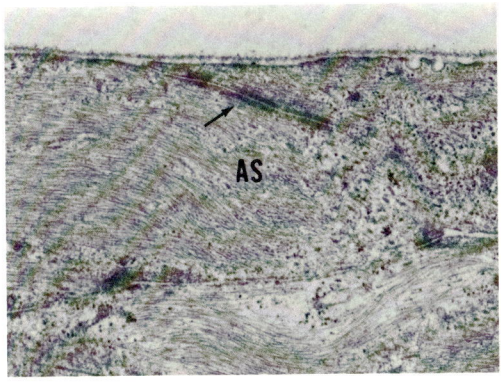

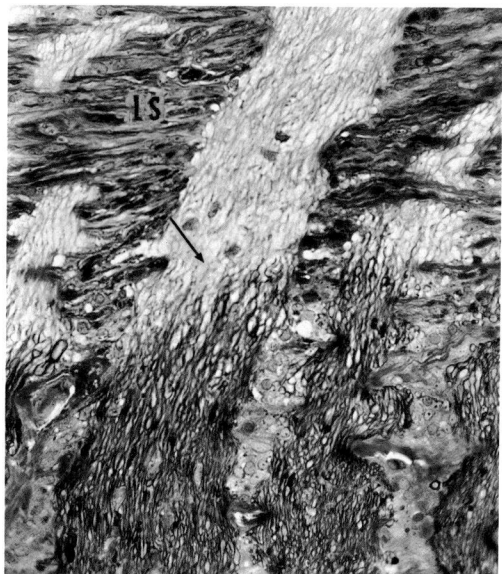

FIG. 12–13. Transition between myelination and nonmyelination (*arrow*) within single bundle of nerve fibers (axons). Collagenous vascular septa of lamina scleralis (*LS*) are prominent. 1.5-μ section, PD, ×360. (AFIP Neg. 69–10516.)

the scleral canal. Myelination doubles the cross-sectional thickness of the nerve so that it rapidly reaches its maximum diameter of ∼ 3 mm at the posterior surface of the sclera (*inset 2*, Fig. 12–3).

In the embryo myelination begins centrally within the brain and proceeds peripherally to cease abruptly at the lamina cribrosa. Myelination may occasionally be observed in portions of the nerve fiber layer of the retina, but even in such instances the region of the lamina cribrosa is typically nonmyelinated (see Chap. 6). Even though myelin may be seen on either side.

Sheaths and Their Spaces

The optic nerve is surrounded by the continuation of the three connective tissue meningeal sheaths that line the cranial cavity and encase the brain (Fig. 12–14): externally, the thick *dura mater*, which fuses distally with the outer layers of the sclera; internally, the thin, vascularized, tightly applied connective tissue sheath, the *pia mater*, covering the nerve proper; and in between, the delicate cellular connective tissue network, the *arachnoid*.

The narrow space between the arachnoid and the dura is but a potential continuation of the subdural space from the cranial cavity, whereas the wide, easily enlarged space between the arachnoid and pia is a true continuation of the subarachnoid space covering the brain. The central retinal artery and vein cross the spaces to enter the optic nerve. Before entering the nerve, the vein frequently travels for some distance between arachnoid and pia in the subarachnoid space, whereas the artery generally takes a shorter and more direct route.

At the optic canal the dural sheath splits into two layers. One layer becomes continu-

FIG. 12–14. Sheaths of optic nerve. Dura (*D*) and arachnoid (*A*) continue from cranial cavity to envelop optic nerve within orbit. *P*, pia mater; *S*, pial septum. H&E, ×70. (AFIP Neg. 70–7288.)

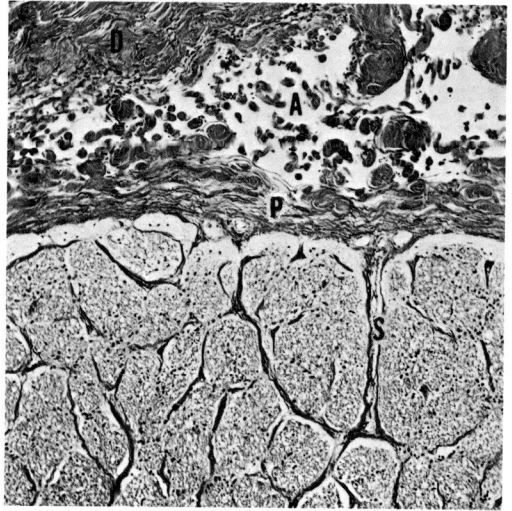

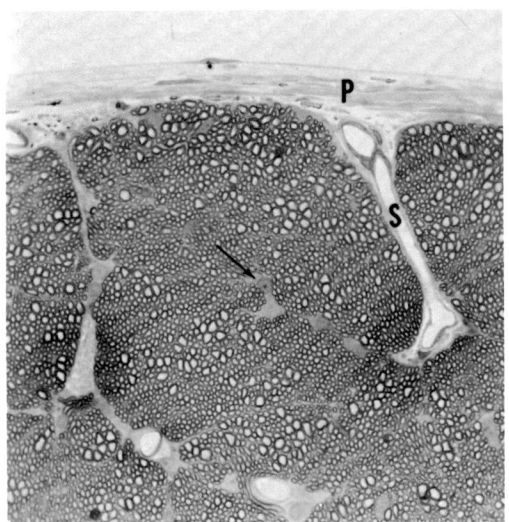

FIG. 12–15. Septa (S) carrying blood vessels into optic nerve from pia (P). Myelinated nerve tissue is separated into nerve fiber bundles. Nucleus and perikaryon of a glial cell (arrow) can be seen among myelinated axons. Rhesus monkey, 1.5-μ section, PD, ×300. (AFIP Neg. 66-8678.)

ous with the periosteum of the orbit, and the other adheres closely to and follows the optic nerve itself, finally becoming indistinguishable from the outer layers of the sclera.

The arachnoid sheath is composed of collagenous tissue and typical meningothelial cells. Delicate fibrous septa cross the subarachnoid space from the arachnoid sheath to the pial sheath, dividing the subarachnoid space into many interconnecting compartments. With aging, laminated calcareous bodies, *corpora arenacea*, may be produced by the meningothelial cells.

Intraneural Components

The collagenous pial sheath (Figs. 12–15 and 12–16) enters the nerve, and the vascularized *pial septa* subdivide the now myelinated nerve fiber bundles into columns. The pial septa blend with the lamina cribrosa (scleralis) anteriorly and with the *central connective tissue* axially around the central vessels of the optic nerve until the latter exit from the nerve inferiorly some 8 to 15 mm behind the globe.

As in the central nervous system, the pial septa are always isolated from the neural elements by a continuous layer of fibrous astrocytes.

In longitudinal section (Fig. 12–17) the myelinated nerve fiber bundles appear to be interrupted. Because the myelinated nerve fibers (axons) continue to their termination in the lateral geniculate body, the apparent interruptions are due to the variability in plane of section and to incomplete encirclement by the collagenous pial septa in any single plane.

Above and below a plane of section the pial septum is continuous around the nerve fiber bundles which are separated from the septal system by a layer of astrocytes.[2,4,13] In

FIG. 12–16. Astrocytic glial processes (G) resting on basement membrane (arrow) separating myelinated nerve fibers (MF) from collagenous pia. Rhesus monkey, ×16,500.

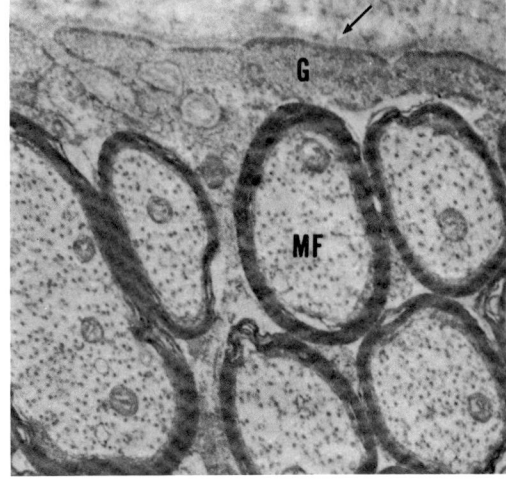

both cross and longitudinal section many nuclei are seen. Those in intimate association with pial septa are probably astrocytic. Those unassociated with pial septa may be either astrocytic or oligodendrocytic.* Distinction between these glia cannot be made from a single conventional microscopic section. It is highly probable that very few of the nuclei observed in any single section belong to oligodendrocytes.

VASCULAR SUPPLY

The central retinal artery and vein penetrate the optic nerve from 8 to 15 mm behind the globe to assume an axial position within the nerve (Fig. 12–18). The artery generally penetrates into the nerve at the same level or anterior to the vein. In its course through the optic nerve, the artery may give off small branches to supply the adjacent nerve fibers. The vascular supply to the optic nerve is extraordinarily rich, especially at its intraocular end where it anastomoses with branches of

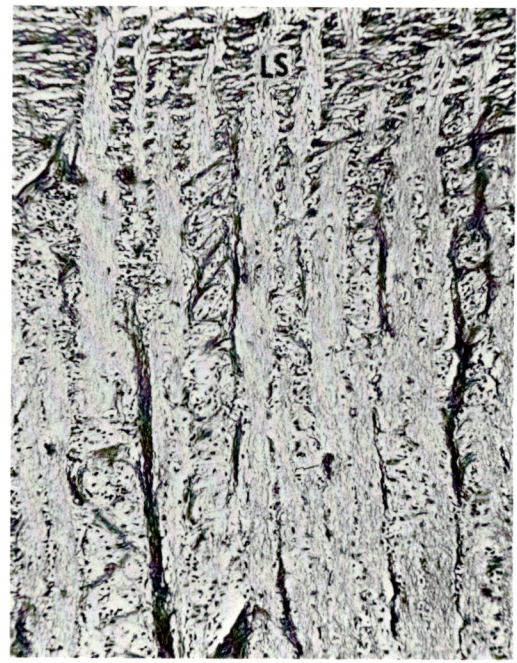

FIG. 12–17. Longitudinal section of orbital portion of optic nerve near lamina scleralis (*LS*). H&E, ×80. (AFIP Neg. 70-7291.)

---

* In recent years electron microscopic identification and the nomenclature of the various types of glial cells (astrocytes, oligodendroglia, and all their variations) have elicited considerable controversy.[9,11] Characterization of these cells in the past has depended upon special "coatings" or "staining" with metallic salts (e.g., silver, gold), either positive or negative, and the subsequent appearance by light microscopy. Attempts to correlate these silver salt impregnations for light microscopy with observations made by electron microscopy generally have been unsatisfactory. At present there is considerable agreement that a glial cell (i.e., nonneuronal cell) containing a *large* number of intracytoplasmic filaments should be designated a "fibrous astrocyte" and that one with a *few* such filaments, but with a large quantity of ribosomes and a dense cytoplasmic ground substance, should be designated an "oligodendrocyte." Other criteria such as the presence of glycogen also assist in identifying an astrocyte. With initial glutaraldehyde fixation, the oligodendrocytic cytoplasm develops microtubules, but most of the cytoplasmic filaments of the astrocyte remain essentially unaltered (see Fig. 12–11). Microglia are currently believed to be monocytes that wander in from the blood stream.

the choroidal system to form the previously mentioned circle of Zinn-Haller.

The afferent blood supply to the nerve head is derived mainly from the central retinal artery in the retinal portion, the arterial circle of Zinn-Haller (lying within the sclera) and the choroidal (or ciliary) vasculature in the choroidal lamina.[6]† Drainage of both retinal and choroidal layers appears to be in great part via the central retinal vein and its branches.[6,7] The blood supply to the choroidal lamina is composed of end vessels meridionally arranged and grouped in sectors. Significant ciliary (or choroidal) contribution to the blood supply of the choroidal lamina is supported by intravenous fluorescein studies which show prompt filling of the vessels along with the adjacent choroidal vasculature long before the retinal arterial phase begins.[6]

† Termed *prelaminar* region by Hayreh.[6]

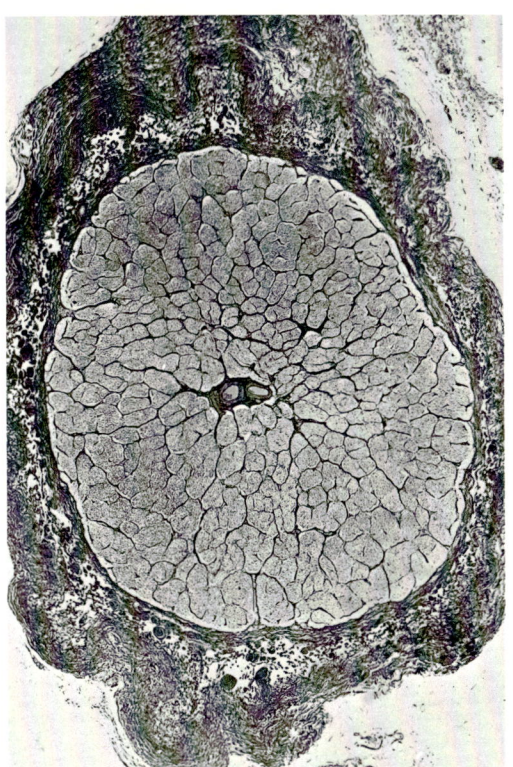

FIG. 12–18. Cross-section of orbital portion of optic nerve within 8 to 15 mm of the globe (before exit of central vessels). H&E, ×20. (AFIP Neg. 70–7290.)

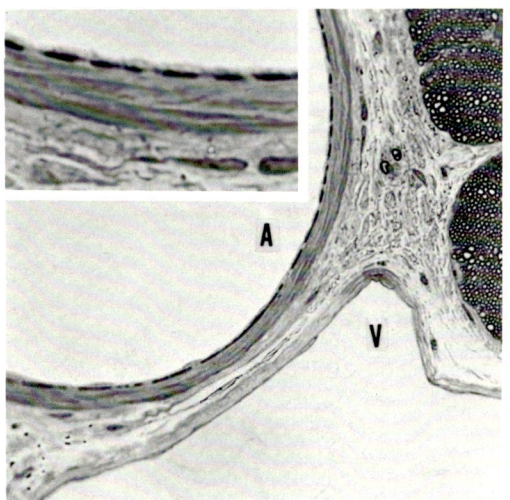

FIG. 12–19. Light micrographs of rhesus monkey central retinal artery (A) and vein (V) in nerve head just behind lamina scleralis. Internal elastic lamina of artery is fenestrated. **Inset** shows discontinuous internal elastic lamina beneath endothelial lining. **Main figure,** PD, ×300 (AFIP Neg. 66–8671); **inset,** ×900 (AFIP Neg. 66–8670).

Filling of the vessels of the choroidal lamina appears most prominently on the temporal side.[6]

The central retinal artery has a fenestrated internal elastic lamina (Figs. 12–19 and 12–20). The intima (i.e., all tissue on the luminal side of the internal elastic lamina) frequently contains a number of muscle-like cells or "myointimal" cells (Fig. 12–20).[3]

Smooth muscle cells not only project through the fenestrated internal elastic lamina of larger arteries such as the aorta, but are also found wholly internal to the membrane (i.e., in the subendothelial space). There is some evidence that these myointimal cells are of importance not only as a possible source for regeneration of endothelial cells[14] but also in the development of atherosclerosis of the vessel wall.[10]

Shortly after the vessels enter the retina the internal elastic lamina is lost entirely (see Fig. 6–66), and the vessels are no longer considered arteries but rather arterioles.

The thin-walled central retinal vein is separated from the surface layer of astrocytes by a thick adventitial layer of collagen (Fig. 12–21). Only a few scattered cells are present within this collagenous layer, and the cells associated with the endothelium are discontinuous and few in number.

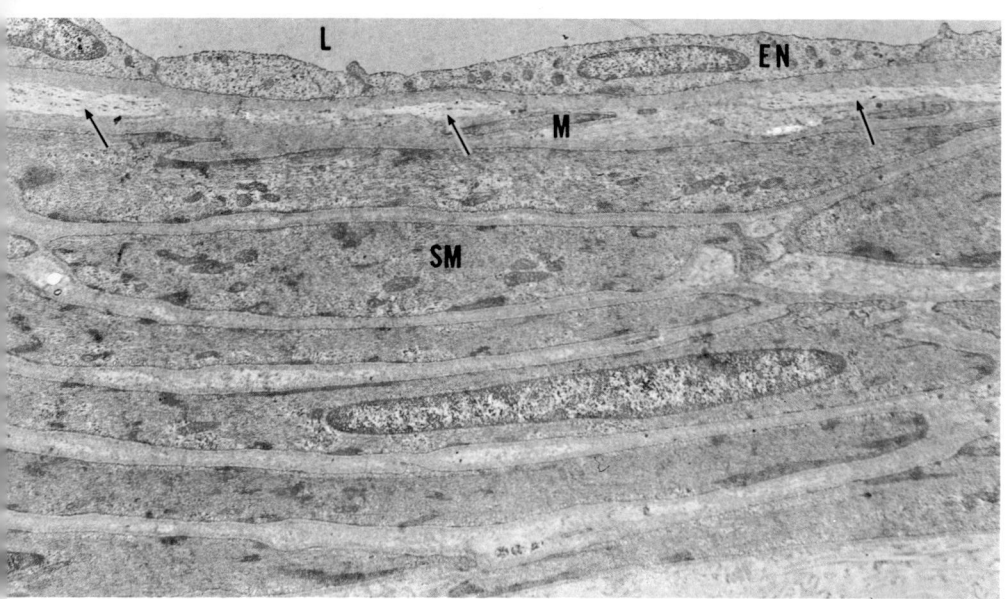

FIG. 12–20. Electron micrograph of portion of artery illustrated in Figure 12–19. Intermittency of internal elastic lamina (arrows) beneath endothelium (EN) is apparent. Fragment of muscle-like cell (M) is seen here in plane of elastica. Typical smooth muscle cells (SM) surrounded by their basement membrane make up remainder of wall of artery. L, lumen. ×4,700.

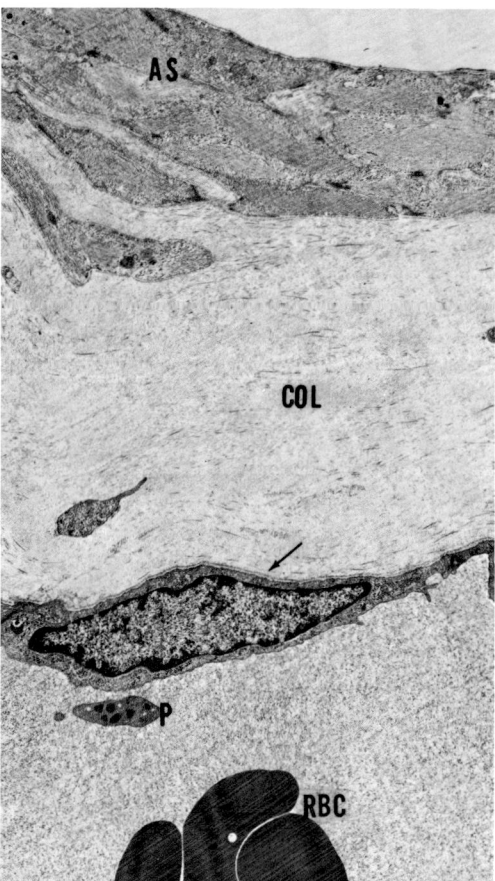

FIG. 12–21. Thin-walled central retinal vein (see Fig. 12–3), containing red blood cells (RBC) and a platelet or fragment of leukocyte (P), separated by its thin basement membrane (arrow) and a thick collagenous adventitia (COL) from surface layer of astrocytes (AS). Thin basement membrane lining free surface of optic cup is almost imperceptible at this magnification. Glutaraldehyde fixation, ×4,800.

REFERENCES

1. Anderson, D. R. Ultrastructure of human and monkey lamina cribrosa and optic nerve head. *Arch Ophthal (Chicago) 82:*800, 1969.
2. Anderson, D. R., and Hoyt, W. F. Ultrastructure of intraorbital portion of human and monkey optic nerve. *Arch Ophthal (Chicago) 82:*506, 1969.
3. Buck, R. C. Intimal thickening after ligature of arteries: An electron microscopic study. *Circ Res 9:*409, 1961.
4. Cohen, A. I. Ultrastructural aspects of the human optic nerve. *Invest Ophthal 6:*294, 1967.
5. Fine, B. S. Observations on the Axoplasm of Neural Elements in the Human Retina. In Titlback, M. (ed.). *Proceedings, Third European Regional Conference on Electron Microscopy.* Prague, Czechoslovak Academy of Sciences, 1964, p. 319.
6. Hayreh, S. S. Blood supply of the optic nerve head and its role in optic atrophy: Glaucoma and oedema of the optic disc. *Brit J Ophthal 53:*721, 1969.
7. Henkind, P., and Levitzsky, M. Angioarchitecture of the optic nerve. *Amer J Ophthal 68:*979, 1969.
8. Lampert, P. W., Vogel, M. H., and Zimmerman, L. E. Pathology of the optic nerve in experimental acute glaucoma: Electron microscopic studies. *Invest Ophthal 7:*199, 1968.
9. Mugnaini, E., and Walberg, F. Ultrastructure of Neuroglia. In *Reviews of Anatomy, Embryology and Cell Biology,* No. 37. Berlin, Springer, 1964.
10. Parker, F., and Odland, G. F. A correlative histochemical, biochemical and electron microscopic study of experimental atherosclerosis in the rabbit aorta with special reference to the myo-intimal cell. *Amer J Path 48:*197, 1966.
11. Peters, A., Palay, S. L., and Webster, H. DeF. *The Fine Structure of the Nervous System: The Cells and Their Processes.* New York, Hoeber, 1970.
12. Peters, A., and Vaughn, J. E. Microtubules and filaments in the axons and astrocytes of early postnatal rat optic nerves. *J Cell Biol 32:*113, 1967.
13. Salzmann, M. *The Anatomy and Histology of the Human Eyeball in the Normal State: Its Development and Senescence.* Tr. by E. V. L. Brown. Chicago, University of Chicago Press, 1912.
14. Ts'ao, C. H. Myo-intimal cells as a possible source of replacement for endothelial cells in the rabbit. *Circ Res 23:*671, 1968.

index

Page numbers in italics refer to illustrations

Acetylcholine, 67
Acid mucopolysaccharide (AMP), 34–35, 36 (table), 57, 61, 62, 64, 111, 113, 121, 142, 148, 170, 193, 197, 198
Adenosine diphosphate (ADP), 19
Adenosine triphosphate (ATP), 19
Adventitia, 88, 94, 238
Agranular endoplasmic reticulum, 17, 18, 50, 51, 51, 57, 60, 72, 219
Alcian blue, 34, 36 (table), 148, 224
Alpha glycogen particles, 22
Aluminum specimen mount, 5, 6
Amacrine cells, 55, 58, 71, 73, 74, 75
Amino acids, 32, 33
Ampulla of choroidal vessels, 187
Angle
  anterior chamber, 214 ff.
  drainage, 215, 217, 219–222, 224, 229 and n., 230
Anterior attachments of vitreous body, 113, 119–120, 121
Anterior chamber angle, 214 ff.
Anterior subretinal cul-de-sac, 103, 104
Apical plasma membrane, 42, 51
Aqueous, 214, 215
Aqueous chambers, 214, 214
Aqueous tubes, 227
Aqueous veins, 162, 222, 224
Arachnoid, 242, 242
Arciform density, 71, 72, 73
Arterioles, 88, 90, 91, 93, 95, 245
Arteriovenous crossing, 94
Artery(ies)
  central retinal, 88, 93, 238, 242, 244, 245, 245, 246
  ciliary
    anterior, 161, 209, 211
    posterior, 160, 161, 185, 209, 211, 241
  cilioretinal, 93
ASCHER, K. W., 222
Astrocytes, 30 n., 87, 92, 94, 238, 238, 239, 239, 240, 241, 244 n., 245, 246
  fibrous, 55, 58, 86, 87, 88, 235, 237, 238, 238, 239, 240, 241, 241, 243, 244 n.
  protoplasmic, 55, 58, 86
  See also Müller cells
Atherosclerosis of central retinal artery, 245

Autophagocytosis, 24
Axenfeld, nerve loop of, 161, 161
Axons, 61, 62, 67, 70, 72, 73, 75, 76, 78, 82, 93, 234, 235, 237, 238, 239, 241, 241, 243
Axoplasm, 69, 76, 82

Banding in collagen fibrils, 33, 33
Basal body, 20, 21, 59 n.
Basement membrane facets, 79, 84
Basement membranes, 34, 35, 37, 37, 38, 38, 42, 42, 73, 235, 237, 246
  and canal of Schlemm, 215, 220, 221
  of cornea, 142, 142, 144, 156, 156
  of retina, 48, 49, 50, 51, 54, 78, 83, 88, 99, 116, 196, 235
  Schwann cells enveloped by, 235
  of uveal tract, 176, 183, 187, 189, 190, 193, 199, 202, 205, 207, 208, 209, 210, 211, 225, 226
  vitreous framework attached to, 116
Basilar plasma membrane, 42, 43, 71
Basophils, 43
Berger, retrolental space of, 120
Bergmeister's papilla, 235
BERMAN, E. R., 105
Beta glycogen particles, 22
Bipolar cells of retina, 10, 10, 58, 71, 72–74, 76
Blood cells, red, 43, 177, 183, 218, 220, 223, 246
Bowman's membrane (layer), 142, 142, 143, 143 and n., 144, 146, 147, 148, 150, 156, 215
Bruch's membrane, 48–49, 50, 188, 189

Cambridge scanning electron microscope, 6
Canal
  of Cloquet, 113, 113, 116, 119, 120, 121, 235
  of Hannover, 125
  of Petit, 125
  of Schlemm, see Schlemm, canal of
Capillaries
  of ciliary crests, 208
  retinal, 88, 89, 95
Capillary bed, 58
Cataract, nuclear, 139
Cationic dyes, 34

Cell(s)
  amacrine, 55, *58*, 71, 73, 74, 75
  apex of, 42
  base of, 42
  bipolar, of retina, 10, *10*, 58, 71, 72–74, *76*
  clump, of Koganei, 175–176, *178*, *179*
  columnar, of nonpigmented epithelial layer of posterior pars plana, 42
  cuboidal, of lens epithelium, 42
  effect of growth and differentiation of, on configuration of cell membrane, *38*, 38–39
  endothelial, 42, 88, *89*
  goblet, in bulbar conjunctival epithelium, 143
  horizontal, 55, *58*, 69, 71, 72, 73
  mast, 43
  mesothelial, 42
  Müller, see Müller cells
  myointimal, 245
  periganglion satellite, *80*
  plasma, 43
  red blood, 43, *177*, *183*, *218*, *220*, *223*, 246
  Schwann, 234, 235
  squamous, of corneal epithelial surface, 42
  wing, of corneal epithelium, 143, *145*
Cement substance, 28
  See also Intercellular cement
Centrioles, 20, 21, *21*
Cerebrosides, 35
Chamber angle, anterior, 214 ff.
Chiasm, 78
Choriocapillaris, 49, *50*, 88, 187, *188*, 188, *189*, 206, *207*
Choroid, 160, *185*, 185–186, *186*
  detachment, 185
  pigmented and nonpigmented cells of, *186*, 186–187, *187*
  stroma of, 185, *186*, 186–187, 188
  vessels of, 187–188, *188*, *189*, *189*
Chromatin aggregation of nucleus, 3, 11
Chrome-osmium fixative, Dalton's 11, 239
Chromium, used in shadow-casting, 5
Cilia, 20, 21, *21*
Ciliary body, 189 and n., 191, *191*, *204*
  components of, 191
  nonpigmented epithelium of, 192, *192*, 193, *193*, *194*, *195*, *196*, 198, *198*, 199, 201, 202, 203, *204*, 205, 206
  and pars plana, 189, 191, *191*, 197, 198, 199, 203, 206, *206*, 210
  and pars plicata, 189, 191, *191*, 199, 201, 202, 203, 206, 210

Ciliary body (*continued*)
  pigment epithelium of, see Pigment epithelium, of ciliary body
  retinal (neuroepithelial) portion of, 192, *192*, 193 ff.
  uveal (mesodermal) portion of, 207–209, 210, 211, *211*
Ciliary crests, 199, *202*, *203*, *205*, 206, 207, 208, *210*, 211
Ciliary muscle, 208, *210*, *211*, 217, 225, 226, *228*
Cilioequatorial zonular fibers, 126
Circle
  major arterial, within uveal tissue, 211
  of Zinn-Haller, 241, 244
Circumferential furrows on posterior iris surface, 180, *182*
Circumpapillary subretinal cul-de-sac, 104, *105*
Citric acid cycle, Krebs, 19
Cloquet, canal of, 113, *113*, 116, 119, 120, 121, 235
Clump cells of Koganei, 175–176, *178*, *179*
Collagen, 32, 33, *33*, 34, 49, 112, 115, 147, *147*, 148, 150, 228, 241, 245
Collagen fibrils, 88, 225
  banding in, 33, *33*
  of choroidal stroma, 188
  of ciliary body, 207
  of iris, 170, *171*, *172*
  scleral, 162, *163*, 215, *216*
Compound granules in pigment epithelial cells, 52, *54*
Compound light microscope, 2, *3*
Condenser lens of transmission electron microscope, 2
Cones of retina, 57, *57*, 59, *59*, 60 n., 61, 62, *62*, 66, 70, 98, *100*, *102*
Connective tissue, 42, *42*, 43, 49, *49*, 51, 243
  juxtacanalicular, 217, *218*, 219, *219*, 220, 221, *221*, *223*, 227
Cornea, 142 ff.
  central, 142, *143*, 156, *156*, 158
  fetal, *152*
  peripheral, 142, 148, 156, *156*, 159
  slit-lamp appearance, 142
Cornea guttata, 158, 225
Corona ciliaris, see Ciliary body, pars plicata
Corpora amylacea, 87, *87*
Corpora arenacea, 243
Cortex, lens, 132, *136*, 137, *137*, 138, *138*, 139
Counterions, 34
Cristae mitochondriales, 19, *19*

Crypts in border layer of iris, 169
Crystal violet, 34
Crystalline materials, 35, 188
Cup
  optic, 48, *48*, 130, *237, 246*
  physiologic, 235, *236*, 238
Cystoid degeneration of peripheral retina, 103, *104*
Cystoid spacing in retina, 98
Cysts
  of iris pigment epithelium, 180, 185
  of pars plana (retina), 104
Cytoplasm, 16
  ganglion-cell, 10, *10*
  of lens cortical cell, 137, *137*, 138
  of Müller cells, *see* Müller cells, cytoplasm of
  vacuolation of, 3
Cytoplasmic inclusions, 22–25, 95
Cytoplasmic organelles, 17–22, 76, 95
Cytosomes, 24

Dalton's chrome-osmium fixative, 11, *239*
Degeneration, cystoid, of peripheral retina, 103, *104*
Dendrite(s)
  of bipolar cells, 62, 67, *71*, 72, *72*, *73*, *74*, *75*, 77
  of ganglion cells, 75, *76*, *78*, *79*
Density, optical, 139
Descemet's membrane, 34, 38, *38*, 43, 142, *142*, *143*, 143 and *n*., 148, 150, *153*, 154, 156, 158, *159*, 214, 215, 217, 222, 223, 225, *225*, *230*, 231
Desmosine, 35
Desmosomes, 21, *28*, 28–29, *29*, 72, 73, 143, 145, 179, 185, *193*, *196*, 198, *239*
Desoxyribonucleic acid (DNA), 18
Diabetes, 88, 91, 208
Dialdehydes (CHO–CHO), 35
Diastase, and glycogen detection, 8, *9*, 22, 35
Dilator muscle of iris, 176, *178*, *179*, *181*, 183
Drusen, 189, *190*
Dura mater, 242, *242*
Dyad configuration, 75 and *n*.
Dystrophy, Fuchs' Combined, 150, 158

Ectoderm, 48, *48*
  ocular epithelia derived from, 42

Ectoplasm, definition of, 16
Ectropion, physiologic, of iris pigment epithelium, 181
Ectropion uveae, 181
Edema
  corneal stroma, 150
  optic nerve head, 237
  retinal, 98
Edinger-Westphal nucleus, 175
Egger's line, 119
Elastic lamina, *see* Lamina elastica
Elastic tissue, 35, 188, 207, *208*
Elastin, 35
Electron-dense structures, 10
Electron-lucent structures, 10
Electron microscopy, 2, *3*, 16, 244 *n*.
  photographic techniques in, 9–13
  special examination techniques in, 4–9
Elschnig, intercalary tissue, 238
Embedding of tissue, 2, 3, 5
Emissaria, scleral, 159–162
Endoplasm, definition of, 16
Endoplasmic reticulum, 17, 17–18, *18*, 43, 50, 51, *51*, 57, 60, 72, 80, 86, 97, 194, *195*, *198*, 219
Endothelium, 42, *42*, 88, 89
  corneal, 142, *143*, 150, *153*, 154, 158, *159*
Eosin, 11 *n*., 16
Eosinophilia, 139
Eosinophils, 43
Ependyma, 57
Episclera, 160, *161*, 162
Epithelium, 42, *42*, 48
  ciliary, 192, *192*, 193, *194*, 199, 206, 208. *See also* Pigment epithelium, of ciliary body
  corneal, 142–143, *143*, *144*, *146*, 150
  lens, 130, *131*, 132, 135, *136*
  pigment, *see* Pigment epithelium
Epon, specimen embedded in, 5, 7, 12
Equatorial plane of eye, 44, *44*
External limiting membrane, 57, *57*, *58*, *59*, 61, 68, *104*, *105*
Extrafoveal cone, and pigment epithelium, 61, 67
Exudates, retinal, 98
Eye
  conventions used in describing, 43–45
  described in terms of three-coat (trilaminar) arrangement, 43–44
  described in terms of three-tissue arrangement, 45

Eye (*continued*)
    equatorial plane of, 44, *44*
    meridional plane of, 44, *44*
    opened, *214*
        aqueous compartment of, 214
        vitreous compartment of, 214
    orientation of, for descriptive purposes, 44–45

Fascia adherens, 30, 193
Fascia occludens, 30, 193
Fenestrated membrane, 29, 30, *104*
Ferritin, *222*, 228
Ferrocyanide, 34
Fibers
    connecting, 61, *68*
    definition of, 33
    Henle, *see* Henle fibers
    Müller, 86
    zonular, of lens, 121, *122*, *123*, *124*, 125, *125*, 126, 134, 135
Fibril, definition of, 33
Fibroblast, 43, 112, 150, 221
Fibrous materials, *32*, 32–34, 43
Filaments
    definition of, 33
    intracellular, 21, *22*, 86, *211*
    intracytoplasmic, 238, 244 *n.*
    vitreous, *84*, 111, 112, *112*, 113, *114*, 115, *115*, 117, 118, 119, *119*, 120, *121*
    of "young" collagen, 33
Fixation of tissue, 3–4, 7, 12
Fluorescein angiography, 52, 244
Formaldehyde, 21
Fovea, 53, 61, 93, 95, *96*, 98–99, *99*, 100, *102*, *111*, 116
    clivus of, *99*, 117
Fuchs' Combined dystrophy, 150, 158
Fuchs' (Michel's) spurs, 175, *177*, *178*

Galactosamine, 36 (table)
Ganglion cell layer (GCL), 57, 76, 95, *96*, 97
Gap junction, 30 *n.*
Glaucoma, 239
Glia, accessory, 86–88
Glial system of neural retina, 53, 55, 58, 86–88
Gliosis, retinal, 86, 87
Globe, *see* Eye

Glucosamine, 36 (table)
Glucuronic acid, 36 (table)
Glutaraldehyde, 4 and *n.*, 11, 12, 21 22, 72, 73, 76, 78, *81*, 86, 238, 244 *n.*
Glycogen, 22, *59*, 67, 86, 88, 201, 244 *n.*
    detection of, 8, *9*, *9*, 22, 35
    glycoprotein differentiated from, 35
Glycolipids, 35
Glycoprotein, 35, 37
    definition of, 34
Gold-palladium in S-E/M, 5
Golgi complex, 20, *20*, 25, 53, 56, 57, *59*, 84, 86, *136*, 137, *194*, 195
Gonioscopy, 170, 215, *230*, 231
Granular endoplasmic reticulum, 17, *18*, 25, 73–74, 80, 86, 97, *194*, 195, 198
Granules
    cytoplasmic, 43
    lipofuscin, 22, 23, 52, *53*, *54*, 98, 201
    matrix, 19, 20
    pigment, 22–23, *23*, 50, 51, 52, *53*, *54*, 55, *125*, 170, 175, *179*, 180, 181, *182*, 183, 201, 203
    secretion, 25
Grids for tissue preparation, *3*, 3
Ground substance, cytoplasmic, 22
Grunert's spurs, 175
Gunn's dots, 79

Halo of synaptic vesicles, 71, *72*, 73
Hannover, canal of, 125
Hassall-Henle, warts of, 156, *159*, *160*
HAYREH, S S., 244 *n.*
Heinrich-Müller, reticulum of, 206, *209*
Hematoxylin, 11 *n.*, 13, 16, 20, 28, 29, 73, *217*, *218*
Hemidesmosomes, 143, *144*, 156
Hemorrhage, 79, 86, 98
Henle fibers, 57, 61, 62 and *n.*, 69, *70*, 72, 74, 75, 95, 98
Heparin, 43
Heterophagocytosis, 24
Heteropolysaccharide, definition of, 34
Histochemical methods in electron microscopy, 8, *9*, 34, 57, 148
Histiocytes, 43
Homopolysaccharide, 35
    definition of, 34
Horizontal cells, 55, 58, 69, 71, 72, 73

Hyalocytes, 121
Hyaloid membrane, 115, 116
Hyaloideocapsular ligament, 119, 125, *125*
Hyaluronic acid, 34–35, 36 (table), *111*
Hyaluronidase, 61, *111*, 170
Hydroxyapatite crystals, 35
Hydroxylysine, 32, 33, 35
Hydroxyproline, 32, 33, 35, 112
Hyperplasias, adenomatous, of nonpigmented ciliary epithelium, 126
Hypertension, 91

Iduronic acid, 36 (table)
Immersion fixation of tissue, 3, 4
Inner nuclear layer (INL), *57*, 71–74
Inner plexiform layer (IPL), 10, *10*, *57*, 62 *n.*, 71, 73, 75, *75*, 77, *79*, *80*, 86, *97*
Inner segments of rods, 57, *57*
Intercalary tissue (of Elschnig), *236*, 238
Intercellular cement, 28, *193*, *195*. *See also* Cement substance
Intercellular spaces, 28, *28*, *30*, 30 and *n.*, 61, 135, *135*, *137*, 138, *196*, *197*, 198, *199*, *205*, 228
Intermediary tissue (of Kuhnt), 237, *238*
Internal limiting membrane, *57*, *58*, 78–79, *81*, *83*, *84*, *91*, 91 and *n.*, *92*, *93*, *94*, *97*, 99, 115, *115*, 116, 118, *118*, 193
Interreceptor mucoid, 53, *60*, *63*, *65*, *67*
Intertrabecular spaces, *223*, *226*, 226, *227*, 227, 228
Iridocorneoscleral angle, 214
Iridodialysis, 168
Iridodonesis, 168
Iris, 168, 211
    anterior border layer of, 168, *169*, 169–170, *171*, *172*, *173*, 180, 183, 184, 185, 231
    blue, 168, *169*, 170, *171*, *173*, *174*, *175*
    clinical appearance, *174*
    clump cells of, *174*, 175–176
    collarette of, 168, *169*, *174*
    color of, 168, *169*, *173*, *174*, *175*
    dilator muscle of, *176*, *178*, *179*, *181*, 183
    retinal (neuroepithelial) portion of, 173 ff.
    root of, 168, *169*, *169*, 180, 181, 201, 214, 222, *225*, 229 and *n.*, 230
    sphincter muscle of, 173, *177*, *178*, 181
    stroma of, 168, *169*, 170, *170*, *171*, *172*, 173, *173*

Iris (*continued*)
    uveal (mesodermal) portion of, 168 ff.
Iris pigment epithelium, 168, *169*, 173, 176, *179*, *181*, 181–182, *182*, *183*, 183–184
    circumferential ridges and furrows, 180, *182*
    longitudinal (radial) furrows, 181, *182*
    pigment seam, 180, *180*, 181
    structural furrows, 181
Iris processes, 170, *230*, 231
Iris trabeculae, *169*, *172*, 173, *174*
Isodesmosine, 35

JAKUS, M. A., 143 *n.*
Janus green, 19
Junctional complex, 30

Keratectomy, 148
Keratocytes, 43, 147, 148, *149*, *150*, *152*
Koganei, clump cells of, 175–176, *178*, *179*
Kolmer crystalloid, 73, *76*, *77*
Krause's end bulbs, 55
Krebs citric acid cycle, 19

Labile appositions, 30 *n.*
Lamellas
    choroidal, 185, *186*, *187*
    continuity of, with surface plasma membrane in cone, 60 *n.*
    corneal stromal, 147, 148, *149*, *152*, *153*
    scleral, *162*, *218*
    synaptic, *see* Synaptic lamellas
Lamina cribrosa, 78, 235, *236*, 239, *240*, *241*, 242
    choroidalis, 239, *240*, *244*, *245*
    scleralis, 239, *240*, 241, *241*, *243*, *244*
Lamina elastica, 49, *49*, *50*, 189, *206*, 207, 245, *245*, *246*
Lamina fusca, *164*, 165, 185
Lamina vitrea, 189
Lateral attachments of vitreous body, *113*, 118–119
Lateral plasma membrane, 42
Lead citrate, tissue sections treated with, 7
Lead hydroxide, tissue sections treated with, 7, 22

Lens
    embryology of, 130, *131*, 132
    slit-lamp appearance, 131
    structure of, 132, *133*, *134*, 134–135, *136*, *137*, 137–138, *138*, 139
Lens capsule, 34, 35, *37*, *38*, 38, 110, 113, *113*, 119, 120, *124*, 125, *125*, 126, 130, 132, *132*, *133*, *134*, 134–135, *135*
Lens shagreen, 134
Leukocytes, polymorphonuclear, 43
Ligamentum nuchae, 35
Limbus, corneoscleral, 215
Lipofuscin granules, 22, 23, 52, *53*, *54*, 98, 201
Lipoidal bodies, 23
Longitudinal furrows
    iris, posterior surface, 181, *182*
    along rod cell outer segments, 61, *64*, *66*
Lymphocytes, 43
Lysosomes, 23–25

Macromolecules, 32, *33*, 113
Macrophages, 43, 88, 175
Macropinocytotic transport mechanism, 219, 220
Macula, 93, 95, 98
    histologic, 95, 98
Macula lutea, 93, 98
Macular stars, 98
Martegiani, area of, *113*, 116, 235
Mast cells, 43
Matrix, intercristal, 19
Meibomian glands, 142
Meissner's corpuscles, 55
Melanin granules, 22–23, *23*, 50, 51, 52, *54*, 55, *125*, *170*, 175, *179*, 180, 181, *182*, 183, 201, 203
Melanocytes, 165, *169*, 175, *176*, 186, *211*
Melanoma, malignant, of uvea, 160
Melanosome, 55
    definition of, 51 *n.*, 52 *n.*
Membrane(s)
    basement, *see* Basement membranes
    Bowman's, 143, *147*
    Bruch's, 48–49, 50, *188*, 189
    Descemet's, *see* Descemet's membrane
    external limiting, *see* External limiting membrane
    fenestrated, 29, 30, *104*
    hyaloid, 115, 116

Membrane(s) (*continued*)
    internal limiting, *see* Internal limiting membrane
    middle limiting, 57, 58, 71, 72, 88, 187
    nuclear double, 17, *18*
    pericapsular, and zonular fibers of lens, 134–135
    plasma, *see* Plasma membrane
    unit, 17, *30*, 30 and *n.*
Meniscus, central supporting tissue (of Kuhnt), 116, 235, *236*, *237*, 238
Meridional plane of eye, 44, *44*
Meshwork
    corneoscleral, *217*, *219*, 221, 223, 224, 227, *229*, *230*
    uveal, *217*, 218, 225, 226, *226*, 227, *230*, 231
Mesoaxon, 234
Mesoderm, 42, 53, 229
Mesothelia, 42, 43, 150, 225
Messenger RNA (m RNA), 18
Metachromasia, 24, 34 and *n.*, 35, 36 (table), 43
Methyl violet, 34
Methylene blue, 13
Michel's (Fuchs') spurs, 175, *177*, *178*
Microbodies, 25
Microfibrils, 35
Microglia, 55, 87, 224 *n.*
Microphages, 43
Micropinocytotic vesicles, 88, 208, *211*, 219, 222
Microscope
    compound light, 2, *3*
    transmission electron, *see* Electron microscopy
Microtomes in tissue preparation, 2–3
Microtubules, 20, 21, 22, *22*, 69, 73, 82, 86, 241, 244 *n.*
Middle limiting membrane, 57, 58, 71, 72, 88, 187
Mitochondria, 17, *18*, 19, *19*, 20, 50, 51, *51*, 60, 64, 65, 71, *71*, 72, *72*, 80, 154
Mittendorf dot, 120
Moll's glands, 142
Monocytes, 43, 244 *n.*
Mounting of tissue, 2, 3, 5, 6
Mucinous materials, 32, 34–35, 43
Müller cells, 39, 55, 58, 59, 61, 66, 68, 70, 71, 73, *73*, 74, 76, *76*, 78, 80, 81, 82, 83, 86, 87, 88, 90, 117, *118*, 235
    cytoplasm of, 10, 11, 63, 69, 82, 86, 91, 91 *n.*, 92, 93, 97, 98, 99
    transition zone of, 86

Müller cells (*continued*)
  villi of, *60, 68*
  See also Astrocytes
Müller fibers, 86
Multilaminar basement membrane, 37
Muscle
  ciliary, 208, *210, 211,* 217, *225, 226, 228*
  dilator, of iris, 176, *178, 179, 181,* 183
  rectus, sclera at attachment of, 164, *164*
Myelin sheath, 234, *234*
Myelination, 78, 241–242, *242,* 243, *243*
Myeloid bodies, 53
Myointimal cells, 245

Negative pressure, and apposition of neural retina to pigment epithelium in vivo, 105
Nerve(s)
  corneal, 146, *148*
  optic, see Optic nerve
Nerve fiber layer (NFL), *57,* 76, 78, *80, 81, 94,* 97
Neurites, 10, 72, 74, 146, *148,* 226
Neuroectoderm, 42, 48, 53
Neuroepithelium, 22–23, 173–174, 180
Neurofilaments, 78
Neuronal system of neural retina, 53, *55, 57, 58*
Neurons, accessory, 73–74
Neurotubules, 78
Nicking, arteriolarvenous, 93
Nissl substance, *18,* 76, 78, *80,* 97
Nodes of Ranvier, 235, *236*
Nucleus, 16
  chromatin aggregation of, 3, 11
  Edinger-Westphal, 175
  of ganglion cell, *80*
  lens, 132, *136, 137,* 139

Objective lens of transmission electron microscope, 2
Occlusion, central retinal artery, 93
Oligodendrocytes, 55, *58,* 234, *234,* 244 and *n.*
Ophthalmoscope, 79
Optic cup, 48, *48,* 130, *237,* 246
Optic disc, 53, 95, 116, 235, *236*
Optic nerve, 44–45, 78, 160, 234–235
  choroidal layer of, 235, *236,* 239, 244
  edema of, 237
  intraneural components of, 243–244

Optic nerve (*continued*)
  intraocular (bulbar) portion of, 235, *236*
  orbital (retrobulbar) portion of, 241–242, 244, *245*
  retinal layer of, 235, *236, 237,* 237–238, *238,* 244
  scleral layer of, 235, *236, 241,* 241–242
  sheaths of, *242,* 242–243
  vascular supply to, 244–245, *245*
Optic vesicle, 48, *48,* 61, 130, 181, 184
Ora serrata, 48, 53, *56,* 61, 93, 103, *104,* 119, *120,* 191, 193, 203
Orbiculoanterior zonular fibers, 125
Orbiculoposterior zonular fibers, 125
Orbiculus ciliaris, see Ciliary body, pars plana
Orcein, 188, 189
Organelles, cytoplasmic, 17–22, 76, 95
Osmium tetroxide fixative, 3, 4, 7, 11, *12,* 22, 76, 238
Outer nuclear layer (ONL), *57,* 61, *70,* 86
Outer plexiform layer (OPL), *57,* 61–62, *69,* 72, 74, *75,* 95, 98
Outer segments of rods, 59, 61, *62, 64, 66, 67*
Outflow, uveoscleral, 229 and *n.*
Ovoid melanin granules, *51, 52*

Pacini's corpuscles, 55
Palladium, used in shadow-casting, 5
Papilla
  Bergmeister's, 235
  optic, 235
Papilledema, 237
Paraphenylenediamine, 11 *n.,* 13
Pedicle, cone, 62, *67, 71,* 71 and *n., 72*
Perfusion fixation of tissue, 3
Pericytes, 88, *89,* 173, *176,* 187
Periganglion satellite cells, *80*
Perikarya, 57, 61, 73, *243*
Periodic acid-Schiff (PAS) reagent, 8, 11 *n.,* 35, 37
Petit, canal of, 125
Phospholipids, 35
Phosphotungstic acid, 8
Photography in electron microscopy, 9–13
Photoreceptor cells, 55, 57, 59, 61, 68, 69, 86, *100*
Physiologic cup, 235, *236,* 238
Pia mater, 242, *242, 243*

Pigment epithelium
  apposition of neural retina to, in vivo, 105
  of ciliary body, 192, *192*, 193, *193*, *195*, 198, 199, 201, 203, *204*, *205*, 206, *206*, 207, 208, *209*, *210*
  and extrafoveal cone, 61, *67*
  iris, *see* Iris pigment epithelium
  retinal, 48–49, *49*, 50, 50–51, *51*, 52, 52–53, *53*, *54*, 55, 56, 57, 61, 67, 90, 98, *102*, 184, 192, 203
Pigment granules, *see* Granules, pigment
Pinocytotic vesicles, 173, 219
Plasma cells, 43
Plasma membrane, 16, *16*, 28, 38, 42, 43, 50, 51, 59, 60 and *n*., 62, 64, 71
  leaflets of, 17
Platelets, 43, *246*
Polyanionic characteristic of AMP, 34
Polymorphonuclear leukocytes, 43
Polysaccharide, definition of, 34
Polysomes, 22
Pores, nuclear, 17–18
PORTER, K. R., 17
Posterior attachments of vitreous body, *113*, 115–116, *116*, 117
Posterior subretinal cul-de-sac, 104, *105*
Premelanosome, 55
  definition of, 51 *n*.
Pressure, negative, and apposition of neural retina to pigment epithelium in vivo, 105
Projector lens of transmission electron microscope, 2
Propylene oxide in S-E/M, 5

Radioautography, 9
Ranvier, nodes of, 235, *236*
Rectus muscle, sclera at attachment of, 164, *164*
Red blood cells, 43, *177*, *183*, *218*, 220, *223*, *246*
Relucency
  diminished, 115
  second zone of, 139
Reticulin, 33–34
Reticulum, endoplasmic, *see* Endoplasmic reticulum
Retina
  bipolar cells of, 10, *10*, *58*, 71, 72–74, 76
  cones of, *see* Cones of retina
  detachment of, 61
  embryology of, 48, *48*, *49*

Retina (*continued*)
  external limiting membrane of, *see* External limiting membrane
  fovea of, *see* Fovea
  ganglion cell layer (GCL) of, *57*, *76*, 95, *96*, 97
  inner nuclear layer (INL) of, *57*, 71–74
  inner plexiform layer (IPL) of, *see* Inner plexiform layer (IPL)
  internal limiting membrane of, *see* Internal limiting membrane
  macula of, *see* Macula
  middle limiting membrane of, *57*, *58*, 71, 72, 88, 187
  nerve fiber layer (NFL) of, *57*, *76*, 78, 80, *81*, 94, 97
  neural (sensory), 53, 55 *ff*., 180, 187
  outer nuclear layer (ONL) of, *57*, 61, 70, 86
  outer plexiform layer (OPL) of, *see* Outer plexiform layer (OPL)
  pigment epithelium of, *see* Pigment epithelium, retinal
  rods of, *see* Rods of retina
  sensory (neural), 53, 55 *ff*., 180, 187
  and subretinal space, 103, 104, 105
Retroillumination, transpupillary, 173
Ribonucleic acid (RNA), 18
Ribosomes, 17, *18*, 22, 50, 51, 57, 60, 244 *n*.
Rods of retina, 55, 57, 59, *59*, 60 *n*., 61, 62, *62*, 66, 70, 98, 99, *100*

Satellite cells, periganglion, *80*
Scanning electron microscopy (S-E/M), 5, *6*, 7, *8*
Schlemm
  canal of, 162, 209, 215, *216*, 217, *218*, 219, *219*, 220, *220*, 221, *221*, 222, 223, 224, 225, 227, 228, *228*, 229, *229*
  canaliculi of, 227
Schwalbe's ring or line, 158, *160*, *161*, 170, 214, 215, 217, 230, 231
Schwann cells, 234, 235
Sclera, 159 *ff*.
  at attachment of rectus muscle, 164, *164*
  lamina fusca of, *164*, 165
  pigment spots of, 160, 161, 162
  "thin" regions of, 164
Scleral roll, *214*, 215, *216*, 217, 223, 225, 227, *228*, 229
Scleral spur, *214*
Sclerosis, arteriolar, 93

Scotoma, 78
Secretion granules, 25
Sectioning of tissue, 2 and *n.*, 3
SELIGMAN, A. M., 8
Septa, pial, 243, 244
Shadow-casting in electron microscopy, 4, 4–5, 6
Sheath
   arachnoid, 243
   myelin, 234, *234*
   pial, 243
SHERRINGTON, C., 69 *n.*
Sialic acid, 35
Silver tetraphenylporphine sulfonate (Ag-TPPS), 9, 207
Space(s)
   intercellular, *see* Intercellular spaces
   interfibrillar, 148
   intermembrane, 19
   intertrabecular, 223, 226, *226*, 227, *227*, *228*
   intracristal, 19
   retrolental, of Berger, 120
   subarachnoid, 242, 243
   subretinal, 103, 104, 105
   suprachoroidal, 185
   transtrabecular, 226, 227, *227*
Spherical melanin granules, 51, 52
Spherule, rod, 62, *71*, 73
Sphincter, iris, 173, *177*, *178*, *181*
Sphingomyelin, 35
Spiral of Tillaux, 45
Staining in electron microscopy, 7, *7*, 9, 11 *n.*, 13
Stroma
   choroidal, 185, *186*, 186–187, 188
   corneal, 142, *143*, 143 and *n.*, 144, 146, 147, 148, *149*, 150, *150*, 152, 153, 225
   iris, 168, *169*, 170, *170*, *171*, *172*, 173, *173*
   scleral, *162*
Subarachnoid space, 242, 243
Subretinal cul-de-sacs, 103, 104, *104*, 105
Subretinal space, 103, 104, 105
Sulcus, scleral, 164, 215, *215*
Suprachoroidal space, 185
Surface replication in electron microscopy, 5
Synapse, definition of, 69 *n.*
Synaptic densities, 69 and *n.*, 72, *72*, 74, 77
Synaptic expansions, 62, 67, *71*, 71, 72, 75, *75*, 77, 78
Synpatic lamellas, 69 and *n.*, *71*, 71 and *n.*, 72, 73, 75, 77, 98

Synaptic ribbons, 69 and *n.*
Synaptic vesicles, 67, *71*, 71, 72, 73, 75, 77

Tear film, 142, *143*, *146*
Tenon's capsule, 162
Terminal bar, 21, *28*, 29, *29*, 30 and *n.*, *50*, 51, 54, *59*, 67, *68*, 73, 88, *158*
Terminal web, 21, *29*
Thiosemicarbazide (TSC) method of glycogen detection, 8, 9, *9*, 22
Tillaux, spiral of, 45
Tissue preparation, 2–4
Toluidine blue, 7, 13, 34, *152*
Tonofibrils, 29, 143 *n.*
Tonofilaments, 143 *n.*
Trabecular meshwork, *see* Meshwork
Transmission electron microscopy, *see* Electron microscopy
Transtrabecular spaces, 226, 227, *227*
Triad configuration, 75 and *n.*
Trichrome, Masson, 11 *n.*
Tropocollagen molecules, 33, 225
Tubes, aqueous, 227
Tubules, intracytoplasmic, 238

Unit membrane, 17, *30*, 30 and *n.*
Uranium, used in shadow-casting, 5
Uranyl acetate, tissue sections treated with, 7
Uveal tract, 168 ff.
   malignant melanomas of, 160
   venous drainage from, 209, 211
Uveitis, 160

Vacuoles
   cytoplasmic, 219
   and secretory cells, 25
Vascular system of neural retina, 53, *58*, 88, *89*, 90, *91*, 91, *92*, 93, *93*
Vein(s)
   anterior ciliary, 161, *162*
   aqueous, 162, 222, 224
   central retinal, 88, 238, 242, 244, 245, *245*, *246*
   vortex, 45, 161, 185, 187, 211
Venules, 88, 91, 93, 95

Verhoeff's stain, 35
Vesicle(s)
  lens, 130, *130*
  micropinocytotic, 88, 208, *211*, 219, 222
  optic, 48, *48*, 61, 130, 181, 184
  pinocytotic, 173, 219
  synaptic, 67, 71, *71*, *72*, *73*, *75*, 77
Vicinal glycol groups (CHOH-CHOH), 35
Villi, 38, *60*, 61, *67*, *68*, 86, 88, 156, 184, *184*, 219
Vitreous body, 110, *110*, *111*, 112, *112*, 193, 198, *199*
  anterior, *113*, 119–120, *121*, 199
  and attachments, *113*, 115–116, 118–120
  boundaries of, 113
  central zone of, 121
  chemical composition of, *111*, 112
  cortical zone of, 115, 118, 120, 121
  embryology of, 110
  face of, 119, 120, *120*, *125*, 134
  fibrous component of, 111, *111*, 112
  mucinous component of, 111, 112–113, 115, *115*
  posterior, *112*, *113*, 115–116, *116*, *117*
  primary, 110
  secondary, 110
  tertiary, 110

Vitreous body (*continued*)
  zones of, 115, 118, 120–121
Vortex veins, 45, 161, 185, 187, 211

Warts of Hassall-Henle, 156, *159*, *160*
Weigert's stain, 35, 49, 50, 189
Wieger, ligament of (hyaloideocapsular ligament), 119, 125, *125*

X-ray diffraction, 32

ZAUBERMAN, H., 105
Zeis's glands, 142
Zinn, zonule of, 121, *122*, *123*, *124*, 125, *125*, 126
Zinn-Haller, circle of, 241, 244
Zonula adherens of terminal bar, 29, *29*, *50*, 51, *52*, 54
Zonula occludens of terminal bar, 29, 29–30, *50*, 51, *52*

Text and cover designed by Maria S. Karkucinski
Composed in Palatino by Brown Bros. Linotypers, Inc.
Printed by Pearl Pressman Liberty

72 73 74 75 9 8 7 6 5 4 3 2 1

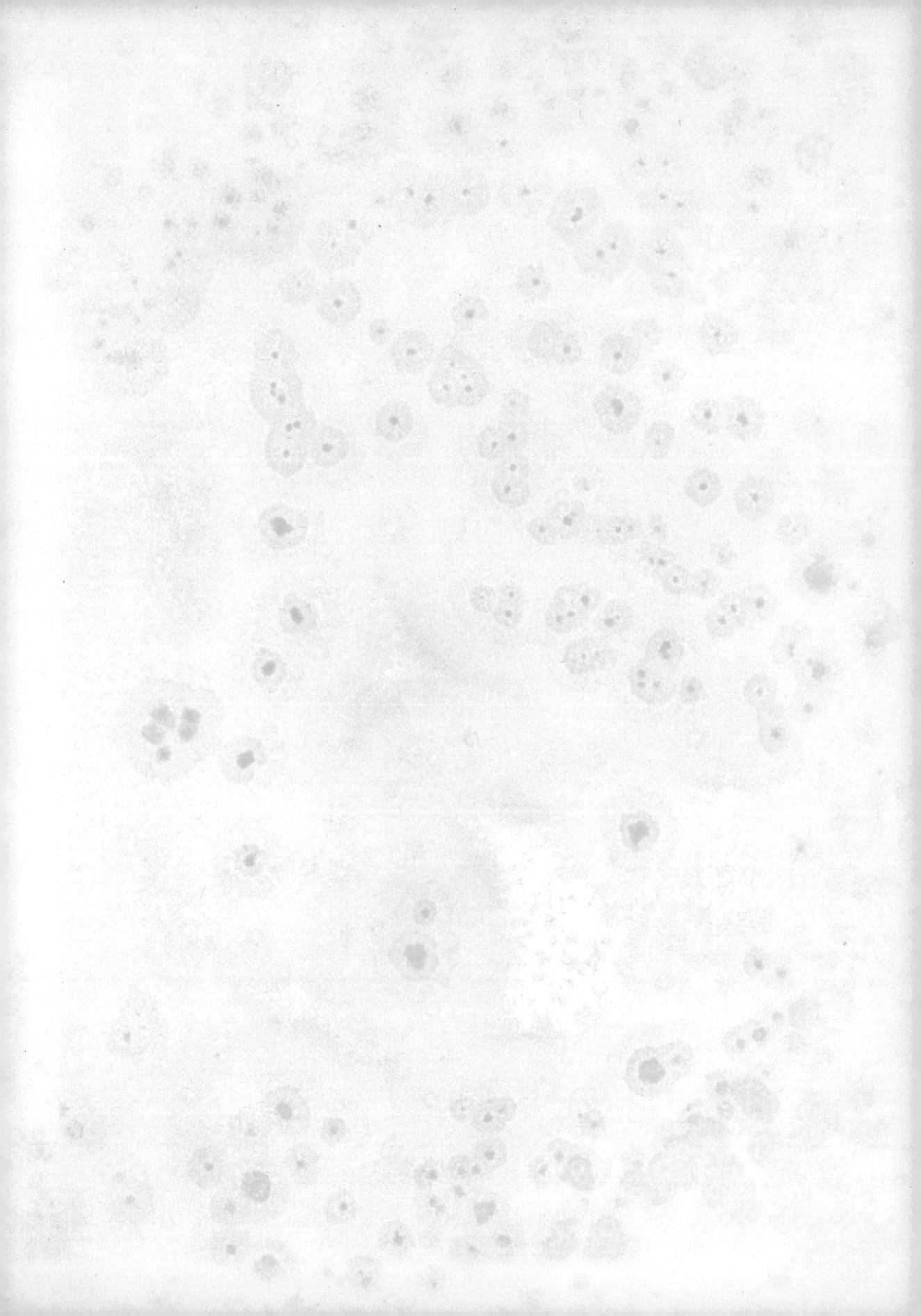